SO-CAH-304

Theoretical and Mathematical Models in Polymer Research

Modern Methods in Polymer Research
and Technology

Series in *Polymers, Interfaces, and Biomaterials*

Series Editor:

Toyoichi Tanaka
Department of Physics
Massachusetts Institute of Technology
Cambridge, MA, USA

Editorial Board:

Sam Safran
Weitzman Institute of Science
Department of Materials
　and Interfaces
Rehovot, Israel

Masao Doi
Applied Physics Department
Faculty of Engineering
Nagoya University
Nagoya, Japan

Charles Knobler
Department of Chemistry
University of California at
　Los Angeles
Los Angeles, CA, USA

Alexander Grosberg
Department of Physics
Massachusetts Institute of
　Technology
Cambridge, MA, USA

Other books in the series:

Kaoru Tsujii, *Chemistry and Physics of Surfactants: Principles, Phenomena, and Applications* (1998).

Teruo Okano, Editor, *Biorelated Polymers and Gels: Controlled Release and Applications in Biomedical Engineering* (1998).

Also available:

Alexander Grosberg and Alexei R. Khokhlov, *Giant Molecules: Here, There, and Everywhere* (1997).

Jacob Israelachvili, *Intermolecular and Surface Forces, Second Edition* (1992).

Theoretical and Mathematical Models in Polymer Research

Modern Methods in Polymer Research and Technology

Edited by

Alexander Grosberg
DEPARTMENT OF PHYSICS
MASSACHUSETTS INSTITUTE OF TECHNOLOGY
CAMBRIDGE, MASSACHUSETTS

ACADEMIC PRESS

Boston   San Diego   New York
London   Sydney   Tokyo   Toronto

This book is printed on acid-free paper. ⊗

Copyright © 1998 by Academic Press
All rights reserved.
No part of this publication may be reproduced or
transmitted in any form or by any means, electronic
or mechanical, including photocopy, recording, or
any information storage and retrievel system, without
permission in writing from the publisher.

ACADEMIC PRESS
525 B Street, Suite 1900, San Diego, CA 92101, USA
1300 Boylston Street, Chestnut Hill, MA 02167, USA
http://www.apnet.com

United Kingdom Edition published by
ACADEMIC PRESS LIMITED
24-28 Oval Road, London NW1 7DX
http://www.hbuk/co.uk/ap/

Library of Congress Cataloging-in-Publication Data

Theoretical and mathematical models in polymer research : modern
   methods in polymer research and technology / edited by Alexander
   Grosberg.
      p.   cm. — (Polymers, interfaces, and biomaterials)
   Includes bibliographical references and index.
   ISBN 0-12-304140-6
   1. Polymers.   I. Grosberg, A.   II. Series.
QD381.7.T485   1998
547′.7—dc21   97-51305
                                                              CIP

Printed in the United States of America
98 99 00 01 02 IC 9 8 7 6 5 4 3 2 1

Q D
381
.7
T485
1998
CHEM

Contents

4   Winding Angle Distributions for Directed Polymers
Barbara Drossel and Mehran Kardar

5   Bulk and Interfacial Polymer Reaction Kinetics
Ben O'Shaughnessy

Contributors

Numbers in parentheses indicate the pages on which the authors' contributions begin.

Barbara Drossel (187), Department of Physics, Massachusetts Institute of Technology, Cambridge, Massachusetts 02139

B.-Y. Ha (1), Department of Chemistry and Biochemistry, University of California at Los Angeles, Los Angeles, California 90024

J.-F. Joanny (37), Institut Charles Sadron, 6 Rue Boussingault, Srasbourg 67083 Cedex, France

A. Johner (37), Institut Charles Sadron, 6 Rue Boussingault, Srasbourg 67083 Cedex, France

Mehran Kardar (187), Department of Physics, Massachusetts Institute of Technology, Cambridge, Massachusetts 02139

Ben O'Shaughnessy (219), Department of Chemical Engineering, Columbia University, New York, New York 10027

S. V. Panyukov (83), Lebedev Physics Institute, Russian Academy of Science, Moscow 117924, Russia

Y. Rabin (83), Department of Physics, Bar-Ilan University, Ramat-Gan 52900, Israel

A. N. Semenov (37), Department of Applied Physics, University of Leeds, Leeds LS2 9JT, United Kingdom

D. Thirumalai (1), Institute for Physical Science and Technology, University of Maryland, College Park, Maryland 20740

Preface by Series Editor

Polymers play a key role in medical, pharmaceutical, chemical, and agriculltural industries. We find polymers in almost every material we use. Our everday life cannot go without polymers. Above all, the most important constituent of our body is polymers. Biopolymers are the bearer of life, responsible for all the biological functions of molecular sensing, homeostasis, cleaning, catalysis, and molecular motions. Synthetic polymers capable of such functions are expected to play a major role in the twenty-first century technologies as a key material for creating high quality life and in solving problems in the global environment.

Scientific and technological communities recognize the importance of polymers and have made strenuous efforts to dramatically advance our understanding of polymers during the last two decades. In particular, novel theoretical concepts and techniques have been developed that allow predictions of many interesting macroscopic behaviors of polymers from the first principles. The predicting power will be of great use for further development of new types of polymers for various applications. We believe that now is the time to review and summarize the state of the art of the modern polymer theories.

Professor Alexander Grosberg, one of the pioneers and leaders of theoretical polymer science, has selected the hottest topics in modern polymer theories and authors who can best teach the field. The topics include statistical mechanics of semiflexible chains, polymer absorption, polymers in fields, topological constraints, polymer reaction kinetics, and replica field theory of polymer networks. They are organized so that readers can survey the scope of the field. From this book, readers should be able to learn the fundamental concepts as well as the techniques that are currently used in the frontier of science.

Toyoichi Tanaka, Series Editor
Massachusetts Institute of Technology
January, 1998

Preface by Editor

There is an interesting question—how many people in the world are doing polymer science? The estimates differ by at least an order of magnitude depending on how one encircles the field. Indeed, there is a variety of views on what polymer science is all about, what is (and what is not) its subject. One of the three most popular approaches associates polymers mainly with materials science; for another approach, the most important aspect is the crucial role the polymers play in molecular biological machinery of the living cell; and there is the third viewpoint, for which the word "polymers" is the generic name for a variety of objects of very different nature that share only the property of being fluctuating strings. This latter viewpoint is far less traditional than the first two, and I would like to comment on that here.

Indeed, while "polymer" literally means something like "many units," this term is now routinely used for such different objects as flux lines in superconductors, rays of light in multiple scattering medium, meandering rivers (on geological time scales), stacks of disc-shaped molecules in certain liquid crystals, some topological defects in crystals and liquid crystals, as well as in liquid helium. "Strings," which are universally considered as the main architectural element of the emerging "theory of everything," are also polymers from that broad point of view. I could have forgotten some examples, but it hardly matters, as the idea is clear: all these objects have little to nothing in common as regards time, distance, and energy scales involved, as well as their underlying "materialistic" nature (what they are made from). One should not forgot that the "traditional" polymers are actually also extremely diverse; for instance, lengths of polymers chains, their rigidities and other basic properties may vary by several orders of magnitude, and the diversity of biological functions of various biopolymer strikes our imagination. Nevertheless, the image of a fluctuating string is something that all these systems have in common, and it is of crucial importance.

As a matter of fact, regular chemical polymers (whose chain-like molecular structure was established by H. Staudinger around 1925) were not the first string-like objects to attract the attention of physicists. As early as 1867, following the pioneering works by H. von Helmholtz on hydrodymanics and vortex lines in an ideal incompressible fluid, W. Thompson came up with the idea that atoms—the greatest puzzle of the time—could be understood in terms of vortex tubes in ether. The idea seemed to explain the diversity of atoms, as each of them could be identified with a certain type of a knot formed by the vortex tube; as there are discretely different knots, so there are many types of atoms, and no way to *continuously* transform one into the other. The stability of atoms would then be attributed to conservation of vorticity in ether, and even atomic spectra seemed to be understandable in terms of oscillations of various knotted tubes. In brief, the idea was really brilliant and attractive. Nowadays, strings generate somewhat similar hopes among quantum field theorists—this fact seems to suggest that every really good idea should eventually be assimilated. The idea of Thompson inspired P. G. Tait to begin systematic inquires into the topology of knots. Even though J. C. Maxwell concluded that atoms do not exhibit sufficient signs of any underlying relation to topology and hydrodynamics, the table of knots and a hypothesis by Tait kept challenging topologists for about a century. Not surprisingly, topological theory of knots currently serves as one of the major working tools in the physics of such an important polymer as DNA.

Thus, with the astonishing variety of both traditional and non-traditional subjects, polymer science in general and polymer theory in particular is mainly concentrated on underlying general concepts, physical insights, and methods. This is why this book is concentrated on *methods*. Indeed, as many people are learning polymers there is a significant demand for pedagogically-oriented texts to supplement reading the original papers. There are very well known books [1, 2, 3, 4, 5, 6, 7, 8, 9, 10, 11] on polymer theory. This book is supposed to be an advanced supplement to allow an interested reader to learn of the new and developing methods in theoretical polymer science. This book is oriented for those, like graduate students or postdoctoral fellows, who are entering the field. It is also oriented for actively working experts in one of the subfields willing to upgrade their theoretical "equipment" to enter another subfield or to cross-disseminate the ideas.

Although this book concentrates on *methods* of polymer theory, it is still unable to cover all methods. Indeed, in the early years of polymer theory, the methods employed [1] were fairly simple by theoretical physics standards, although we should not forget that using these methods P. Flory was able to compute an excellent approximation for the non-trivial critical exponent—decades before this problem was addressed by the inventors of renormalization group and ϵ-expansion. It was an achievement when matrix methods of one-dimensional Ising model were systematically applied to polymers [2, 3, 4]. This was the way to tackle *linear* interactions between the *neighboring monomers along the chain,* including mechanisms of chain flexibility. As regards *volume* interactions between monomers that come close to each other in space due to chain wiggling, the first systematic treatment for them was of a perturbative nature [5]. This situation changed radically when, in the middle of the 1960s, S. F. Edwards and I. M. Lifshitz (independently) discovered the analogy between polymer statistics and Feynman formulation of quantum mechanics. Indeed, both polymer partition function and Feynman path integral are just the sums over all possible spatial curvilinear shapes. Soon after that, another remarkable discovery by P.-G. de Gennes came—the one of $n = 0$ polymer-magnetic analogy. These two seminal achievements completely revolutionized the field and brought it into the realm of theoretical physics. In the last quarter of the century, the methodical arsenal of polymer theory have matured, and it is now just as broad and sophisticated as that of theoretical physics in general; there is hardly any general method which is not of use in polymers. This is why the idea to review *all* theoretical and mathematical methods in polymers is hopeless and useless. Instead, this book concentrates on several specific topics in which recent methodical achievements have yielded important new insights. There are five main contributions in the book.

• The work "Statistical Mechanics of Semiflexible Chains: A Mean Field Variational Approach," by D. Thirumalai and B.-Y. Ha revisits the classical polymer theory single-chain problem. Although this has been extensively studied many years ago (see, for instance, [3]), there is now a renewed interest due to the completely new experimental techniques of manipulations with *single* DNA molecules. Ingenious combination of molecular-biological manipulations with DNA and the use of optical tweezers allows one to measure directly, say, elongation of a single DNA which is mechanically pulled by its ends. Understanding these kinds of experiments required completely new levels of analysis; new works in this direction are under way, attempting to incorporate the effects of twist, heterogeneity, etc. The work by D. Thirumalai and B.-Y. Ha will serve as a valuable introduction in this subfield.

• The article "Polymer Adsorption: Mean Field Theory and Ground State Dominance Approximation," by A. N. Semenov, J. F. Joanny, and A. Johner scrutinizes another classical problem of polymer theory. The problem of chain adsorption has attracted a great deal of attention for decades both because of its direct experimental relevance and as a simple theoretical model of a broad class of collapse phase transition, in the case of adsorption collapse as being driven by an external field (attraction to the absorber). These problems have been addressed in [6, 7, 9, 11], but the present authors found that the treatment of chain ends was not completely satisfactory, they re-examine the problem, and indeed their article sheds an important new light on it.

• One of the hottest topics in modern condensed matter physics and statistical mechanics is the problem of disorder and disordered systems. Although spin glasses is the example of disordered system beloved by theorists in the field, it seems that disordered polymer materials of various sorts are very important for technological and biological applications. "Replica Field Theory Methods in Physics of Polymer Networks," by S. V. Panyukov and Y. Rabin is the most sophisticated and systematic presentation of the replica and field theory methods as applied to the gels with frozen disordered pattern of cross-links. This work is based on the seminal contribution by the same authors and supplements the text of the original presentation [12] with stress on methods involved.

• "Winding Angle Distributions for Directed Polymers," by B. Drossel and M. Kardar deals with yet another classical problem of topological entanglements between a fluctuating polymer and an external rigid obstacle, such as a rod. Directly in the polymer context, this problem was first addressed about 30 years ago in the work by S. F. Edwards and in the work by S. Prager and H. Frisch, but in pure mathematical context the problem has been known even earlier. This is, of course, the simplest among the problems related to entanglements, knots, and other topological constraints in polymers. Its advantage is in its tractability, and the present authors demonstrate how this problem can be addressed in a variety of circumstances, including finite non-zero size of the obstacle, disorder in the medium, chiraltiy, etc.

• Kinetics of polymers has always been particularly challenging for researchers. In this book, the subfield is presented by the article "Bulk and Interfacial Polymer Reaction Kinetics," by B. O'Shaughnessy. This is the most comprehensive and up-to-date treatment of theoretical methods employed in modern studies of chemical reactions in a variety of diffusion controlled regimes.

Clearly, all the articles in this book are largely independent and could be arbitrarily permuted.

On behalf of all authors, I express the hope that this book will be useful for the reader. We wish you, the reader, a success.

Alexander Grosberg
Brookline, MA
January, 1998

References

[1] P. Flory, *Principles of Polymer Chemistry,* Cornell University Press, 1953.
[2] M. V. Volkenstein, *Configurational Statistics of Macromolecules,* Interscience, NY, 1963.

[3] T. M. Birshtein, O. B. Ptitsyn, *Conformations of Macromolecules,* Interscience, NY 1966.

[4] P. Flory, *Statistical Mechanics of Chain Molecules,* Interscience, NY 1969.

[5] H. Yamakawa, *Modern Theory of Polymer Solutions,* Harper and Row, NY, 1971.

[6] P. G. de Gennes, *Scaling Concpts in Polymer Physics,* Cornell Univ. Press, Ithaca, NY, 1979.

[7] M. Doi, S. F. Edwards, *The Theory of Polymer Dynamics,* Oxford University Press, Oxford 1986.

[8] K. F. Freed, *Renormalization Group Theory of Macromolecules,* J. Wiley, NY, 1987.

[9] J. des Cloizeaux, G. Jannink *Polymers in Solution: their Modelling and Structure,* Oxford Univ. Press, 1990.

[10] A. V. Vologodskii *Topology and Physics of Circular DNA,* CRC Press, Boca Raton, 1992.

[11] A. Yu. Grosberg, A. R. Khokhlov *Statistical Physics of Macromolecules,* AIP Press, NY 1994.

[12] S. Panyukov, Y. Rabin *Statistical Physics of Polymer Gels,* Physics Reports, v. 269, p. 1, 1996.

Theoretical and Mathematical Models in Polymer Research

Modern Methods in Polymer Research
and Technology

Chapter 1 | Statistical Mechanics of Semiflexible Chains: A Mean Field Variational Approach

D. Thirumalai

Institute for Physical Science and Technology
University of Maryland, College Park, Maryland

B.-Y. Ha

Department of Chemistry and Biochemistry
University of California at Los Angeles
Los Angeles, California

1.1 Introduction

The random walk model for neutral polymers is perhaps the simplest mathematical representation for long flexible chains [1]. The tremendous progress made in the theoretical understanding of conformational and dynamical properties of flexible polymer systems becomes possible because systematic calculations using the random walk model in conjunction with the inclusion of excluded volume interactions can be carried out, at least in principle [2]. The resulting model, referred to as the Edwards model, is the minimal representation of real polymers that adequately describes the global properties of several polymeric systems. The random walk model views the flexible polymer chains as a Brownian curve. In the discrete representation, a flexible chain can be modeled as one for which angles between successive chain segments are not correlated. Since the orientations of chain segments are independent, the segment vectors have the Markovian property so that the mean squared end-to-end distance is proportional to the number N of segments of size a in the chain. In the continuum limit this chain becomes a Brownian curve.

In contrast, there are a number of issues concerning the behavior of semiflexible chains that are not satisfactorily solved. Many polymeric molecules have internal stiffness and hence can not be modeled as freely jointed chains [3]. This is especially the case in several biopolymers such as actin, DNA, and microtubules [4]. A measure of the stiffness of a polymer is in terms of the so-called persistence length, l_p, which gives an estimate of the length scale over which the tan-

1

THEORETICAL AND MATHEMATICAL
MODELS IN POLYMER RESEARCH

Copyright © 1998 by Academic Press.
All rights of reproduction in any form reserved.
ISBN 0-12-304140-6/$25.00.

gent vectors along the contour of the chain backbone are correlated. The typical values of l_p of biopolymers lie in the range of several nm to a few mm. This range is several orders of magnitude larger than the persistence length of flexible chains. If the contour length, L, of such molecules is of the same order of magnitude as l_p, then it is imperative to include bending rigidity to describe the conformations of the chain. In these situations the chain behavior is wormlike; the appropriate model for this behavior was introduced some time ago [3]. It has been known for a long time that the wormlike model provides a good starting point in the theoretical description of chains with internal stiffness. Inspired by several recent experiments [5], which have probed a number of properties of stiff biological molecules, there has been a renewed interest in understanding their shapes. The purpose of this chapter is to provide a simple way of calculating a number of interesting properties of semiflexble chains using a mean field variational approach.

The effect of excluded volume profoundly changes the properties of flexible chains. Although historically the importance of excluded volume was recognized some time ago, the introduction of the Edwards model to represent the effect of this short-range interaction made possible systematic calculation of various static and dynamic properties using field theoretical methods [2]. In this chapter, in which we focus on the properties of semiflexible chains, the excluded volume effects will be ignored. This is physically reasonable for relatively stiff chains for which deviations from rodlike conformations are negligible.

A simple way to account for the stiffness of a semiflexible chain is to constrain the angles between two successive segments θ to be fixed. The value of θ depends on the local stiffness of the chain. This prescription leads to the freely rotating chain model. If we describe the configurations of a polymer chain by the set of position vectors $\{\mathbf{r}_n\} = (\mathbf{r}_0, \ldots, \mathbf{r}_N)$ or, alternatively, by the set of segment vectors $\{\Delta\mathbf{r}_n\} = (\mathbf{r}_1 - \mathbf{r}_0, \ldots, \mathbf{r}_N - \mathbf{r}_{N-1})$, then the spatial correlation, $\langle \Delta\mathbf{r}_n \cdot \Delta\mathbf{r}_{n-1}\rangle$, in the freely rotating chain has the assigned value $a^2 \cos\theta$. In the continuous limit ($a \to 0$, $\theta \to 0$, $N \to \infty$, $Na = L$), the freely rotating chain becomes the so-called wormlike chain [3]. In this case, the ratio $2a/\theta^2$ defines the *persistence length, l_p*, which is the typical length scale over which the chain changes its direction appreciably. Other conformational properties of such a model are well known in the literature [3, 6–21].

The spatial correlations $\langle \Delta\mathbf{r}_n \cdot \Delta\mathbf{r}_m\rangle$, which characterize the properties of a semiflexible chain, decay exponentially as $\exp[-a|n - m|/l_p]$. Thus, the conformational properties of a semiflexible chain beyond the length scale l_p reduce to those of flexible chains; i.e., one can view the stiff chain as being made up of several rigid segments of length l_p that are freely joined. However, because of the intrinsic skeletal stiffness of many synthetic polymers as well as biopolymers, a model that explicitly builds effects due to chain bending is needed. The chain stiffness turns out to be a relevant parameter in the description of the isotropic–nematic transition condition in liquid–crystalline polymers [15]. Even for an isolated chain, the chain stiffness should be taken into account to describe the local properties of stiff polymer chains. This is especially important in polyelectrolytes. The scaling behavior of the electrostatic persistence length l_e is known to depend on the rigidity of the chain [16, 17]. Many biological molecules and short chains of otherwise flexible chains also belong to the class for which the chain stiffness plays an important role.

A number of theoretical models have been introduced in the literature to account for chain stiffness. The earliest model for stiff chains is the wormlike chain (also known as the Kratky-Porod model) in which the angles between successive chains are constrained [3]. Although physically reasonable, this model has not yielded analytically tractable results for equilibrium and dynamical properties. Harris and Hearst introduced a "simplified model" of stiff chains in which the tangent vector $\mathbf{u}(s) = \partial \mathbf{r}/\partial s$ was allowed to fluctuate as opposed to having the constraint $\mathbf{u}^2(s) = 1$ for all s [7]. However, it has been noted that the resulting model does not satisfy the spatial homogeneity of stiff chains. More recently, a model that does not suffer from this restriction and that uses a functional integral formalism was proposed by Lagowski et al. [18]. These authors showed that the resulting model yielded the mean squared end-to-end distance in agreement with Kratky-Porod. The spatial correlations decay exponentially with a slightly shorter value of the persistence length.

In this chapter we show that the model for stiff chains proposed by Lagowski et al. (LNN) [18] results from a stationary phase evaluation of certain functional integrals that occur in an appropriate field theory for stiff chains. Our approach is systematic and can be applied to many diverse problems involving semiflexible chains. We should note that Winkler et al. [19] have also obtained a model for stiff chains using the maximum entropy principle. These authors did not notice that their model in the continuum limit is identical to that of LNN. Furthermore, their method appears more cumbersome than the standard functional integral approach presented here.

The chapter is organized as follows. In Section 1.2 we present the basic strategy behind the mean field variational approach. The resulting model, as mentioned earlier, leads to the LNN representation of wormlike chains. The theoretical ideas are used to calculate the distribution of end-to-end distance in semiflexible chains in Section 1.3. The application of the theory to the problem of semiflexible chains under tension with forces or the stretching of DNA is developed in Section 1.4.

The possible limitations of the theory are illustrated in Section 1.5 through the study of the behavior of semiflexible chains in a stretching nematic field. Section 1.6 concludes the chapter with a few additional remarks.

1.2   Mean Field Model

1.2.1   FLEXIBLE CHAINS

The basic methodology can be illustrated using the simpler example of a flexible chain. This is the limiting case of a stiff chain as the rigidity vanishes. The probability function for the flexible chain conformations without excluded volume interactions can be written as

$$\Psi\{\mathbf{r}_n\} = \prod_{n=1}^{N} \psi(\Delta\mathbf{r}_n) \tag{1}$$

where $\psi = \delta(|\Delta\mathbf{r}| - a)/4\pi a^2$ denotes the random distribution of a segment vector of length a. Eq. (1) accounts for the chain connectivity. We can rewrite the probability weight in Eq. (1) by introducing auxiliary fields λ_n as

$$\Psi\{\mathbf{r}_n\} \propto \int_{-i\infty}^{i\infty} \prod_{n=1}^{N} d\lambda_n \exp\left[-\sum_{n=1}^{N} \frac{\lambda_n}{a}((\Delta\mathbf{r}_n)^2 - a^2)\right]. \tag{2}$$

We now show that a stationary phase evaluation of the free energy of the chain described by the above weight leads to the probability weight for the Brownian chain. This approximation amounts to relaxing the locally enforced constraint of $(\Delta\mathbf{r}_n)^2 = a^2$ to a global one, $\langle(\Delta\mathbf{r}_n)^2\rangle = a^2$, and the validity of the approximation can be justified *a posteriori* as $a \to 0$. The free energy F of a noninteracting flexible chain can be written as

$$\exp(-F/k_B T) = \text{const} \int_{-i\infty}^{i\infty} \prod_{n=1}^{N} d\lambda_n \exp(-\mathcal{F}\{\lambda_n\}) \tag{3}$$

where the free energy functional $\mathcal{F}\{\lambda_n\}$ is defined by

$$\mathcal{F}\{\lambda_n\} \equiv -\ln\left\{\int \prod_{n=1}^{N} d\mathbf{r}_n \exp\left[-a^{-1}\sum_{n=1}^{N} \lambda_n \mathbf{r}_n^2\right]\right\} - a\sum_{n=1}^{N} \lambda_n$$

$$= \sum_{n=1}^{N}\left(\frac{3}{2}\ln\lambda_n - \lambda_n a\right) + \text{const}. \tag{4}$$

In the above equation, the order of the $\mathbf{r}_n$ and λ_n integrations is interchanged. If we denote the trajectory λ_n along which the integrand in Eq. (3) has its maximum value by λ_n^{cl}, then the free energy can be expanded around this stationary phase trajectory, λ_n^{cl}. In the following discussion, the superscript cl will be omitted.

In the mean field theory for which the constraint is imposed only on an average, we retain only the leading term in this expansion. By setting the partial derivative of the free energy functional $\mathcal{F}\{\lambda_n\}$ with respect to λ_n to zero, we get the stationary phase condition

$$\frac{\partial}{\partial \lambda_n} \mathcal{F}\{\lambda_n\} = 0 \Rightarrow \lambda_n = \frac{3}{2a}, \quad 0 \leq n \leq N. \tag{5}$$

The independence of λ_n on n reflects the symmetry of the problem of an ideal flexible chain. Since the delta function can be also represented as

$$\delta(\mathbf{r}) = \lim_{a \to 0} \left(\frac{3}{2\pi a^2} \right)^{3/2} \exp(-r^2/2a^2), \tag{6}$$

the stationary phase evaluation becomes very accurate in the continuum limit, $a \to 0$. Thus, long flexible chains—i.e., $N \gg 1$—can be well described by the following weight in the continuum limit

$$\Psi_{\mathrm{MF}}[\mathbf{r}(s)] \propto \exp\left[-\frac{3}{2a} \int_0^L ds \left(\frac{\partial \mathbf{r}}{\partial s} \right)^2 \right] \tag{7}$$

where $\Psi[\mathbf{r}(s)]$ is written in the functional integral notation and is the Wiener measure.

By treating the random fields $\lambda(s)$ at the mean field level, the microscopic constraints conjugate to the fields $\lambda(s)$—which ensure that the chain segments are connected but otherwise randomly distributed—are relaxed to the global ones. This results in the expected probability weight given in Eq. (7) for a long flexible chain and is the Wiener measure obtained in the path integral description of a diffusion equation.

1.2.2 LINEAR STIFF CHAINS

The approach described above can be extended to semiflexible chains. In this calculation we assume that the stretching of two connected chain segments is not important so that the coupling between this degree of freedom and the bending degree of freedom can be ignored [20]. In this case, the weight in Eq. (1) needs to be modified so that it yields the nonvanishing correlations $\langle \Delta \mathbf{r}_n \cdot \Delta \mathbf{r}_{n-1} \rangle = a^2 \theta^2 = 2a^3/l_\mathrm{p}$. This can be achieved if we multiply the weight in Eq. (1) by the Boltzmann weight $\exp(l_\mathrm{p} a^{-3} \sum_{n=1}^{N-1} \Delta \mathbf{r}_{n+1} \cdot \Delta \mathbf{r}_n)$ corresponding to the local interactions between adjacent segments. This term favors parallel alignment of adjacent segments over bent configurations. In the λ_n representation of the probability weight, this can be rewritten as $\exp[-\frac{1}{2}l_p a^{-3}(\Delta\mathbf{r}_{n+1} - \Delta\mathbf{r}_n)^2]$ with a redefinition of λ_n. Then the weight associated with a particular configuration of a semiflexible chain becomes

$$\Psi\{\mathbf{r}_n\} \propto \int_{-i\infty}^{i\infty} \prod_{n=1}^{N} d\lambda_n \, \exp\left[-\sum_{n=1}^{N} \frac{\lambda_n}{a}((\Delta\mathbf{r}_n)^{2-a^2}) - \frac{l_p}{2a^3} \sum_{n=1}^{N-1}(\Delta\mathbf{r}_{n+1} - \Delta\mathbf{r}_n)^2\right]. \quad (8)$$

In the continuum limit, this can be written as the functional integral

$$\Psi[\mathbf{u}(s)] \propto \exp\left[\frac{-l_p}{2} \int_0^L ds \left(\frac{\partial\mathbf{u}}{\partial s}\right)^2\right] \prod_{0 \leq s \leq L} \delta(\mathbf{u}^2(s) - 1) \quad (9)$$

where $\mathbf{u}(s) \equiv \partial\mathbf{r}(s)/\partial s$ is a unit tangent vector. The properties associated with the weight $\Psi[\mathbf{u}(s)]$ are well known in the literature [2–3, 10]. The random variable $\mathbf{u}(s)$ describes the rotational Brownian motion on a unit sphere, $\mathbf{u}^2 = 1$. If we let $P(\mathbf{u}_s, \mathbf{u}_{s'}; s', s)$ be the probability that $\mathbf{u}(s') = \mathbf{u}_{s'}$ when $\mathbf{u}(s) = \mathbf{u}_s$, then this function obeys a diffusion equation on the unit sphere. The solution of the diffusion equation can be expanded in terms of spherical harmonics. This enables us to compute the correlation

$$\langle\mathbf{u}(s') \cdot \mathbf{u}(s)\rangle = \exp(-|s' - s|/l_p). \quad (10)$$

This correlation, along with the Markovian property of $\mathbf{u}$, leads to the mean squared end-to-end distance given by

$$\langle R^2 \rangle = \int_0^L \int_0^L ds \, ds' \langle\mathbf{u}(s') \cdot \mathbf{u}(s)\rangle$$

$$= 2l_p L - 2l_p^2(1 - e^{-L/l_p}). \quad (11)$$

Even though the results given in Eq. (10) and Eq. (11) are exact, the use of Eq. (9) to describe nonideal semiflexible chains turns out to be quite formidable. The major difficulty arises because of the constraint $\mathbf{u}^2(s) = 1$. One encounters similar difficulty in other physical systems described by the nonlinear σ model [22] for which the magnitude of a spin $\mathbf{S}$ is held fixed, $\mathbf{S}^2 = \text{const}$. Thus, it is of practical interest to obtain a tractable model for such constrained systems. We will extend the stationary phase approach adopted for the flexible chain to obtain a tractable mean field model for a semiflexible chain.

In our stationary phase approach, the field λ_n is treated as a parameter to be determined. The dependence of λ_n on n depends on the problem under consideration. The free energy functional for an ideal semiflexible chain can be written as

$$\mathcal{F}\{\lambda_n\} = -\ln \int \prod_{n=1}^{N} d\mathbf{r}_n \exp\left[-\frac{E}{k_{\mathrm{B}}T} + a \sum_{n=1}^{N} \lambda_n\right] \tag{12}$$

where E in a matrix form is given by

$$\frac{Ea}{k_{\mathrm{B}}T} = \zeta^T Q \zeta \tag{13}$$

with $\zeta \equiv \{\mathbf{r}_1, \ldots, \mathbf{r}_N\}^T$. The $3N \times 3N$ matrix Q is defined by

$$Q_{nm} = \lambda_n \delta_{nm} - \frac{l_{\mathrm{p}}}{2a^2}(1 + \delta_{nm\pm1}). \tag{14}$$

Then the free energy $\mathcal{F}$ is given by

$$\mathcal{F}\{\lambda_n\} = \frac{3}{2} \ln(\det Q) - a \sum_{n=1}^{N} \lambda_n + \text{const.} \tag{15}$$

The stationary phase evaluation of λ_n amounts to minimizing the free energy with respect to λ_n; i.e.,

$$\frac{\partial}{\partial \lambda_n} \mathcal{F}\{\lambda_n\} = 0 \Rightarrow \frac{3}{2} \frac{\partial \ln(\det Q)}{\partial \lambda_n} = a, \quad 1 \leq n \leq N. \tag{16}$$

It can be easily shown that the minimization condition in Eq. (16) amounts to requiring $\langle \mathbf{u}^2 \rangle = 1$ in the continuous limit. This follows because Eq. (16) can be rewritten as

$$\frac{\partial}{\partial \lambda_n} \mathcal{F} = \langle (\Delta \mathbf{r}_n)^2 / a^2 \rangle - 1.$$

This is a simultaneous equation for the unknown parameters λ_n for which we can not find an analytical solution. However, an examination of the structure of the matrix Q leads to the following properties of λ_n, which satisfy the above equation: $\lambda_1 = \lambda_N \neq \lambda_2 = , \ldots, = \lambda_{N-1}$. For our purposes, it suffices if λ_n can be chosen so that $\langle \mathbf{u}^2(s) \rangle = 1$ and other conformational properties are reproduced. If all λ_n are equal to each other, as is the case for the flexible chain, then the chain described by the probability weight in Eq. (8) shows inhomogeneity; i.e., the chain fluctuates more strongly at both ends than elsewhere. Having recognized the translational asymmetry in the problem of a semiflexible chain, it is convenient to rewrite λ_n as follows: $\lambda_1 = \lambda_N = \lambda + \delta/a$, $\lambda_n = \lambda(2 \leq n \leq N-1)$. With these simplifications, the weight for the semiflexible chain at the level of a stationary phase approximation becomes

$$\Psi_{MF}[\mathbf{u}(s)] \propto \exp\left[-\lambda \int_0^L ds\mathbf{u}^2(s) - \frac{l_p}{2} \int_0^L ds\left(\frac{\partial \mathbf{u}}{\partial s}\right)^2 - \delta(\mathbf{u}_0^2 + \mathbf{u}_L^2)\right]. \quad (17)$$

The functional in Eq. (17) is identical in form to that proposed by LNN. The explicit expression for det Q and thus the stationary point conditions for λ and δ can be obtained by setting a recursion relation in N. Alternatively, we can exploit an analogy between the path integral in Eq. (17) and the harmonic oscillator in quantum mechanics [23]. The propagator of a harmonic oscillator of a mass l_p and a frequency $\Omega = \sqrt{2\lambda/l_p}$, denoted by $Z(\mathbf{u}_0, \mathbf{u}_L; L)$, is given by

$$Z(\mathbf{u}_0, \mathbf{u}_L; L) = \left(\frac{2\pi \sinh \Omega L}{\Omega l_p}\right)^{3/2} \exp\left[-\frac{(\mathbf{u}_L^2 + \mathbf{u}_0^2)\cosh \Omega L - 2\mathbf{u}_0 \cdot \mathbf{u}_L}{\Omega l_p/2 \cdot \sinh \Omega L}\right]. \quad (18)$$

We can thus rewrite the free energy as

$$\mathcal{F}[\lambda, \delta] = -\ln \int d\mathbf{u}_0\, d\mathbf{u}_L\, e^{-\delta(\mathbf{u}_2^0 + \mathbf{u}_2^L)}Z(\mathbf{u}_0, \mathbf{u}_L; L) - (L\lambda + 2\delta) + \text{const}$$

$$= \frac{3}{2}\ln\left[\left(\delta \sinh L\sqrt{\frac{2\lambda}{l_p}} + \sqrt{\frac{\lambda l_p}{2}}\cosh L\sqrt{\frac{2\lambda}{l_p}}\right)^2 - \frac{\lambda l_p}{2}\right]$$

$$-\frac{3}{2}\ln\sqrt{\frac{\lambda l_p}{2}} - \frac{3}{2}\ln\left(\sinh L\sqrt{\frac{2\lambda}{l_p}}\right) - (L\lambda + 2\delta) + \text{const} \quad (19)$$

where we have used

$$\int_{-\infty}^{\infty}\int_{-\infty}^{\infty} dx\, dy\, e^{-p^2(x^2+y^2)\pm qxy} = \frac{\pi}{\sqrt{p^4 - \frac{1}{4}q^2}}. \quad (20)$$

A little algebra leads to the following stationary phase conditions for λ and δ, which require $(\partial/\partial\lambda)\mathcal{F} = 0 = (\partial/\partial\delta)\mathcal{F}$:

$$\sqrt{\frac{\lambda l_p}{2}} = \delta = \frac{3}{4}. \quad (21)$$

Note here that the values of λ and δ do not depend on the contour length L of a chain. Since the stationary phase condition is imposed on λ, only one of l_p or λ is independent.

To understand the features implied by the weight in Eq. (17), let us compute the correlation $\langle\mathbf{u}(s) \cdot \mathbf{u}(s')\rangle$. Using the Markovian property of $\mathbf{u}$, the correlation can be computed as

$$\langle\mathbf{u}(s) \cdot \mathbf{u}(s')\rangle = \mathcal{N}^{-1} \int \mathcal{D}[\mathbf{u}]\mathbf{u}(s) \cdot \mathbf{u}(s')\Psi_{MF}[\mathbf{u}]$$

$$= \mathcal{N}^{-1} \int d\mathbf{u}_0\, d\mathbf{u}_s\, d\mathbf{u}_{s'}\, d\mathbf{u}_L\, e^{-\delta(\mathbf{u}_0^2 + \mathbf{u}_L^2)}Z(\mathbf{u}_L, \mathbf{u}_{s'}; L - s')$$

$$\times Z(\mathbf{u}_{s'}, \mathbf{u}_s; s' - s)\mathbf{u}(s) \cdot \mathbf{u}(s')Z(\mathbf{u}_s, \mathbf{u}_0; s)$$

$$= \frac{\partial}{\partial\alpha}\bigg|_{\alpha=0} \ln[\int d\mathbf{u}_s \, \mathbf{u}_{s'} e^{-\delta(\mathbf{u}_s^2 + \mathbf{u}_{s'}^2) + \alpha \mathbf{u}_s \cdot \mathbf{u}_{s'}} Z(\mathbf{u}_{s'}, \mathbf{u}_s; s' - s)]$$

$$= \exp(-|s' - s|/l_0) \tag{22}$$

where $\mathcal{N}$ is the normalization constant given by $\mathcal{N} = \int \mathcal{D}[\mathbf{u}] \, \Psi_{\mathrm{MF}}[\mathbf{u}]$ and $l_0 \equiv \frac{2}{3}l_p$. A direct consequence of the above correlation is $\langle \mathbf{u}^2 \rangle = 1$, and thus the constraint $\mathbf{u}^2 = 1$ is enforced only on an average in the mean field model of semiflexible chains with the weight given in Eq. (17). The above correlation can be compared with the one in Eq. (10) obtained with the exact weight. A comparison of Eq. (10) and Eq. (22) shows that the persistence length for the approximate model for stiff chains (cf. Eq. (17)) is smaller by a factor of $\frac{2}{3}$. The plausible reason for this is the following: In the original model the constraint condition is $\mathbf{u}^2(s) = 1$ for all values of s. The model obtained by enforcing the global condition $\langle \mathbf{u}^2 \rangle = 1$ allows for unrestricted (restricted only on an average) fluctuations in $\mathbf{u}$, thus allowing for configurations that would be prohibited by the restricted condition $\mathbf{u}^2(s) = 1$. Thus we would expect l_0 to be less than l_p.

Winkler *et al.* [19] have obtained exactly the same result (see their Eq. (4.18)) using the maximum entropy principle. Since they also only enforce the constraint on an average, the resulting theory (as described here) should be viewed as mean-field-like. If l_0 is understood as a new definition of the persistence length, then the stationary phase weight in Eq. (17) predicts the same conformational behavior as the exact one in Eq. (9). If dimension d is large enough, we expect the effect of fluctuations around the mean value to become minimal. It is known that the stationary phase approach described here becomes exact as $d \to \infty$ [21]. Note that for a chain described by this weight with $\delta = 0$, the above correlation holds only for $0 \ll s, s' \ll L$. This is because of the excess end fluctuations in this case.

1.2.3 CLOSED STIFF CHAINS

For practical purposes it is more convenient to use a translationally symmetric model for a semiflexible chain, i.e., the one described by $\Psi_{\mathrm{MF}}[\mathbf{u}(s)]$ in Eq. (17) with $\delta = 0$. For ring polymers for which the periodic condition $\mathbf{u}(0) = \mathbf{u}(L)$ and the closure relation $\int_0^L \mathbf{u}(s)ds = 0$ are imposed, we expect all the $\lambda(s)$'s to be equal or $\lambda(s)$ to be independent of s. The free energy in this case can be written as

$$\mathcal{F}[\lambda] = -\ln \int_{\mathbf{u}_0 = \mathbf{u}_L} \mathcal{D}[\mathbf{u}(s)]\delta(\int_0^L ds \; \mathbf{u}(s)) \Psi_{\mathrm{MF}}[\mathbf{u}(s)]|_{\delta = 0} - L\lambda. \qquad (23)$$

To compute this integral, it is convenient to introduce the Fourier transform

$$\mathbf{u}(s) = \sum_{n=-\infty}^{\infty} \mathbf{u}_n \exp\left[i\frac{2\pi s n}{L}\right]. \qquad (24)$$

Completeness of this expansion is reflected in the closure condition

$$\frac{1}{L}\int ds \exp\left[i\frac{2\pi s}{L}(n' - n)\right] = \delta_{nn'}. \qquad (25)$$

In terms of the new variables $\mathbf{u}_n$, the weight can be written as

$$\Psi_{\mathrm{MF}}\{\mathbf{u}_n\} = \exp\left\{-\sum_{n\neq 0}\left[\frac{l_p}{2}\frac{(2\pi n)^2}{L} + \lambda L\right]\mathbf{u}_n \cdot \mathbf{u}_{-n}\right\} \qquad (26)$$

where the contribution from $n = 0$ is excluded to incorporate the closure condition $\int_0^L \mathbf{u}(s)ds = 0$. Consequently, free energy becomes

$$\mathcal{F}[\lambda] = -\ln \int \prod_{n\neq 0} d\mathbf{u}_n \left|\det \frac{\partial \mathbf{u}(s)}{\partial \mathbf{u}_n}\right| \Psi_{\mathrm{MF}}\{\mathbf{u}_n\} - L\lambda. \qquad (27)$$

Since the transformation in Eq. (24) is linear, the Jacobi determinant $|\det (\partial \mathbf{u}(s)/\partial \mathbf{u}_n)|$ is independent of $\mathbf{u}_n$. Furthermore, it does not depend on λ. Its value is thus unimportant and will give rise to an additive constant to the free energy:

$$\mathcal{F}[\lambda] = -\ln \int \prod_n d\mathbf{u}_n \Psi_{\mathrm{MF}}\{\mathbf{u}_n\} - L\lambda + \mathrm{const.}$$

$$= \frac{3}{2}\sum_{n\neq 0}\ln\left(1 + \frac{\lambda L^2}{2l_p n^2 \pi^2}\right) + \mathrm{const.}$$

$$= 3\ln\left(\sinh L\sqrt{\frac{\lambda}{2l_p}}\right) - \frac{3}{2}\ln(L\lambda) - L\lambda + \mathrm{const.} \qquad (28)$$

where const refers to λ-independent terms. In the last step of the above equation, we have used

$$\ln \sinh x = \ln x + \sum_{n=1}^{\infty}\ln\left(1 + \frac{x^2}{n^2\pi^2}\right). \qquad (29)$$

Now the stationary phase condition reads

$$\sqrt{\frac{l_p \lambda}{2}} = \frac{3}{4}\left(\coth L\sqrt{\frac{\lambda}{2l_p}} - \frac{1}{L}\sqrt{\frac{2l_p}{\lambda}}\right). \tag{30}$$

Here we have an L-dependent condition for λ because only paths that satisfy the cyclic conditions, $\mathbf{u}(L) = \mathbf{u}(0)$ and $\int_0^L \mathbf{u}(s)ds$, contribute to the free energy $\mathcal{F}$. The cyclic conditions are also incorporated in the correlation of $\mathbf{u}(s)$; i.e.,

$$\langle \mathbf{u}(s') \cdot \mathbf{u}(s)\rangle = \sum_{n,n' \neq 0} \langle \mathbf{u}_n \cdot \mathbf{u}_{n'}\rangle \exp\left[i\frac{2\pi}{L}(s'n' + sn)\right]$$

$$= \frac{3}{2}\sum_{n \neq 0} \frac{\exp\left[i\frac{2\pi n}{L}(s' - s)\right]}{\lambda L + \frac{l_p}{2}\frac{(2\pi n)^2}{L}} \tag{31}$$

where $\langle \ldots \rangle$ is defined by

$$\langle \mathbf{u}_n \cdot \mathbf{u}_{n'}\rangle = \frac{\int \prod_{n \neq 0} d\mathbf{u}_n \, \mathbf{u}_n \cdot \mathbf{u}_{n'} \Psi_{\mathrm{MF}}\{\mathbf{u}_n\}}{\int \prod_{n \neq 0} d\mathbf{u}_n \Psi_{\mathrm{MF}}\{\mathbf{u}_n\}}. \tag{32}$$

Using the identity

$$\sum_{n=1}^{\infty} \frac{\cos nx}{n^2 + \alpha^2} = \frac{\pi}{2\alpha} \cdot \frac{\cosh \alpha(\pi - x)}{\sinh \alpha\pi} - \frac{1}{2\alpha^2} \tag{33}$$

we can rewrite the correlation of the unit tangent vectors in a closed form

$$\langle \mathbf{u}(s') \cdot \mathbf{u}(s)\rangle = \frac{\cosh[\Omega(L - 2|s' - s|)/2] - 2/\Omega L \cdot \sinh(\Omega L/2)}{\cosh(\Omega L/2) - 2/\Omega L \cdot \sinh(\Omega L/2)}. \tag{34}$$

Note here that $\langle \mathbf{u}(s + L) \cdot \mathbf{u}(s)\rangle = \langle \mathbf{u}^2(s)\rangle = 1$. In the limit of $L \to \infty$, however, the cyclic conditions are irrelevant. That is, the stationary phase condition and the correlation given above reduce to those of open chains. This can be checked by taking the limit $L \to \infty$ in the above equations for the ring polymers. An alternative method (without derivation) for ring segments that involves modifying the original Harris-Hearst model for open chains has been proposed [24]. Huber *et al.* have computed quasi-elastic scattering for a modified version of this model [25].

1.2.4   *LINEAR PERIODIC CHAINS*

The ring polymer described by Eq. (26) was introduced to circumvent difficulties associated with inhomogeneity in the linear chain with a uniform stationary phase value λ. According to the correlation function in Eq. (31), however, we can assume a uniform λ even for the chains with only periodic boundary conditions without violating the homogeneity of the chain. In this case, we should include the $n = 0$ contribution in the Fourier modes; the free energy is thus given by

$$\mathcal{F}[\lambda] = 3 \ln\left(\sin L \sqrt{\frac{\lambda}{2l_p}} \right) - L\lambda + \text{const.} \tag{35}$$

This leads to the stationary phase condition

$$\sqrt{\frac{l_p\lambda}{2}} = \frac{3}{4} \coth L \sqrt{\frac{\lambda}{2l_p}}. \tag{36}$$

The periodic boundary condition is also incorporated in the correlation function, which reads

$$\langle \mathbf{u}(s) \cdot \mathbf{u}(s') \rangle = \frac{\cosh[(L - 2|s' - s|)\, \Omega/2]}{\cosh(\Omega L/2)}. \tag{37}$$

This correlation shows that the chain with a periodic boundary condition is homogeneous; $\langle \mathbf{u}^2(s) \rangle = 1$ for all s, $0 \leq s \leq L$. As $L \to \infty$, the periodic boundary condition is irrelevant as in the ring polymers.

An alternative and useful presentation of this calculation can be made writing the curvature term in a symmetric way. Due to the periodic boundary conditions, i.e., $\mathbf{u}(0) = \mathbf{u}(L)$ and $(\partial/\partial s)\mathbf{u}(0) = (\partial/\partial s)\mathbf{u}(L)$, the free energy functional for the semiflexible chain can be written as

$$\mathcal{F}[\lambda] = -\ln \int \mathcal{D}\mathbf{u}(s) \exp\left[-\frac{1}{2} \int_0^L \int_0^L ds\, ds'\, \mathbf{u}(s)Q(s, s')\mathbf{u}(s) \right] - \int ds\, \lambda(s) \tag{38}$$

where the operator Q is defined by

$$Q(s, s') = \left[-l_p\left(\frac{\partial}{\partial s} \right)^2 + 2\lambda(s) \right] \delta(s' - s). \tag{39}$$

Integration with respect to $\mathbf{u}(s)$ leads to

$$\mathcal{F}[\lambda] = \frac{3}{2} \operatorname{tr} \ln Q - \int ds\, \lambda(s). \tag{40}$$

The stationary phase condition can be obtained by requiring $(\partial/\partial\lambda(s))\mathcal{F} = 0$:

$$1 = \frac{3}{2}\left(\frac{2}{Q}\right)_{s,s} = \frac{3}{2}\left(\frac{1}{-\frac{l_p}{2}\frac{\partial^2}{\partial s'^2} + \lambda(s')}\right)_{s,s}. \tag{41}$$

If a uniform stationary phase value λ is assumed, the above equation can be Fourier transformed. To this end, it is convenient to define a complete set of eigenstates $\{|s\rangle\}$, with s a curvilinear space label, and a complete set $\{|n\rangle\}$, Fourier-conjugate to this, such that

$$\frac{\partial}{\partial s}|n\rangle = i\frac{2\pi n}{L}|n\rangle.$$

Thus, $|s\rangle$ and $|n\rangle$ are related with each other through

$$\langle n|s\rangle = \frac{1}{\sqrt{L}}\exp\left[i\frac{2\pi ns}{L}\right].$$

With the aid of these, the right-hand side of Eq. (41) can be written as

$$\frac{3}{2}\left(\frac{1}{-\frac{l_p}{2}\frac{\partial^2}{\partial s'^2} + \lambda}\right)_{s,s} = \frac{3}{2}\langle s|\left(\frac{1}{-\frac{l_p}{2}\frac{\partial^2}{\partial s'^2} + \lambda}\right)|s\rangle$$

$$= \frac{3}{2}\sum_{n=-\infty}^{\infty}\langle s|\left(\frac{1}{-\frac{l_p}{2}\frac{\partial^2}{\partial s'^2} + \lambda}\right)|n\rangle\langle n|s\rangle$$

$$= \frac{3}{2}\sum_{n=-\infty}^{\infty}\frac{1}{\lambda L + \frac{l_p}{2}\frac{(2\pi n)^2}{L}}. \tag{42}$$

Combined with the identity in Eq. (33), the stationary phase condition relation can be rewritten as

$$\sqrt{\frac{l_p\lambda}{2}} = \frac{3}{4}\coth L\sqrt{\frac{\lambda}{2l_p}}. \tag{43}$$

This condition becomes the same one as in Eq. (21). The model described by Eq. (35) with $L \to \infty$ leads to the same result for the stationary phase condition and correlation of $\mathbf{u}(s)$ as the one in Eq. (17). In Section 1.4, we adopt this approach to examine the elastic response of a semiflexible chain under tension.

1.2.5   OPERATOR REPRESENTATIONS

For periodic chains, we can exploit the analogy with quantum mechanics and use an operator representation of the free energy. This is especially useful as $L \to \infty$, which allows the use of ground state dominance approximation. If we interpret $\hat{\mathbf{p}} \equiv l_p (\partial/\partial s)\mathbf{u}$ as the momentum operator and $\hat{\mathbf{u}}$ as the position operator such that they satisfy the commutation relation $[\hat{p}_j, \hat{u}_k] = -i\delta_{jk}$, the free energy of the periodic chain can be written as [26]

$$\mathcal{F}[\lambda] = -\ln \text{tr } e^{-L\hat{\mathcal{H}}[\lambda]} - L\lambda \qquad (44)$$

where the Hamiltonian $\hat{\mathcal{H}}$ is defined by

$$\hat{\mathcal{H}}[\lambda] = \hat{\mathbf{p}}^2\backslash 2l_p + \lambda \hat{\mathbf{u}}^2. \qquad (45)$$

This problem is equivalent to a three-dimensional quantum harmonic oscillator moving in imaginary time $s = -it$; if we denote the energy eigenvalues of the Hamiltonian in Eq. (45) by $E_\mathbf{n}$ with $\mathbf{n} = (n_1, n_2, n_3)$ and $n_i = 0, 1, 2, \ldots$, then

$$\text{tr } e^{-L\hat{\mathcal{H}}[\lambda]} = \sum_{n_1,n_2,n_3=0}^{\infty} e^{-LE_\mathbf{n}}$$

$$= \left(\sum_{0}^{\infty} e^{-L\Omega(n+1/2)} \right)^3$$

$$= \left(\frac{1}{2 \sinh\left(\frac{1}{2}\Omega L\right)} \right)^3. \qquad (46)$$

This calculation gives the same stationary phase condition as before.

This approach is especially useful as $L \to \infty$. For large L, we can use the following formula [26]

$$e^{-L\hat{\mathcal{H}}} = e^{-LE_0}[|0\rangle\langle 0| + O(e^{-L(E_1-E_0)})] \qquad (47)$$

where $E_1 = \frac{3}{2}\Omega$ and $E_1 = \frac{5}{2}\Omega$. The ground state denoted by $|0\rangle$ corresponds to the lowest eigenvalue E_0 of $\hat{\mathcal{H}}$ and is assumed to be unique and separated by a gap from E_1. As $L \to \infty$, this expression is dominated by the ground state. The term tr $\exp(-L\hat{\mathcal{H}})$ can be easily computed to yield the stationary phase condition, i.e., $\sqrt{\lambda l_p/2} = \frac{3}{4}$. Introducing $\hat{\mathbf{U}}(s)$, the interaction representation of the operator $\hat{\mathbf{u}}$,

$$\hat{\mathbf{U}}(s) = e^{-s\hat{\mathcal{H}}} \hat{\mathbf{u}} e^{s\hat{\mathcal{H}}} \qquad (48)$$

we can express the correlation of $\mathbf{u}$ as

$$\langle \mathbf{u}(s) \cdot \mathbf{u}(s') \rangle = \frac{\text{tr}[\hat{\mathbf{U}}(s) \cdot \hat{\mathbf{U}}(s') c e^{-L\hat{\mathcal{H}}}]}{\text{tr } e^{-L\hat{\mathcal{H}}}}, \quad 0 \le s \le s' \le L. \tag{49}$$

As $L \to \infty$, the ground state is dominant in the above expression contribution:

$$\langle \mathbf{u}(s) \cdot \mathbf{u}(s') \rangle = \langle 0 | \mathcal{T} \hat{\mathbf{U}}(s) \cdot \hat{\mathbf{U}}(s') | 0 \rangle \tag{50}$$

where $\mathcal{T}$ is a time-ordering operator. In general, the correlation functions of the statistical analog correspond with the time-ordered products of the corresponding quantum fields. If we use the basis $|\mathbf{n}\rangle$, in which $\hat{\mathcal{H}}$ is diagonal, the correlation can be further simplified to yield

$$\langle \mathbf{u}(s') \cdot \mathbf{u}(s) \rangle = \langle 0 | \hat{\mathbf{u}} e^{-|s'-s|\hat{\mathcal{H}}} \mathbf{u} e^{|s'-s|\hat{\mathcal{H}}} | 0 \rangle$$

$$= \sum_{\mathbf{n}} |\langle 0 | \hat{\mathbf{u}} | \mathbf{n} \rangle|^2 \, e^{-(E_{\mathbf{n}} - E_0)|s'-s|}$$

$$= e^{-(E_1 - E_0)|s'-s|}$$

$$= e^{-|s'-s|/l_0}. \tag{51}$$

As we have seen, the operator representation of the free energy is especially useful for a very long chain. In general, we can make a ground state dominance approximation for a long chain as long as the Hamiltonian of the statistical analog $\hat{\mathcal{H}}$ is Hermitian bounded below and has a discrete energy spectrum. In this case, the free energy per unit length is approximately equal to the ground state energy of $\hat{\mathcal{H}}$. The calculation of the correlation function reduces to the calculation of transition amplitudes $\langle 0 | \hat{\mathbf{u}} | \mathbf{n} \rangle$. For an ideal semiflexible chain, $\hat{\mathbf{u}}$ connects $\langle 0 |$ with the first excited state with nonvanishing transition amplitude, resulting in the correlation function given in Eq. (51).

1.3   End-to-end Distribution Function

There have been several studies over the years that have focused on the distribution of end-to-end distance of semiflexible chains. Most of these studies have started by considering the chains near the rod limit and have computed corrections in powers of t^{-1}, where t is the ratio of the bare persistence length to the contour length. In general, the calculations have been done only to low orders in t^{-1}. These calculations are quite involved; more important, they do not provide reliable results in the interesting cases, such as when t is on the order of unity. In a recent paper, Wilhelm and Frey [27] have reported analytical and numerical (Monte Carlo) calculations for the radial distribution function for a range of values t. The analytical expressions are given in terms of an infinite series involving the second Hermite polynomials for chains in three dimensions. The mean-field-like theory presented in the previous section can be used to derive a very simple expression for the distribution function of end-to-end distance.

The distribution function of the end-to-end distance R is

$$G(R; L) = \left\langle \delta \left(\mathbf{R} - \int_0^L \mathbf{u}(s) ds \right) \right\rangle \tag{52}$$

where the average is evaluated as

$$\langle \ldots \rangle = \frac{\int \mathcal{D}[\mathbf{u}(s)] \ldots \Psi_{\mathrm{MF}}[\mathbf{u}(s)]}{\int \mathcal{D}[\mathbf{u}(s)] \Psi_{\mathrm{MF}}[\mathbf{u}(s)]}. \tag{53}$$

The weight $\Psi[\mathbf{u}(s)]$ is (see Eq. (9))

$$\Psi[\mathbf{u}(s)] \propto \exp \left[-\frac{l_\mathrm{p}}{2} \int_0^L ds \left(\frac{\partial \mathbf{u}}{\partial s} \right)^2 \right] \prod_{0 \leq s \leq L} \delta(\mathbf{u}^2(s) - 1). \tag{54}$$

Following the previous section, at the level of the stationary phase approximation, the weight can be replaced by

$$\Psi_{\mathrm{MF}}[\mathbf{u}(s)] \propto \exp \left[-\frac{l_\mathrm{p}}{2} \int_0^L ds \left(\frac{\partial \mathbf{u}}{\partial s} \right)^2 - \lambda \int_0^L ds \, \mathbf{u}^2(s) - \delta(\mathbf{u}_0^2 + \mathbf{u}_L^2) \right] \tag{55}$$

where $\mathbf{u}_0 = \mathbf{u}(0)$ and $\mathbf{u}_L = \mathbf{u}(L)$. The parameters λ and δ, which are used to enforce the constraint $\mathbf{u}^2(s) = 1$ by a global constraint $\langle \mathbf{u}^2(s) \rangle = 1$, are determined variationally. From the result in Section 1.2.2, it is clear that due to chain fluctuations at the ends the global constraint $\langle \mathbf{u}^2(s) \rangle$ cannot be imposed using only one parameter as suggested by others [21]. The variational solutions for λ and δ depend on the problem under consideration. For the calculation of $G(R; L)$ the optimal values of λ and δ are different from those obtained in Section 1.2.

We calculate $G(R; L)$ by replacing the true weight in Eq. (54) by $\Psi_{\mathrm{MF}}[\mathbf{u}(s)]$ as given in Eq. (55). The equation for $G(R; L)$ with the weight given by $\Psi_{\mathrm{MF}}[\mathbf{u}(s)]$ is

$$G(R; L) = \mathcal{N}^{-1} \int_{-i\infty}^{i\infty} \frac{d\mathbf{k}}{(2\pi)^3} \int d\lambda \, d\delta \int \mathcal{D}[\mathbf{u}(s)] \, e^{-\mathbf{k} \cdot (\mathbf{R} - \int_0^L ds \, \mathbf{u}(s))} \Psi_{\mathrm{MF}}[\mathbf{u}(s)] \tag{56}$$

where $\mathcal{N}$ is an appropriate normalization constant. The functional integral over $\mathbf{u}(s)$ in Eq. (56) is done by replacing $\mathbf{u}(s)$ by $\mathbf{u} - (\mathbf{k}/2\lambda)$. The resulting path integral corresponds to a harmonic oscillator that makes a transition from $\mathbf{u}_0 -$

$(\mathbf{k}/2\lambda)$ to $\mathbf{u}_L - (\mathbf{k}/2\lambda)$ in "time" L. Using the standard result for a harmonic oscillator propagator, the distribution function $G(R; L)$ becomes

$$G(R; L) = \int d\lambda \int d\delta \exp \{-\mathcal{F}[\lambda, \delta]\} \tag{57}$$

where

$$\mathcal{F}[\lambda, \delta] = \frac{3}{2}\left[\ln\left(\frac{L\lambda + 2\delta}{4\lambda^2}\right) + \ln\left(\frac{\sinh \Omega L}{\Omega L}\right) + \ln(\alpha\beta)\right] - \lambda L - 2\delta + \frac{R^2\lambda^2}{\lambda L - 2\delta} \cdot \frac{\gamma}{\beta}, \tag{58}$$

with

$$\alpha = \delta + l_{\mathrm{p}}\Omega \coth\left(\frac{\Omega L}{2}\right), \tag{59}$$

$$\beta = \delta + \frac{l_{\mathrm{p}}\Omega}{2} \tanh\left(\frac{\Omega L}{2}\right) + \frac{\delta^2}{\lambda L - 2\delta}, \tag{60a}$$

$$\gamma = \delta + \frac{l_{\mathrm{p}}\Omega}{2} \tanh\left(\frac{\Omega L}{2}\right), \tag{60b}$$

and

$$\Omega = \sqrt{\frac{2\lambda}{l_{\mathrm{p}}}}. \tag{61}$$

We evaluate the integrals over λ and δ (see Eq. (57)) by the stationary phase approximation. This approximation replaces the constraint $\mathbf{u}^2(s) = 1$ by a global constraint $\langle \mathbf{u}^2(s)\rangle = 1$. If the global constraint is enforced using only one variational parameter, as suggested elsewhere [21], then one simply obtains the incorrect Gaussian expression for $G(R; L)$. The presence of the parameter δ accounts for the suppression of the fluctuations of the ends of the chain. The stationarity condition

$$\frac{\partial}{\partial \lambda} \mathcal{F}[\lambda, \delta] = 0 \tag{62}$$

gives, after some algebra,

$$\sqrt{\frac{\lambda l_{\mathrm{p}}}{2}} = \frac{3}{4}\left(\frac{1}{1 - \dfrac{R^2}{L^2}}\right) \tag{63}$$

and we find that δ does not show any significant variation with R. With the stationary values of λ and δ, the distribution of the end-to-end distance for the semiflexible chain in the large L limit becomes

$$G(R; L) = \text{const} \frac{1}{\left(1 - \dfrac{R^2}{L^2}\right)^{9/2}} \exp\left[-\frac{9L}{8l_p} \frac{1}{\left(1 - \dfrac{R^2}{L^2}\right)}\right]. \tag{64}$$

We showed in the previous section that within the stationary phase approach, which enforces the constraint only globally, the persistence length is reduced to $l_0 = \frac{2}{3}l_p$ instead of l_p. If we let $t = (L/l_0)$, the radial probability density in three dimensions for semiflexible chains is given by the simple expression

$$P(r; t) = 4\pi C \frac{r^2}{(1 - r^2)^{9/2}} \exp\left[-\frac{3t}{4} \frac{1}{(1 - r^2)}\right] \tag{65}$$

where $r = R/L$. The normalization constant C is determined using the condition

$$\int_0^1 dr\, P(r; t) = 1 \tag{66}$$

and is given by

$$C = \frac{1}{\pi^{3/2} e^{-\alpha} \alpha^{-3/2}\left(1 + 3\alpha^{-1} + \dfrac{15}{4}\alpha^{-2}\right)} \tag{67}$$

where $\alpha = 3t/4$.

The distribution function, $P(r; t)$, goes as r^2 as $r \to 0$ and vanishes at $r = 1$. The peak of the distribution function occurs at

$$r_{\max} = \sqrt{\frac{\eta + \sqrt{\eta^2 + 14}}{7}} \tag{68}$$

where $\eta = \frac{5}{2} - \frac{3}{4}t$. In Fig. 1.1 we plot $P(r; t)$ for the five values of t for which Wilhelm and Frey [27] have presented simulation data. We find that our simple expression in Eq. (64) almost quantitatively reproduces the data with the maximum deviation of about 10% at the peak for $t = 0.5$.

We have also calculated the first two moments of the distribution function and find that they reproduce the exact results in both the random coil limit and the rod limit. The distribution function $P(r; t)$ also has the correct limiting behavior $(\sim \delta(1 - r))$ as $t \to 0$. Thus, the mean field variational approach yields a simple

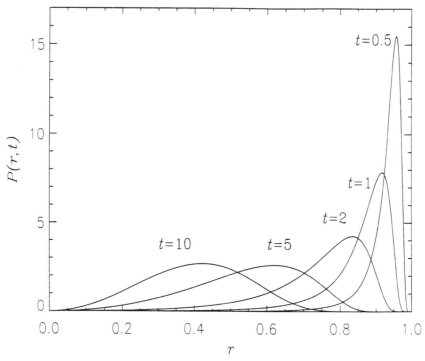

Figure 1.1   Plot of $P(r; t)$ (see Eq. (65)) as a function of r for a number of values of t. The values of t are labeled. The results of the theory are in excellent agreement with the simulation results of Wilhelm and Frey [27].

expression for $P(r; t)$ that is in quantitative agreement with simulation, and hence, can form the basis to analyze experiments.

1.4   Semiflexible Chains Under Tension

The mean field theory described in the previous section is especially useful when one encounters semiflexible chains in the presence of fields for which tractable calculations are difficult. In this section we use the variational theory to investigate the external field on the conformational properties of a semiflexible chain. Even though the analogous problems in flexible chains have been extensively studied, the corresponding problems in semiflexible chains have only recently attracted much attention [27–33]. The earliest theoretical paper dealing with this problem was initiated by Fixman and Kovacs [28]. These authors used a modified version of the Gaussian model for stiff chains and provided expressions for the stretching as a function of applied force. Their treatment is only valid when the applied force is small, and significant deviation from these predictions are observed at sufficiently large values of the external force. Marko and Siggia [32, 33] have calculated the extension as a function of force for wormlike chains and found that their results fit the experimental data very well. Some aspects of this theory have also been considered by Odijk [29], who also discusses the competition between entropically dominated effects and elasticity effects. The mean field approach adopted here is, we believe, more general than those adopted in the literature.

Gaussian chains can be arbitrarily extended under tension f as implied by Hook's law of elastic response, i.e., $R \sim L/f$; real chains cannot be extended beyond the contour length L. As the magnitude of tension exerted on both ends of a semiflexible chain increases, one expects an interesting crossover to occur from Hook's limit to the fully extended limit ($R = L$). Thus, the problem of stretched semiflexible chains entails a competition between entropy-dominated effects and ordering due to the external field $\mathbf{f}$. We will show that the mean field treatment quantitatively reproduces experiments on DNA [5], thus providing further evidence for the validity of a mean field approach.

1.4.1 STATIONARY PHASE CONDITION

For mathematical convenience, we will adopt the translationally symmetric chain model introduced in Eq. (38). Each segment in a semiflexible chain is assumed to be stretched by the tangential force $\mathbf{f}(s)$, which tends to suppress chain fluctuations. The weight for the case of semiflexible chains in the presence of the stretching is

$$\Psi[\mathbf{u}(s), \lambda(s)] \propto \exp\left[-\frac{1}{2}\int_0^L\int_0^L ds\,ds'\,\mathbf{u}(s)Q(s,\,s')\mathbf{u}(s') + \int_0^L \mathbf{u}(s)\cdot\mathbf{f}(s)ds\right] \quad (69)$$

where Q is the same operator as introduced in Eq. (39), i.e.,

$$Q(s,\,s') = \left[-l_p\left(\frac{\partial}{\partial s}\right)^2 + 2\lambda\right]\delta(s' - s). \quad (70)$$

The last term in the exponent is the energy penalty for chain conformations that are not parallel to $\mathbf{f}(s)$. The free energy can be written as

$$\exp(-F/k_B T) \propto \int_{-i\infty}^{i\infty} \mathcal{D}[\lambda(s)] \int \mathcal{D}[\mathbf{u}(s)] \, e^{\int_0^L \lambda(s)ds} \, \Psi[\mathbf{u}(s), \lambda(s)]$$

$$\propto \int_{-i\infty}^{i\infty} \mathcal{D}[\lambda(s)] \exp\{-\mathcal{F}[\lambda(s), \mathbf{f}(s)]\} \tag{71}$$

where $\mathcal{F}$ is the free energy functional given by

$$\mathcal{F}[\lambda(s), \mathbf{f}(s)] = -\ln \int \mathcal{D}[\mathbf{u}(s)] \Psi[\mathbf{u}(s), \lambda(s)] - \int_0^L \lambda(s)ds + \text{const.} \tag{72}$$

Integration with respect to $\mathbf{u}(s)$ yields

$$\mathcal{F}[\lambda(s), \mathbf{h}(s)] = \frac{3}{2} \operatorname{tr} \ln Q - \frac{1}{2} \int_0^L \int_0^L \mathbf{f}(s) \, Q^{-1}(s, s') \mathbf{f}(s') - \int_0^L \lambda(s)ds + \text{const.} \tag{73}$$

Note that $\mathcal{F}[\lambda(s), \mathbf{f}(s)]$ is the generating functional for the connected correlation function defined by

$$\langle \mathbf{u}(s) \cdot \mathbf{u}(s') \rangle_c = -\frac{\delta}{\delta \mathbf{f}(s)} \cdot \frac{\delta \mathcal{F}}{\delta \mathbf{f}(s')}$$

$$= \frac{\partial}{\delta \mathbf{f}(s)} \cdot \langle \mathbf{u}(s') \rangle$$

$$= \langle \mathbf{u}(s) \cdot \mathbf{u}(s') \rangle - \langle \mathbf{u}(s) \rangle \cdot \langle \mathbf{u}(s') \rangle. \tag{74}$$

As $\mathbf{f}$ goes to zero, $\langle \mathbf{u}(s) \rangle$ vanishes. In this case, the distinction between the connected correlation function and the correlation function disappears. Performing the functional differentiations in Eq. (46), we get

$$\langle \mathbf{u}(s) \cdot \mathbf{u}(s') \rangle_c = 3Q^{-1}(s, s') \tag{75}$$

and

$$\langle \mathbf{u}(s) \rangle = -\frac{\delta \mathcal{F}}{\delta \mathbf{f}(s)} = \int_0^L Q^{-1}(s, s') \mathbf{f}(s'). \tag{76}$$

Integrations of each term in Eq. (74) with respect to s and s' lead to the following expression for ΔR^2:

$$\Delta R^2 = \langle R^2 \rangle - \langle R \rangle^2$$

$$= \int_0^L \int_0^L \langle \mathbf{u}(s) \cdot \mathbf{u}(s') \rangle \, ds \, ds' - \int_0^L \int_0^L \langle \mathbf{u}(s) \rangle \cdot \langle \mathbf{u}(s') \rangle ds \, ds'$$

$$= \int_0^L \int_0^L \frac{\delta}{\delta f(s)} \cdot \frac{\delta \mathcal{F}}{\delta f(s')} ds\, ds'. \tag{77}$$

All relevant quantities can be thus expressed in terms of the free energy functional $\mathcal{F}[\lambda(s), f(s)]$.

Following the general formalism described in Section 1.2, in the case of ideal semiflexible chains the free energy is approximated by the minimum value of the corresponding free energy functional that occurs along the stationary phase trajectory λ. The stationary phase condition is obtained by requiring $[\delta/\delta\lambda(s)]\mathcal{F} = 0$. To this end, we first note that

$$\frac{\delta}{\delta\lambda(s)} \operatorname{tr} \ln Q = \frac{\delta}{\delta\lambda(s)} \sum_{s'} (\ln Q)(s', s')$$

$$= 2Q^{-1}(s, s) \tag{78}$$

and

$$\frac{\delta}{\delta\lambda(s)} Q^{-1}(s', s'') = 2Q^{-1}(s', s)Q^{-1}(s, s''). \tag{79}$$

The stationary phase condition leads to

$$1 = \frac{3}{2}\left(\frac{2}{Q}\right)_{s,s} + \int_0^L \int_0^L ds'\, ds''\, f(s')\, Q^{-1}(s', s)Q^{-1}(s, s'')f(s'').$$

The stationary phase value $\lambda(s)$ thus depends on the value of the external force $f(s)$. The simplest but nontrivial case corresponds to the case of $f(s) = f = $ const. It is the case of stretching of semiflexible chains under a constant value of f that is appropriate to the recent experiments on DNA [5]. If f is uniform, then so is λ. In this case, the stationary phase condition reduces to

$$1 = \frac{3}{2}\left(\frac{1}{-\frac{l_p}{2}\left(\frac{\partial}{\partial s'}\right)^2 + \lambda}\right)_{s,s'} + \frac{f^2}{4\lambda^2}$$

$$= \frac{3}{2}\sum_{-\infty}^{\infty} \frac{1}{\lambda L + \frac{l_p}{2}\frac{(2\pi n)^2}{L}} + \frac{f^2}{4\lambda^2}. \tag{81}$$

With the aid of Eq. (33), this can be written in a closed form

$$1 = \frac{3}{4}\sqrt{\frac{2}{l_p\lambda}} \coth\left(\frac{1}{2}\Omega L\right) + \frac{f^2}{4\lambda^2} \tag{82}$$

where $\Omega = \sqrt{2\lambda/l_p}$. As $L \to \infty$, this equation can be further simplified to yield

$$1 - \frac{3}{4}\sqrt{\frac{2}{l_p\lambda}} = \frac{f^2}{4\lambda^2}. \tag{83}$$

As $f \to 0$, this equation reduces to the earlier one in Eq. (21), as expected.

1.4.2 CONFORMATION OF A SEMIFLEXIBLE CHAIN UNDER TENSION

In the case of uniform $\mathbf{f}$, we obtain a simple expression for the correlation function:

$$\langle \mathbf{u}(s) \cdot \mathbf{u}(s') \rangle = 3Q^{-1}(s, s') + \frac{f^2}{4\lambda^2}$$

$$= \frac{3}{2} \sum_{-\infty}^{\infty} \frac{\exp\left[i\dfrac{2\pi n}{L}|s' - s|\right]}{\lambda L + \dfrac{l_p}{2}\dfrac{(2\pi n)^2}{L}} + \frac{f^2}{4\lambda^2}$$

$$= \frac{3}{4}\sqrt{\frac{2}{l_p\lambda}} \frac{\cosh[(L - 2|s' - s|)\Omega/2]}{\sinh(\Omega L/2)} + \frac{f^2}{4\lambda^2} \tag{84}$$

where $\Omega = \sqrt{2\lambda/l_p}$. To derive the last step from the previous one, we used Eq. (33). As $L \to \infty$, the above equation can be further simplified to yield

$$\langle \mathbf{u}(s) \cdot \mathbf{u}(s') \rangle = \frac{3}{4}\sqrt{\frac{2}{l_p\lambda}}\, e^{-|s'-s|\Omega} + \frac{f^2}{4\lambda^2}. \tag{85}$$

If we set $|s' - s|$ to zero, this equation reduces to the stationary phase condition. Thus, the free energy minimization condition in this case also amounts to requiring $\mathbf{u}^2(s) = 1$. As $\mathbf{f} \to 0$, the coefficient in the exponential function becomes 1, resulting in the same expression as in Eq. (51).

Using the correlation function in Eq. (85), the mean squared internal distance of the chain under tension can be obtained.

$$\langle |\mathbf{r}(s') - \mathbf{r}(s)|^2 \rangle = \int_s^{s'} \int_s^{s'} \langle \mathbf{u}(s_1) \cdot \mathbf{u}(s_2) \rangle ds_1\, ds_2$$

$$= \frac{3}{4}\sqrt{\frac{2}{l_p\lambda}}\left[\frac{2}{\Omega}|s' - s| - \frac{2}{\Omega^2}(1 - e^{-\Omega|s'-s|})\right] + \frac{f^2}{4\lambda^2}(s' - s)^2. \tag{86}$$

We can easily see that, for small $|s' - s|$, the first two terms on the right-hand side of Eq. (86) are dominant. As $|s' - s|$ becomes larger, the last term is more important. At large length scales, the chain conformation is thus mainly determined by the orienting field. We can now introduce a crossover length S_f at which these two length scales coincide. If the crossover occurs at S_f, which is somewhat larger than Ω^{-1}, then we can get a simple expression for S_f; this can be obtained by balancing the first term with the last one in the right-hand side of Eq. (86)

$$S_f = \frac{3}{4}\sqrt{\frac{2}{l_p\lambda}}\left(\frac{4}{f^2}\frac{\lambda^2}{\Omega}\right). \tag{87}$$

The condition $S_f \gg \Omega^{-1}$ is equivalent to

$$\frac{f^2}{4\lambda^2} \ll \frac{3}{4}\sqrt{\frac{2}{l_p\lambda}} = O(1)$$

or $fl_p \ll 1$. Under this condition, the stationary phase condition coincides with that for an ideal semiflexible chain. This results in

$$S_f \sim \frac{1}{f^2 l_p}. \tag{88}$$

The longitudinal size corresponding to S_f is called a *tensile screening length* [34] and is denoted by ξ_f. By noting that parts of the chain within the length scale S_f resemble an ideal chain, we can have

$$\xi_f \sim f^{-1}. \tag{89}$$

The condition $(f^2/4\lambda^2) \ll 1$ for the above expression to be valid in turn amounts to requiring $fl_p \ll 1$ or, equivalently, $\xi_f \gg l_p$.

It is interesting to note that the results for S_f and ξ_f given above are the same as those for a Gaussian chain under tension. This can be understood as follows: The condition for the above results to be valid, $\xi_f \gg l_p$, is equivalent to saying that the tensile screening length ξ_f contains a large number of chain segments of length l_p. The chain stiffness becomes marginal in determining the shape of parts of chains within the length scale ξ_f. Since the effect of $\mathbf{f}$ is important only beyond this length scale, the chain stiffness is not "coupled" to the external field $\mathbf{f}$.

As $f^2/4\lambda^2$ becomes much larger than $\frac{3}{4}\sqrt{(2/l_p\lambda)}$ or, equivalently, $fl_p \to \infty$, S_f approaches 0. This implies that, at any length scale beyond $S_f \approx 0$, the chain conformation is governed by the interaction term $-\int \mathbf{f} \cdot \mathbf{u}(s)$. As we will see below, as

$l_p f \to \infty$, the entropical contribution is not balanced by the energy penalty for crumpled conformation. That is, the chain segments are aligned with $\mathbf{f}$ as expected. These quantitative considerations, which were already implicit in the calculations of Fixman and Kovacs [28] can be made quantitative using our general formalism.

1.4.3 ELASTIC RESPONSE OF A SEMIFLEXIBLE CHAIN

Let us now consider the average elongation z defined by

$$z = \left\langle \int_0^L ds\, \mathbf{u}(s) \cdot \frac{\mathbf{f}}{f} \right\rangle. \tag{90}$$

This quantity, which is experimentally measurable, shows how the chain responds to the tangential field. The statistical average in Eq. (90) can be conveniently expressed in terms of the free energy functional $\mathcal{F}$ (see Eq. (73)). For the case of uniform $\mathbf{f}$, we obtain the simple expression

$$z = \frac{\mathbf{f}}{f} \cdot \frac{\delta \mathcal{F}}{\delta \mathbf{f}} = f \int_0^L \left[-l_p \left(\frac{\delta}{\delta s} \right)^2 + 2\lambda \right]^{-1} ds$$

$$= f \frac{L}{2\lambda(f)} \tag{91}$$

where the argument in $\lambda(f)$ is introduced to emphasize the dependence of λ on f. This equation and the stationary phase condition Eq. (83) provide the average elongation with respect to f. To obtain analytical expressions, we first take two limiting cases of $l_p \to 0$ and $f \to \infty$. As $l_p \to 0$ or, more precisely, $l_p f \to 0$, the problem reduces to that of a stretched Gaussian chain (as mentioned in Section 1.4.1). In this case, we can get the elastic response relation

$$z = f \frac{R_0^2}{3} \tag{92}$$

where $R_0^2 = 2l_0 L = 2(\frac{2}{3}l_p)L$ is the size of the corresponding ideal semiflexible chain. This is the elastic response relation of a Gaussian chain under tension [35]. For the case of $f l_p \ll 1$, the stationary phase condition can be expanded in terms of f^2. Up to f^2, we get

$$\lambda(h) \approx \lambda + \frac{f^2}{2\lambda} \tag{93}$$

where λ is the stationary phase condition for $f = 0$. Thus the average elongation is given by

$$z \approx f \frac{R_0^2}{3}\left[1 - \frac{8}{9}f^2 l_0^2\right].$$ (94)

Note here that the control parameter in this expansion is $f l_p$ (or $f l_0$).

As l_p increases, the term $f^2/4\lambda^2$ in the stationary phase condition becomes important. If the value of l_p reaches the same order of magnitude as that of ξ_f, we expect the effect of the orienting field to be important over all length scales. In other words, chain fluctuations become frozen as $l_p f$ becomes large. As $f \rightarrow \infty$, the stationary phase condition has a solution at $\lambda^{-1} = 0$, resulting in $\lambda = \frac{1}{2}f$. The average elongation is thus given by $z = L$. Chain fluctuations are totally suppressed in this case. For finite but much larger f than l_p, $\lambda(f)$ can be approximated as

$$\lambda(f) = \frac{1}{2}f\left(1 + \frac{3}{4}\sqrt{\frac{1}{l_p f}}\right).$$ (95)

Accordingly, z for finite f is smaller than L, which is due to chain fluctuations.

$$z \approx L\left(1 - \frac{3}{4}\sqrt{\frac{1}{l_p f}}\right).$$ (96)

A similar result with a slightly larger numerical factor was obtained by Odijk [29], who considered a semiflexible chain near the rod limit.

Since the chain under strong elongation as described by Eq. (96) is nearly parallel to the orienting force $\mathbf{f}$, we can approximate z as

$$z = \left\langle \int_0^L \mathbf{u}(s)ds \cdot \frac{\mathbf{f}}{f} \right\rangle$$

$$= \left\langle \int_0^L \cos \theta_f(s)ds \right\rangle$$

$$= L\left[1 - \frac{1}{2}\langle \theta_f^2(s) \rangle\right]$$ (97)

where $\theta(s)$ is the angle between $\mathbf{f}$ and $\mathbf{u}(s)$. For a uniform field, we expect $\langle \theta_f^2(s) \rangle$ not to depend on s for a long chain, i.e., $\langle \theta_f^2(s) \rangle = \langle \theta_f^2 \rangle$. Comparing Eqs. (96) and (97), we can obtain the expression for this quantity:

$$\langle \theta_f^2 \rangle = \frac{3}{2}\sqrt{\frac{1}{l_p f}}.$$ (98)

This result again differs in numerical factor from the corresponding expression for $\langle \theta_f^2 \rangle$ derived by Odijk [29].

From the above analysis we see that the stationary phase approach reproduces results consistent with the literature in limiting cases. To test our theory further, we solved the stationary phase condition numerically. To compare with experimental data obtained by Smith *et al.* [5] on the stretching of DNA (see also Fig. 2 in ref. [32]), l_0 and L were chosen to be 53.4 nm and 32.80 μm, respectively. In Fig. 1.2, z/L is plotted against $\ln f$. Our numerical solution (as continuous curve) is in good quantitative agreement with the experimental data [5].

In the previous analysis, we have seen that a strong orienting field tends to suppress chain fluctuations as is implied by the correlation and the elastic response relation. This is also manifested in the ratio $\Delta R^2/\langle R^2 \rangle$. Using Eq. (77), we get

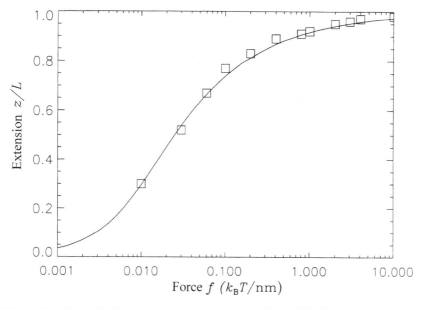

Figure 1.2   The solid line is the force-extension curve for semiflexible chains under tension obtained by solving Eqs. (83) and (91). We have taken the parameters to correspond with the experiments on DNA stretching by Smith *et al.* [5]: $L = 32.9$ μm and $l_0 = 53$ nm. The squares are experimental points as reported in ref. [5].

$$\Delta R^2 = \frac{3}{4}\sqrt{\frac{2}{l_p\lambda}}\left[\frac{2}{\Omega}L - \frac{2}{\Omega^2}(1 - e^{-\Omega L})\right]$$ (99)

and thus,

$$\frac{\Delta R^2}{\langle R^2\rangle} = \frac{\Delta R^2}{\Delta R^2 + \dfrac{f^2L^2}{4\lambda^2}}.$$ (100)

For a short chain, $\Delta R^2/\langle R^2\rangle = O(1)$ while it approaches zero as $L \to \infty$. It can be easily seen that the value of this quantity crosses over at the contour length $L \approx S_f$ from the fluctuation dominated limit ($\Delta R^2/\langle R^2\rangle = O(1)$) to the fluctuation suppressed limit ($\Delta R^2/\langle R^2\rangle \approx 0$). As $l_p f \to \infty$, S_f approaches zero. Thus, in this case, chain fluctuation becomes totally suppressed for a chain with contour length $L \gg S_f \approx 0$.

1.4.4   OPERATOR REPRESENTATION

Just as in the case of an ideal semiflexible chain, it is instructive to use the operator representation of the free energy functional $\mathcal{F}[\lambda, \mathbf{f}]$. The corresponding Hamiltonian in this case describes a quantum harmonic oscillator in the linear force field described by $\mathbf{f}$:

$$\hat{\mathcal{H}}[\lambda,\mathbf{f}] = \hat{\mathcal{H}}[\lambda] + \Delta\hat{\mathcal{H}}$$

$$\equiv \frac{\hat{\mathbf{p}}^2}{2l_p} + \lambda\hat{\mathbf{u}}^2 - \hat{\mathbf{u}}\cdot\mathbf{f}$$

$$= \frac{\hat{\mathbf{p}}^2}{2l_p} + \lambda\left(\hat{\mathbf{u}} - \frac{\mathbf{f}}{2\lambda}\right)^2 - \frac{f^2}{4\lambda}.$$ (101)

Since $[\hat{\mathbf{p}}, \mathbf{f}] = 0$ and thus the commutator $[\hat{p}_j, \hat{u}_k] = -i\delta_{jk}$ is invariant under $\hat{\mathbf{u}} \to \hat{\mathbf{u}} - \mathbf{f}/2\lambda$, the external field just shifts eigenvalues of $\hat{\mathcal{H}}[\lambda, \mathbf{f}]$ by the same amount $-f^2/4\lambda$. This results in

$$\mathrm{tr}\, e^{-L\hat{\mathcal{H}}[\lambda,\mathbf{f}]} = \left(\frac{1}{2\sinh\left(\dfrac{1}{2}\Omega L\right)}\right)^3 \exp\left(\frac{f^2}{4\lambda}L\right).$$ (102)

This leads to the same stationary phase condition as in Eq. (83).

The correlation function of $\mathbf{u}(s)$ also can be computed following steps similar to those leading to Eq. (51).

$$\langle \mathbf{u}(s) \cdot \mathbf{u}(s') \rangle = \sum_{\mathbf{n}} \left| \left\langle 0 \left| \left(\mathbf{u} + \frac{\mathbf{f}}{2\lambda} \right) \right| \mathbf{n} \right\rangle \right|^2 e^{-(E_{\mathbf{n}} - E_0)|s' - s|}$$

$$= |\langle 0 | \hat{\mathbf{u}} | \mathbf{n} \rangle|^2 e^{-(E_1 - E_0)|s' - s|} + \frac{f^2}{4\lambda^2}$$

$$= \frac{3}{4} \sqrt{\frac{2}{l_p \lambda}} e^{-|s' - s|\Omega} + \frac{f^2}{4\lambda^2}. \tag{103}$$

The operator representation again produces the same result as the functional integral formalism. One of the advantages of using the operator representation is that we can use ground state dominance approximation, provided there is an energy gap.

1.5   Semiflexible Chains in a Nematic Field

As a final example of the utility of our approach, we consider a semiflexible chain in a nematic environment. These calculations expose conditions under which the stationary phase approach is not successful. The interaction energy of a chain in a nematic field along $\mathbf{n}$ is assumed to be $-g \int_0^L [\mathbf{u}(s) \cdot \mathbf{n}]^2$, where the coupling constant g measures the strength of the nematic field. Thus, conformations of nematic polymers either parallel or antiparallel to the nematic field are equally energetically favorable. This is in contrast to the case of a stretched chain. The Hamiltonian of a single chain in the present case is given by [9, 36]

$$\frac{\mathcal{H}[g]}{k_B T} = \int_0^L ds \left\{ \frac{l_p}{2} \left(\frac{\partial \mathbf{u}}{\partial s} \right)^2 - g[\mathbf{u}(s) \cdot \mathbf{n}]^2 \right\} \tag{104}$$

where $\mathbf{u}(s)$ is defined only on a unit sphere of $|\mathbf{u}(s)| = 1$. At the mean field level this takes the form

$$\frac{\mathcal{H}[\lambda, g]}{k_B T} = \frac{1}{2} \int_0^L \int_0^L ds \, ds' \, \mathbf{u}(s) Q(s, s') \mathbf{u}(s') - g \int_0^L ds \, \mathbf{u}_z^2(s) \tag{105}$$

such that the weight is given by

$$\Psi_{MF}[\mathbf{u}(s), g] \propto \exp \left\{ \frac{-\mathcal{H}[\lambda, g]}{k_B T} \right\} \tag{106}$$

where the direction of the nematic field is chosen to be parallel to the z-axis. With the convention $\mathbf{u}_\perp = (u_x, u_y, 0)$, this can be rewritten as

$$\frac{\mathcal{H}[\mathbf{u}(s)]}{k_B T} = \frac{1}{2} \int_0^L \int_0^L ds \, ds' \, [\mathbf{u}_\perp(s)Q(s, s')\mathbf{u}_\perp(s') + u_z(s)Q_1(s, s')u_z(s')] \quad (107)$$

where

$$Q(s, s') = \left[-l_p \left(\frac{\partial}{\partial s}\right)^2 + 2\lambda\right]\delta(s' - s) \quad (108)$$

and

$$Q_1(s, s') = \left[-l_p \left(\frac{\partial}{\partial s}\right)^2 + 2(\lambda - g)\right]\delta(s' - s). \quad (109)$$

The corresponding generating functional $\mathcal{F}[\lambda, g, \mathbf{f}(s)]$, up to an additive constant, now has the form

$$\mathcal{F}[\lambda, g] = \text{tr} \ln Q - \frac{1}{2} \int_0^L \int_0^L ds \, ds' \mathbf{f}_\perp(s) Q^{-1}(s, s') \, \mathbf{f}_\perp(s')$$

$$+ \frac{1}{2} \, \text{tr} \ln Q_1 - \frac{1}{2} \int_0^L \int_0^L ds \, ds' f_z(s) \, Q_1^{-1}(s, s') f_z(s') - \int_0^L \lambda(s) ds \quad (110)$$

from which we can obtain

$$\langle \mathbf{u}_\perp(s) \cdot \mathbf{u}_\perp(s') \rangle = -\frac{\delta}{\delta \mathbf{f}_\perp(s)} \cdot \frac{\delta \mathcal{F}}{\delta \mathbf{f}_\perp(s')} + \langle \mathbf{u}_\perp(s) \rangle \cdot \langle \mathbf{u}_\perp(s') \rangle$$

$$= \frac{1}{2} \sqrt{\frac{2}{l_p \lambda}} \, e^{-|s' - s|\Omega} \quad (111)$$

and

$$\langle u_z(s) \cdot u_z(s') \rangle = -\frac{\delta}{\delta f_z(s)} \cdot \frac{\delta \mathcal{F}}{\delta f_z(s')} + \langle u_z(s) \rangle \cdot \langle u_z(s') \rangle$$

$$= \frac{1}{4} \sqrt{\frac{2}{l_p(\lambda - g)}} \, e^{-|s' - s|\Omega_1} \quad (112)$$

with $\Omega = \sqrt{2\lambda/l_p}$ and $\Omega_1 = \sqrt{2(\lambda - g)/l_p}$. In the last steps of the above equations, we set $\mathbf{f}(s)$ to zero. The stationary phase value of λ now satisfies the equality

$$1 = \frac{1}{2} \sqrt{\frac{2}{l_p \lambda}} + \frac{1}{4} \sqrt{\frac{2}{l_p(\lambda - g)}} \quad (113)$$

which ensures

$$1 = \langle \mathbf{u}_{\perp}^2(s) \rangle + \langle u_z(s) \rangle = \langle \mathbf{u}^2(s) \rangle. \tag{114}$$

It can be easily seen that the stationary phase value of λ in this case is larger than that for the case of free chains. But $(\lambda - g)$ is smaller than that for the latter case. This implies that the nematic field induces ordering along the field but decreases the persistence length perpendicular to it. That is,

$$\langle R_{\perp}^2 \rangle = \int_0^L \int_0^L \langle \mathbf{u}_{\perp}(s) \cdot \mathbf{u}_{\perp}(s') \rangle ds \, ds'$$

$$= \frac{2}{3}(2l_{\text{eff}}^{\perp} L)$$

$$< \frac{2}{3}(2l_0 L) \tag{115}$$

and

$$\langle R_z^2 \rangle = \int_0^L \int_0^L \langle u_z(s) \cdot u_z(s') \rangle ds \, ds'$$

$$= \frac{2}{3}(2 \, l_{\text{eff}}^z L)$$

$$< \frac{1}{3}(2l_0 L) \tag{116}$$

where $l_0 = \frac{2}{3}l_p$ is the persistence length of a free chain. The size of chains in the nematic field does not grow as L as $L \to \infty$, as is the case for the stretched chain within the mean field variational theory. To be more precise, let us consider the stationary phase solutions for two extreme cases. In the following derivations any quantity with (without) argument g corresponds to a nematic polymer (noninteracting polymer).

(a) *The weak nematic limit, $g \to 0$.*

For small g, we can write $\lambda(g) \approx \lambda(1 + \varepsilon)$. By assuming both ε and g small, we can have

$$\lambda(g) \approx \lambda\left(1 + \frac{2}{3}gl_0\right) \tag{117}$$

and thus,

$$\Omega(g) = \sqrt{2\lambda(g)/l_p} \approx \sqrt{2\lambda/l_p}\left(1 + \frac{2}{3}gl_0\right) \tag{118}$$

and

$$\Omega'(g) = \sqrt{2(\lambda(g) - g)/l_p} \approx \sqrt{2\lambda/l_p}\left(1 - \frac{1}{3}gl_0\right). \tag{119}$$

These lead to

$$\frac{\langle R_\perp^2 \rangle}{\frac{2}{3}(2l_0 L)} \approx 1 - \frac{1}{3}gl_0 \tag{120}$$

and

$$\frac{\langle R_z^2 \rangle}{\frac{1}{3}(2l_0 L)} \approx 1 + \frac{1}{3}gl_0. \tag{121}$$

Thus, the persistence length along the nematic field is increased by the factor of $(1 + \frac{1}{3}gl_0)$, which can be compared with $(1 + \frac{2}{3}gl_0)$ (in our notation) obtained by Warner *et al.* [9].

(b) *The strong nematic limit, $g \to \infty$.*

If we simply let $g \to \infty$ in the stationarity condition in Eq. (113), then $\lambda(g)$ and thus $\Omega(g)$ should diverge so that $(\lambda(g) - g)$ remain positive. This results in

$$\langle \mathbf{u}_\perp(s) \cdot \mathbf{u}_\perp(s') \rangle = 0$$

and

$$\langle u_z(s) \cdot u_z(s') \rangle = \exp(-|s' - s|/2l_p). \tag{123}$$

The diverging nematic field totally suppresses chain fluctuations perpendicular to it but increases the persistence length parallel to it by the factor of 2. The increase of the persistence length along the nematic field by only numerical factor, however, is *not* correct. Several authors predicted an effective persistence that grows exponentially with $\sqrt{gl_p}$ [9, 12, 36, 37]. Following ref. [36], this can be explained as follows: Since the interaction energy $-gu_z^2$, which is quadratic in u_z, has two minima for $u_z = \pm 1$, the "tunneling" probability from one minimum to the next is very small for large g. This results in the length scale that varies exponentially as gl_p over which the chain is parallel to the nematic field. (For more details, see ref. [36].)

The stationary phase free energy in Eq. (105) also contains the quadratic interaction term $-gu_z^2$. This term if added to λu_z^2 becomes $(\lambda - g)u_z^2$ with positive $(\lambda - g)$, which has only *one* minimum at $u_z = 0$. Thus the stationary phase approach does not account for the instanton solution, resulting in much smaller persistence length than the exact results. This problem illustrates the possible limitations of the stationary phase approach; if there is a possibility of broken symmetry in the problem, then the mean field variational theory should be used with caution.

1.6   Conclusions

In this chapter we have described a simple mean field variational approach to study a number of properties of intrinsically stiff chains that are appropriate models for a large class of biopolymers. The exact statistical mechanics of such systems are complicated by the constraint that the tangent vector being unity has to be enforced at all points along the contour. In the mean field approach this local constraint, which is difficult to impose, is replaced by a global one. The global constraint ensures that the condition of the tangent vector being unity is obeyed on an average. We have described the calculation of the distribution of end-to-end distance and the elastic response of stiff chains under tension using this basic methodology. In the former example, we find that the simple expression almost quantitatively fits the results of simulation. For the case of the stiff chain under tension we recover analytically all the known limits. Furthermore, we obtain quantitative agreement with recent experiments on the stretching of DNA using the total contour length and the persistence length as adjustable parameters. The limitations of our approach become obvious in situations that involve broken symmetry—such as the case of a stiff chain in a strong nematic potential. Nevertheless, it is clear that the mean field variational approach lays the foundation for systematic studies of more complicated problems involving semiflexible chains.

References

1. See, for example, P. J. Flory, *Statistical Mechanics of Chain Molecules,* Interscience, New York, 1969; K. F. Freed, *Adv. Chem. Phys.* **22:** 1 (1972); M. Doi and S. F. Edwards, *The Theory of Polymer Dynamics,* Oxford University Press, Oxford, 1986.

2. See, for example, J. des Cloiaeaux and G. Jannink, *Polymers in Solution, Their Modelling and Structure,* Oxford University Press, Oxford, 1990; M. Doi and S. F. Edwards, *The Theory of Polymer Dynamics,* Oxford University Press, Oxford, 1986.
3. H. Yamakawa, *Modern Theory of Polymer Solutions,* Harper and Row, New York, 1971.
4. K. Kroy and E. Frey, *Phys. Rev. Lett.* **77:** 306 (1996) and references therein.
5. S. B. Smith, L. Finzi, and C. Bustamante, *Science* **258:** 1122 (1992).
6. N. Saito, J. Takahashi, and Y. Yunoki, *J. Phys. Soc. Jpn.* **22:** 219 (1967).
7. R. A. Harris and J. E. Hearst, *J. Chem. Phys.* **44:** 2595 (1966).
8. K. F. Freed, *J. Chem. Phys.* **54:** 1453 (1971); M. G. Bawendi and K. F. Freed, *J. Chem. Phys.* **83:** 2491 (1985).
9. M. Warner, J. M. F. Gunn, and A. B. Baumgartner, *J. Phys. A* **18:** 3007 (1965); M. Warner, J. M. F. Gunn, and A. B. Baumgartner, *J. Phys. A* **19:** 2215 (1986).
10. For a review of functional integral methods in polymers, see K. F. Freed, *Adv. Chem. Phys.* **22:** 1 (1972).
11. V. N. Tsvetkov, *Rigid Chain Polymers,* Consultants Bureau, New York, 1989.
12. G. J. Vroege and T. Odijk, *Macromolecules* **21:** 2848 (1988).
13. T. Odijk, *Macromolecules* **16:** 1340 (1983); T. Odijk, *Macromolecules* **19:** 2313 (1986).
14. B.-Y. Ha and D. Thirumalai, *J. Chem. Phys.* **103:** 9408 (1995).
15. A. M. Gupta and S. F. Edwards, *J. Chem. Phys.* **98:** 1588 (1993); Z. Y. Chen, *Phys. Rev. Lett.* **71:** 93 (1993).
16. B.-Y. Ha and D. Thirumalai, *Macromolecules* **28:** 577 (1995).
17. J. L. Barrat and J. F. Joanny, *Europhys. Lett.* **24:** 333 (1993).
18. J. B. Lagowski, J. Noolandi, and B. Nickel, *J. Chem. Phys.* **95:** 1266 (1991).
19. R. G. Winkler, P. Reineker, and L. Harnau, *J. Chem. Phys.* **101:** 8119 (1994).
20. K. Soda, *J. Phys. Soc. Jpn.* **35:** 866 (1973).
21. M. Otto, J. Eckert, and T. A. Vilgis, *Macromol. Theory Simul.* **3:** 543 (1994).
22. D. J. Amit, *Field Theory, the Renormalization Group, and Critical Phenomena,* World Scientific, Singapore, 1978.
23. R. P. Feynman and A. R. Hibbs, *Quantum Mechanics and Path Integral,* McGraw-Hill, New York, 1965.
24. O. G. Berg, *Biopolymer* **18:** 2861 (1979).
25. K. Huber, W. H. Stockmayer, and K. Soda, *Polymer J.***31:** 1811 (1990).
26. J. Zinn-Justin, *Quantum Field Theory and Critical Phenomena,* Oxford University Press, Oxford, 1989.
27. W. F. J. Wilhelm and E. Frey, *Phys. Rev. Lett.* **77:** 2581 (1996).
28. M. Fixman and J. Kovacs, *J. Chem. Phys.* **58:** 1564 (1973).
29. T. Odijk, *Macromolecules* **28:** 7016 (1995).
30. B.-Y. Ha and D. Thirumalai, *J. Chem. Phys.* **106:** 4243 (1997).
31. A. Kholondenko and T. Vilgis, *Phys. Rev. E* **50:** 1257 (1994).
32. J. F. Marko and E. D. Siggia, *Science* **265:** 1599 (1994).

33. J. F. Marko and E. D. Siggia, *Macromolecules* **28:** 8759 (1995).

34. P. Pincus, *Macromolecules* **9:** 386 (1976).

35. P.-G. de Gennes, *Scaling Concepts in Polymer Physics,* Cornell University Press, Ithaca and London, 1979.

36. K. D. Kamien, P. L. Doussal, and D. R. Nelson, *Phys. Rev. A* **45:** 8727 (1992).

37. P.-G. de Gennes, in *Polymer Liquid Crystals,* A. Ciferri, W. R. Kringbaum, and R. B. Meyer Eds., Academic Press, New York, 1982.

Chapter 2 | Polymer Adsorption: Mean Field Theory and Ground State Dominance Approximation

A. N. Semenov

Department of Applied Mathematics
University of Leeds, Leeds, United Kingdom

J.-F. Joanny and A. Johner

Institut Charles Sadron
Strasbourg, France

Abstract

We review a mean field theory that we have recently developed to describe the adsorption of a polymer solution on an interface. The configuration of the chains is described by two order parameters associated to the central monomers of the chains (loops) and the end monomers of the chains (tails), respectively. This order parameter approach gives the large molecular asymptotic limit of the mean field theory studied numerically by Scheutjens and Fleer.

Our presentation emphasizes the role of the chain end points. We describe the properties of adsorbed polymer layers in equilibrium with a solution that can be dilute, semidilute, or concentrated. We also discuss the interaction between adsorbed polymer layers in both limits of reversible or irreversible adsorption (fixed adsorbed polymer amount).

In addition to the scaling behavior obtained for the various physical quantities in the limit of large molecular weights, our approach also predicts some new, nonclassical behaviors—for example, for the adsorbed polymer amount or for the equilibrium interaction between adsorbed polymer layers.

2.1 Introduction

A versatile way to monitor the interactions between colloidal particles in solution is to add a polymer to the solution [1]. If the polymer is repelled by the surface of the particles, it induces a depletion attraction between particles. The reverse case where the polymer adsorbs is far more subtle and the very sign of the polymer-induced interaction depends on the precise conformation of the polymer chains on the surface and even, in certain cases, on the history of the solution.

THEORETICAL AND MATHEMATICAL
MODELS IN POLYMER RESEARCH

Copyright © 1998 by Academic Press.
All rights of reproduction in any form reserved.
ISBN 0-12-304140-6/$25.00.

This important issue and various other applications have prompted very early theoretical studies of the behavior of polymer solutions in the vicinity of an interface [2–4].

Over the last few years, the structure of an adsorbed polymer layer (Figure 2.1) has been a very controversial issue [5]. The role of the chain ends, pointed out several years ago in the work of Scheutjens and Fleer, has only been clarified recently [6]. The early theoretical treatment of a polymer solution in the vicinity of a solid surface was a mean field theory based on the so-called ground state dominance approximation for the polymer partition function. This model turns out to be equivalent to a van der Waals free energy functional approach where the free energy density depends on a single order parameter ψ related to the local monomer concentration by $c = \psi^2$. The approximations involved ignore the chain end effects and consider the polymer chains as infinite. This model captures several important properties of adsorbed polymer layers. For example, the monomer concentration profile decays according to the power law from the adsorbing surface and the adsorbed polymer layer can thus be described as a transient self-similar network with a mesh size decreasing toward the adsorbing surface. Also, the adsorbed polymer forms a thick layer with a characteristic size of the order of magnitude of the correlation length of the polymer solution, i.e., much larger than the microscopic monomer size [7]. De Gennes [8] has calculated the interaction between two parallel layers within the framework of this model, taking into account the reversibility of the adsorption. De Gennes was also able to extend the model beyond the mean field approximation, using scaling arguments and a rescaled density functional theory that enforces nonclassical exponents for

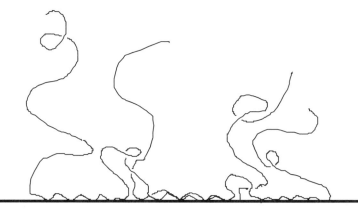

Figure 2. 1   Schematic picture of loops and tails in an adsorbed polymer layer

the polymer behavior. This scaling theory based on the self-similar structure of the adsorbed layer [9] gives a good description of many of the experiments performed on adsorbed polymer layers, such as the concentration profile as measured by neutron scattering [10] or neutron reflectivity [11, 12], the layer thickness obtained by optical [13] or hydrodynamic [14] measurements, and the interaction between adsorbed polymer layers as measured by a surface force apparatus [15]. The theory has also been generalized to various solvency conditions [16–19] and to the calculation of the interactions between undersaturated polymer layers [20].

On the other hand, Scheutjens and Fleer and their coworkers have questioned the validity of the ground state dominance approximation and have solved numerically the full mean field equations for the conformation of a polymer chain in an adsorbed layer [5]. They have shown unambiguously that the chain end points play a major role in the structure of an adsorbed polymer layer. An adsorbed polymer chain must be considered as a succession of loops starting from the surface and coming back to the surface and of two tails going from the last monomers in contact with the adsorbing surface to the chain ends in the solution. In the vicinity of the adsorbing surface, the concentration is dominated by loops and most monomers belong to loops. In the outer parts of the adsorbed polymer layers, the chains form long tails that fully dominate the concentration profile. Most monomers belong to short loops, but a significant portion of the adsorbed layer contains mostly tail monomers. The numerical scheme has been implemented in a broad variety of situations [21, 22], and in many cases it gives as good a description of the experiments as can be expected for not-too-long chains when the excluded volume correlations are not important [23]. The method has also been applied to various polymer architectures, to various solvency conditions, and to various types of monomer–monomer interactions.

Recently, we have proposed a modified version of the mean field theory that goes beyond the ground state dominance approximation [6] and which should constitute the large molecular weight limit of the Scheutjens-Fleer theory. This approach allows the detailed study of some asymptotic limits, provides simple analytic laws for the variation of the relevant physical parameters, and gives some new insights into the conformation of a polymer chain in an adsorbed layer and into the nature of tail-induced effects (such as end-induced repulsion between adsorbed layers [24–26]). For some of the problems, we were also able to go beyond the mean field approximation using the scaling approach proposed by de Gennes [27, 6].

The aim of this chapter is to review in detail this new mean field theory and to show how it can be applied to the adsorption of a polymer solution onto a single surface when the bulk monomer concentration is increased and to the study of interactions between adsorbed polymer layers. In the next section, we present the general theoretical background of the new mean field theory. The adsorption from a dilute solution is studied in Section 2.3 and the adsorption from a semidilute and a concentrated solution in Section 2.4. Sections 2.5 and 2.6 are devoted to the derivation of the free energy of the polymer solution and to the problem of the interactions between adsorbed polymer layers—first in the case of concentrated solutions and then in the case of dilute or semidilute solutions. In the last section, we discuss the limitations of the theory and possible extensions beyond the mean field approximation.

2.2   Mean Field Theory and Ground State Dominance Approximation

In this section we present the general mean field theory that we use to describe polymer adsorption. In the mean field approach, the monomers feel an average potential proportional to the local concentration; if the local concentration at a distance z from the wall is $c(z)$, we write the mean field potential as $U(z) = v(c(z) - c_b)$, where v is the excluded volume parameter (which is positive in a good solvent) and c_b is the monomer bulk concentration far away from the adsorbing surface. (The thermal energy $k_B T$ is assumed to be the energy unit throughout the chapter.) In the following, it is convenient to use an effective volume fraction $\phi = vc$ rather than the concentration c. The mean field potential then reads $U(z) = \phi(z) - \phi_b$. The structure of the adsorbed layer can be determined from the partition function of a chain in the potential $U(z)$.

2.2.1   PARTITION FUNCTIONS AND CONCENTRATION PROFILES

In the presence of an adsorbing surface, the total potential seen by the monomers is the sum of the mean field potential U and the wall potential $U_s(z)$. We consider here that the wall potential is short range and has a weakly bound state (in the quantum mechanical sense). Similar results would be obtained for more realistic potentials with a sufficiently fast decay at large distances, such as the van der Waals interaction between the wall and the monomers (decaying as z^{-3}). We describe the statistics of a chain of N monomers by the partition function $Z_N(z)$ with the constraint that one chain end is fixed at a distance z from the wall, all the other degrees of freedom being integrated out. This partition function is calculated in an external potential equal to the total potential $U + U_s$. It is known to satisfy the Schrödinger-like equation [28–30]

$$\frac{-\partial Z_n}{\partial n} = -\frac{\partial^2 Z_n}{\partial z^2} + (U + U_s)Z_n. \tag{1}$$

In this equation, the unit length has been defined as $a/6^{1/2}$, where a is the statistical length (monomer size). In the following, we are not interested in the details of the concentration in the very vicinity of the wall where it depends explicitly on the shape of the wall potential; the wall potential, being sufficiently short range, can be replaced by an effective boundary condition for the partition function at short distances:

$$\frac{-1}{Z_n} \frac{\partial Z_n}{\partial z}\Big|_{z=0} = \frac{1}{b}. \tag{2}$$

The extrapolation length b measures the strength of the adsorption and varies roughly as the inverse of the free energy gained by a monomer upon adsorption. We assume here a rather weak adsorption limit where b is larger than the monomer size but remains a molecular length. The partition function then satisfies

$$\frac{-\partial Z_n}{\partial n} = -\frac{\partial^2 Z_n}{\partial z^2} + UZ_n \tag{3}$$

with the boundary condition (2). We choose the normalization of the partition function to be one when the chain reduces to a monomer, $Z_0(z) = 1$. Far away from the surface, the chains do not feel the wall attraction and their partition function is $Z_N = 1$.

The probability to find monomer n of the chain at a distance z from the surface is proportional to the product of two partition functions of the subchains starting from monomer n, $Z_n(z)Z_{N-n}(z)$. The total monomer volume fraction is thus

$$\phi(z) = \frac{\phi_b}{N} \int_0^N dn\, Z_n(z)Z_{N-n}(z). \tag{4}$$

The normalization has been chosen here by imposing that very far away from the surface, the concentration is the imposed bulk concentration.

We now divide the chain conformations into two sets: The free states are those where none of the monomers are adsorbed and thus where no monomer touches the surface; the adsorbed states are those where at least one of the monomers is adsorbed. The total partition function can be written as $Z_N = Z_N^a + Z_N^f$. The adsorbed chain partition function Z_N^a and the free chain partition function Z_N^f satisfy the same equation (3) as the total partition function but with different boundary conditions. The free chain partition function vanishes on the wall; in the limit of very small chains it has the same behavior as the total partition function $Z_0^f = 1$; far away from the adsorbing wall, the chains are not adsorbed and the free chain partition function is equal to the total partition function $Z_N^f = 1$. The adsorbed chain partition function has the same boundary condition at the wall (2) as the total partition function.[1] In the limit of very small chains the adsorbed chain partition function vanishes $Z_0^a = 0$; far away from the adsorbing wall the adsorbed chain partition function also vanishes since all the conformations are free.

Using the adsorbed and free chain partition functions, we can split the total volume fraction into several contributions. If two subchains starting from monomer n are in an adsorbed state, the monomer belongs to a loop. The volume fraction of monomers belonging to loops is thus

$$\phi_l(z) = \frac{\phi_b}{N} \int_0^N dn \, Z_n^a(z) Z_{N-n}^a(z). \tag{5}$$

If one of the subchains is in a free state and the other one is in an adsorbed state, monomer n belongs to a tail. The concentration of monomers belonging to tails is therefore

$$\phi_t(z) = \frac{2\phi_b}{N} \int_0^N dn \, Z_n^f(z) Z_{N-n}^a(z). \tag{6}$$

One can similarly define a concentration of free chains. We will in a first step neglect the free chains in the adsorbed layer and then briefly consider them at the end of this section.

The volume fraction of chain end points can also be obtained from the partition functions

$$\phi_e(z) = \frac{2\phi_b}{N} Z_N(z) \tag{7}$$

where the factor 2 accounts for the existence of two end points per chain. The volume fraction of the end points of the adsorbed chains can similarly be written as

$$\phi_{ea}(z) = \frac{2\phi_b}{N} Z_N^a(z). \tag{8}$$

---

[1]This statement is approximately true if the bound state is dominant (see below).

2.2.2   GROUND STATE DOMINANCE APPROXIMATION
AND ORDER PARAMETERS

The partition function of a polymer chain in a solution adsorbing onto a wall can be decomposed onto the eigenstates of the Schrödinger-like equation (3). We label the energy levels with an index i and call $\psi_i(z)$ the eigenstate corresponding to an energy E_i; the eigenstates are given by

$$E_i\psi_i = -\frac{\partial^2 \psi_i}{\partial z^2} + U\psi_i. \tag{9}$$

At the wall, the boundary condition for the eigenfunctions is the same as that of the partition function:

$$\frac{-1}{\psi_i}\frac{\partial \psi_i}{\partial z}\bigg|_{z=0} = \frac{1}{b}.$$

The eigenstates form an orthonormal set of functions so that $\int_0^\infty \psi_i \psi_j\, dz = \delta_{ij}$. The total partition function $Z_N(z)$ is written as

$$Z_N(z) = \sum_i K_i\psi_i(z)e^{-E_i N} \tag{10}$$

where all eigenfunctions $\psi_i(z)$ are assumed to be real. The amplitudes K_i are fixed by the normalization of the partition function $Z_0(z) = 1$, $K_i = \int_0^\infty \psi_i(z)dz$. We consider here the case when the adsorption is weak enough so that there is only one bound state of energy $E_0 = -\epsilon$ (in fact this is true for all the cases of adsorption considered so far). All the other states are unbound states with a positive energy.

The ground state dominance approximation amounts to considering the limit of very large molecular weights, $\epsilon N \gg 1$, when the sum giving the partition function is dominated by the first term $Z_N(z) = Z_N^a(z) = K_0\psi_0(z)e^{\epsilon N}$. Clearly this implies that all the conformations of a chain are adsorbed and this approximation ignores the tail contribution to the concentration profile. In this approximation, the monomer volume fraction is proportional to the square of the ground state eigenfunction $\phi(z) \simeq \text{const } \psi_0(z)^2$, where const is proportional to the total number of chains. We want here to go beyond this approximation and consider the effect of the free chain conformations—i.e., of the tails of the chains.

To describe quantitatively the effect of the tails, we separate the ground state in the partition function and write

$$Z_N(z) = e^{\epsilon N}(K_0\psi_0(z) + r_N(z)). \tag{11}$$

In the calculation of the total concentration, three terms appear. In the limit where we neglect the free chains, it is consistent to ignore the last term proportional to $r_n r_{N-n}$. The monomer volume fraction then reads

$$\phi(z) = \phi_b e^{\epsilon N} K_0^2 \left(\psi_0^2(z) + \frac{2}{NK_0} \psi_0(z) r(z) \right). \tag{12}$$

We have defined here the rest function as $r(z) = \int_0^N dn\ r_n(z) = \int_0^\infty dn(e^{-\epsilon n} Z_n - K_0 \psi_0)$. The integral defining r can be to a good approximation extended to infinity since the integrated function decreases exponentially at large values of n; physically this is due to the fact that the tails of the chains have a finite maximum length of order $1/\epsilon$. By direct derivation, we can check that the rest function $r(z)$ is a solution of

$$1 - K_0 \psi_0(z) = -\frac{\partial^2 r}{\partial z^2} + (U + \epsilon) r. \tag{13}$$

In the expression (12) of the concentration, it is tempting to consider the first term (equivalent to the ground state dominance approximation) as the contribution of the loop monomers and the second term as the contribution of the tail monomers. This is approximately true but the second term also includes a small correction to the loop concentration. This can be seen by calculating directly the tail concentration from Eq. (6). The adsorbed chain partition function in this equation can be approximated by the ground state term:

$$Z_{N-n}^a = K_0 \psi_0 e^{\epsilon(N-n)}. \tag{14}$$

We obtain

$$\phi_t(z) = \frac{2\phi_b}{N} e^{\epsilon N} K_0 \psi_0(z) \varphi(z). \tag{15}$$

The order parameter $\varphi(z)$ is defined here as

$$\varphi(z) = \int_0^\infty dn\ e^{-\epsilon n} Z_n^f(z). \tag{16}$$

The equation satisfied by φ is also obtained directly from the Schrödinger equation for Z_n^f

$$1 = -\frac{\partial^2 \varphi}{\partial z^2} + (U + \epsilon) \varphi \tag{17}$$

with the boundary condition $\varphi(0) = 0$.

The connection between two functions $r(z)$ and $\varphi(z)$ has been discussed in detail in the appendix of ref. [25]. We give here a more qualitative discussion. At distances from the adsorbing wall larger than the extrapolation length b, the term $K_0\psi_0$ turns out to be negligible compared to 1. The two functions r and φ are then solutions of the same linear differential equation and are equal at large distances. The ground state eigenfunction ψ_0 being the solution of the corresponding homogeneous equation, we can approximately write $r(z) = \varphi(z) + A\psi_0(z)$, where A is a constant. The constant can be determined by imposing that the functions r and ψ_0 are orthogonal ($\int_0^\infty r(z)\psi_0(z)dz = 0$). This gives $A = -\int_0^\infty dz\ \varphi(z)\psi_0(z)$. The total volume fraction can thus be written as the sum of the loop and tail volume fractions: $\phi(z) = \phi_l(z) + \phi_t(z)$ where the tail volume fraction is given by Eq. (15) and the loop volume fraction is given by

$$\phi_l(z) = \phi_b e^{\epsilon N} K_0^2 \psi_0^2(z)\left(1 - \frac{2}{NK_0}\int_0^\infty dz'\ \varphi(z')\psi_0(z')\right). \tag{18}$$

The leading term is equivalent to the ground state dominance approximation but there is a small negative $1/N$ correction to the ground state dominance approximation that slightly overestimates the loop concentration. In the following, this correction will often be neglected. Note also that there is another (smaller) correction to the loop concentration due to the small term $K_0\psi_0$ neglected on the left-hand side of Eq. (13).

The volume fraction of chain end points of adsorbed chains can be calculated from the ground state dominance approximation using Eqs. (8) and (14).

$$\phi_{ea}(z) = \frac{2\phi_b}{N} e^{\epsilon N} K_0 \psi_0(z). \tag{19}$$

2.2.3 FREE CHAINS

So far, we have neglected the free chains that do not adsorb on the solid surface. The volume fraction of the free chains $\phi_f(z) = (\phi_b/N)\int_0^N dn\ Z_n^f(z)Z_{N-n}^f(z)$ can be calculated from the free chain partition function $Z_N^f(z)$ that satisfies the Schrödinger-like equation

$$\frac{-\partial Z_n^f}{\partial n} = -\frac{\partial^2 Z_n^f}{\partial z^2} + UZ_n^f$$

with the boundary condition $Z_N^f(z=0) = 0$. It is convenient to expand the free chain partition function on the set of eigenfunctions, $\chi_j(z)$, corresponding to "energy" levels E_j', which satisfy the condition $\chi_j(0) = 0$. The eigenstate equation is

$$E_j'\chi_j = -\frac{\partial^2\chi_j}{\partial z^2} + U\chi_j.$$

The energy spectrum is here continuous and the lowest eigenvalue is $E'_j = 0$. It is useful to introduce the free chain order parameter $\psi_f(z) = \chi_0(z)/\chi_0(\infty)$. This order parameter is obtained from the equation

$$0 = -\frac{\partial^2 \psi_f}{\partial z^2} + U\psi_f \tag{20}$$

with the boundary conditions $\psi_f(0) = 0$ and $\psi_f(\infty) = 1$. The expansion of the free chain partition function reads

$$Z_N^f(z) = \psi_f(z) + \sum_{j \neq 0} L_j \chi_j(z) e^{-E_j N} \tag{21}$$

where $L_j = \int_0^\infty dz\, \chi_j(z)$. In the limit where the thickness of the adsorbed layer λ is small compared to the chain end-to-end distance $R = N^{1/2}$, the partition function is dominated by the ground state. We then calculate the volume fraction of free chains in a way similar to that used for adsorbed chains to obtain Eq. (12). The volume fraction of free chains is written as

$$\phi_f(z) = \phi_b\left(\psi_f^2(z) + \frac{2}{N}\psi_f r_f(z)\right) \tag{22}$$

where the rest function is defined as $r_f(z) = \int_0^\infty dn(Z_n^f(z) - \psi_f(z))$. The rest function is a solution of

$$1 - \psi_f(z) = -\frac{\partial^2 r_f}{\partial z^2} + Ur_f \tag{23}$$

with the boundary conditions $r_f(0) = r_f(\infty) = 0$.

When the first term is dominant in the right-hand side of Eq. (22), the free chains penetrate the adsorbed layers via central monomers; we call this mechanism hairpin penetration. On the other hand, if the second term is dominant, the free chains penetrate the adsorbed layer by their end point and we call this mechanism end penetration.

2.2.4 PROPERTIES OF ADSORBED POLYMER LAYERS

The relevant physical properties of polymer layers adsorbed on a wall are the total amount of adsorbed polymer, the adsorbance Γ, the monomer concentration profile and the relative contributions of tails and loops, and eventually the contribution of the adsorbed polymer to the surface tension. Ignoring the contribution of the free chains, the adsorbance is directly obtained by integration of the volume fraction $\phi(z) = \phi_l(z) + \phi_t(z)$ given by [15, 18]

$$\Gamma = \phi_b e^{\epsilon N} K_0^2. \tag{24}$$

The same result is obtained by integrating the end points of the adsorbed chains and noting that there are $2\Gamma/N$ end points per unit area or by equilibrating the chemical potential of the adsorbed chains to the chemical potential of the bulk chains.

The concentration profile of the adsorbed chains can be obtained by solving the equations for the order parameters φ and ψ_0 with the self-consistency condition that the mean field potential is $U = \phi - \phi_b$. It is often useful to change the normalization of the ground state eigenfunction and to define $\psi = \Gamma^{1/2}\psi_0$. The order parameter ψ is obtained by solving

$$0 = -\frac{\partial^2 \psi}{\partial z^2} + (U + \epsilon)\psi \tag{25}$$

with the boundary condition

$$\frac{-1}{\psi} \frac{\partial \psi}{\partial z}\bigg|_{z=0} = \frac{1}{b}.$$

The total volume fraction can then be written as

$$\phi(z) = \psi^2(z) + B\psi(z)\varphi(z) \tag{26}$$

where we have introduced $B = 2\Gamma/(N\int_0^\infty \psi(z)dz)$. Note that we ignore here the small correction to the loop volume fraction found in Eq. (18) and the free chain contribution. The contributions to the adsorbance due to the different terms in Eq. (26) are shown in Figure 2.2.

The concentration profiles are then given by the simultaneous resolution of the coupled Eqs. (26), (25), and (17). A detailed study of the concentration profiles as the bulk volume fraction is varied is given in the following sections.

It is interesting to note that these equations have a first integral

$$\left(\frac{\partial \psi}{\partial z}\right)^2 + B\frac{\partial \psi}{\partial z}\frac{\partial \varphi}{\partial z} + \phi_{ea}(z) - \frac{1}{2}(\phi - \phi_b)^2 - \epsilon(\phi - \phi_b) = 0. \tag{27}$$

This first integral can for example be used to calculate the surface volume fraction $\phi_s = \phi(0)$. In the limit of large molecular weights, the tails and the end

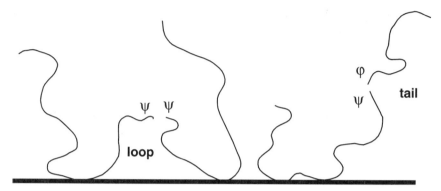

Figure 2.2 Loops and tails contributions to the adsorbance. The loops contribution is proportional to ψ^2 and the tails contribution is proportional to $\psi\varphi$.

points have a negligible contribution close to the adsorbing wall. The bulk concentration is also much smaller than the surface concentration. Using the boundary condition for the order parameter ψ, we get

$$\phi_s = \frac{2}{b^2} + 2(\phi_b - \epsilon). \qquad (28)$$

The surface concentration is indeed large ($b \simeq 1$) and increases with the bulk concentration (as shown below, ϵ decreases with the concentration).

The contribution of the polymer to the interfacial energy γ between the solution and the adsorbing wall can be determined from the surface concentration. The effective boundary condition that we use for the partition function is equivalent to a local interaction $-1/b$ between the adsorbed monomers and the wall. When the strength of this interaction is changed by $d(1/b)$, the change in the interfacial tension is $d\gamma = -\phi_s d(1/b)$. The integration of this differential relation gives the interfacial tension

$$\gamma = \frac{-2}{3b^3} - \frac{2}{b}(\phi_b - \epsilon). \qquad (29)$$

Here we take into account the fact that ϵ almost does not depend on ϕ_b or b: Eq. (24) shows a weak logarithmic dependence. As expected, the interfacial tension of the adsorbed polymer solution is negative and decreases with the bulk concentration. The adsorbance can be obtained from the surface tension by using Gibbs law, $\Gamma = -(\partial\gamma/\partial\mu)$, where μ is the monomer chemical potential.

As shown by Eq. (24), if the variation of the adsorbance with the concentration is weak, the chemical potential is up to a constant $\mu = (\phi_b - \epsilon)$. The adsorbance is thus

$$\Gamma = \frac{2}{b}. \tag{30}$$

This is the same result as obtained from the ground state dominance approximation. The determination of the corrections to this asymptotic value due to the finite size of the chains requires a detailed knowledge of the concentration profiles.

Up to now we have only considered the adsorption of a polymer solution onto a single wall. To study the interactions between colloidal particles mediated by a polymer solution, we need to consider the adsorption of a polymer solution onto two identical parallel adsorbing walls separated by a distance $2h$. The formalism presented here can be used to study this problem. The boundary conditions at infinity must then be replaced by boundary conditions at the midplane (which are easily obtained by symmetry arguments), and the integrations over space should be made over half the space between the planes (between 0 and h). The formalism allows a calculation of the concentration profiles, but the determination of the interaction force requires a precise calculation of the free energy of the polymer solution (to be given in Section 2.5).

2.3   Polymer Adsorption from Dilute Solution

In this section, we discuss polymer adsorption from a dilute solution ($\phi_b \ll 1/N$). An adsorbed layer is described by four order parameters: ψ, φ related to the adsorbed polymer conformations and ψ_f, r_f related to the free conformations. The layer acts as a repulsive potential for the free configurations. Well inside the layer, free chains are not expected to play any role and the order parameters ψ and φ can be determined from Eqs. (25) and (17), which we rewrite below

$$0 = -\frac{d^2\psi}{dz^2} + (U(z) + \epsilon)\psi(z)$$

$$1 = -\frac{d^2\varphi}{dz^2} + (U(z) + \epsilon)\varphi(z). \tag{31}$$

The molecular potential increases linearly with the local concentration $U(z) = \psi^2 + B\varphi\psi - \phi_b$. These order parameter equations must be solved with the boundary conditions

$$\frac{1}{\psi}\frac{d\psi}{dz}\bigg|_w = -\frac{1}{b}, \qquad \varphi|_w = 0, \qquad \lim_{z\to\infty}\psi(z) = 0. \tag{32}$$

When the solution is dilute, the contribution of the bulk concentration ϕ_b to the molecular potential U can be neglected inside the layer. Three regions can then be identified:

- the proximal region closest to the wall within a distance b (comparable to the monomer size in the limit of strong adsorption).
- the central region in the intermediate range ($b < z < \lambda = \epsilon^{-1/2}$) where the concentration is high and thus U dominates over ϵ.
- the distal region ($z > \lambda$) where ϵ dominates over U.

2.3.1   CROSSOVER LENGTH AND CONCENTRATION PROFILES IN THE CENTRAL REGIME

In the central regime the order parameter equations are written in a dimensionless form as (note that here both ϕ_b and ϵ are much smaller than U):

$$0 = -\frac{d^2\tilde{\psi}}{d\zeta^2} + \tilde{U}\tilde{\psi}$$

$$1 = -\frac{d^2\tilde{\varphi}}{d\zeta^2} + \tilde{U}\tilde{\varphi} \tag{33}$$

with $\tilde{U} = \tilde{\psi}\tilde{\psi} + \tilde{\varphi}\tilde{\psi}$. The reduced variables are defined as follows:

$$l = \left(\frac{1}{B}\right)^{1/3}, \quad \zeta = z/l, \quad \tilde{\psi}(\zeta) = l\psi(z), \quad \tilde{\varphi}(\zeta) = l^{-2}\varphi(z), \quad \tilde{U}(\zeta) = l^2 U(z). \tag{34}$$

The concentration inside the central region obeys the scaling law $\phi(z) = l^{-2}\tilde{U}(z/l)$. The existence of the characteristic length l is in contradiction with the ansatz made by de Gennes that there is no intermediate length scale, which leads to the self-similar construction of the concentration profile in an adsorbed polymer layer [9]. Nonetheless, the ansatz that there is no intermediate length scale can be made in both asymptotic limits $z \ll l$ and $z \gg l$. In these limits a z^{-2} power law decay of the concentration is thus expected.

Close to the wall, ($z \ll l$) the loop contribution dominates over the tail contribution, and we find the following asymptotic behavior:

$$\psi = \frac{\sqrt{2}}{z}, \qquad \varphi = \frac{1}{3}z^2 \log\left(\frac{l}{z}\right), \qquad U(z) = \frac{2}{z^2}. \tag{35}$$

Further away from the wall ($z \gg l$), the tail contribution dominates over the loop contribution, and inside the central region ($\lambda > z \gg l$) we find

$$\psi = 360 l^3/z^4, \qquad \varphi = \frac{1}{18}z^2, \qquad U(z) = \frac{20}{z^2}. \tag{36}$$

The large crossover between the asymptotic behaviors has been obtained by numerical solution of Eq. (33) and is shown in Figure 2.3. The concentrations of monomers belonging to loops and tails cross at a distance $z^* = 1.43l$ from the wall, whereas the tail contribution reaches its maximum at $z = 1.29l$.

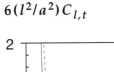

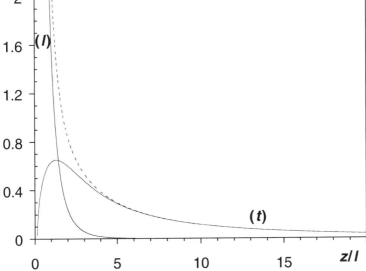

Figure 2.3 Loops C_l and tails C_t concentration profiles in the central regime ($b < z < \lambda$): The distance to the wall is reduced by l and the concentrations are reduced by l^{-2}. The total concentration is also shown (dashed curve).

The asymptotic analysis allows for the calculation of $B = 2\Gamma/(N\int_0^\infty \psi(z)dz)$ and of the crossover length l; with a logarithmic accuracy, we find

$$l = \frac{1}{3^{1/3}2^{1/2}}(Nb)^{1/3}\left(\log\left(\frac{N}{b^2}\right)\right)^{1/3}. \tag{37}$$

2.3.2   THICKNESS OF THE ADSORBED LAYER AND EQUILIBRIUM WITH A BULK SOLUTION

The equilibrium between bulk chains and adsorbed chains imposes that $\Gamma = \phi_b\int_0^\infty dz\, Z^a$ and determines the value ϵ of the ground state energy

$$\epsilon \approx \frac{1}{N}\log\left(\frac{1}{\phi_b b^2}\right) \tag{38}$$

with logarithmic accuracy. In the distal regime ($z > \lambda$), the tail contribution to the concentration still dominates, but the mean field potential U is negligible compared to ϵ. In this regime the natural unit length is

$$\lambda = \epsilon^{-1/2} \tag{39}$$

and the concentration of adsorbed polymer decays exponentially:

$$\psi \simeq \frac{l^3}{\lambda^4}\exp-(z/\lambda), \quad \varphi = \lambda^2, \quad \phi_t \simeq \frac{1}{\lambda^2}\exp-z/\lambda, \quad \phi_l \simeq \frac{l^6}{\lambda^8}\exp-(2z/\lambda). \tag{40}$$

The decay length for the loop monomer concentration is half that for the tail monomer concentration. In the sublayer beyond l ($z \gg l$), the loop contribution to the molecular potential U is negligible. Choosing λ as a unit length, the concentration of loops and tails can be written in the scaled form

$$\phi_l = \frac{l^6}{\lambda^8}f_l(z/\lambda), \quad \phi_t = \frac{1}{\lambda^2}f_t(z/\lambda)$$

with $f_l = \tilde{\psi}_\lambda^2, f_t = \tilde{\psi}_\lambda\tilde{\varphi}_\lambda$ where the reduced order parameters obey

$$0 = -\frac{d^2\tilde{\psi}_\lambda}{d\zeta_\lambda^2} + (\tilde{\psi}_\lambda\tilde{\varphi}_\lambda + 1)\tilde{\psi}_\lambda, \qquad \tilde{\psi}_\lambda(0) = \infty$$

$$1 = -\frac{d^2\tilde{\varphi}_\lambda}{d\zeta_\lambda^2} + (\tilde{\psi}_\lambda\tilde{\varphi}_\lambda + 1)\tilde{\varphi}_\lambda, \qquad \tilde{\varphi}_\lambda(0) = 0. \tag{41}$$

The scaled concentrations $f_{l,t}$ are plotted in Figure 2.4. The various concentration profiles in the dilute regime are sketched in Figure 2.5(a).

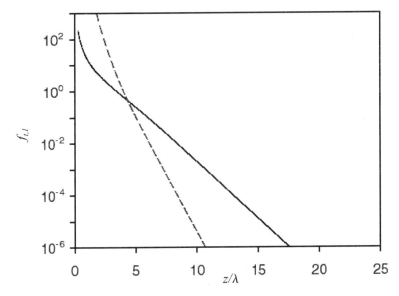

Figure 2.4 Loops and tails concentration profiles in the distal regime ($\lambda < z < R$): The scaled loops concentration $f_l(x)$ and tails concentration $f_t(x)$ are plotted against $x = z/\lambda$. The function f_l given by the dashed line is decaying faster.

2.3.3 FREE CHAIN DEPLETION EFFECTS

The free chains are repelled by the adsorbed chains and feel the adsorbed layer essentially as a hard wall; they are depleted from the vicinity of the wall. Outside the layer, the free chain concentration profile thus is identical to the depletion profile of polymer chains near a hard wall: $\phi_f = \phi_b P(z/R)$, where $P(x) = 1 - 8i^2 \operatorname{erfc}(x/2) + 4i^2 \operatorname{erfc}(x)$, with $i^n \operatorname{erfc}(x)$ a repeated integral of the complementary error function $\operatorname{erfc}(x)$. Close to the layer, at distances smaller than the radius of gyration $R = \sqrt{N}$, $\phi_f \simeq \phi_b (z^2/N)$, whereas at distances larger than R, $\phi_f \simeq \phi_b[1 - (16/\sqrt{\pi})(z/R)^{-3} \exp(-z^2/4R^2)]$. At the edge of the layer $z \simeq \lambda$, the concentration of free chains is small:

$$\phi_f(\lambda) \simeq \frac{\phi_b}{\epsilon N}. \tag{42}$$

The total concentration profile outside the layer ($z \gg \lambda$) reads

$$\phi(z) = \text{const } \epsilon \exp(-z/\lambda) + \phi_b P(z/R). \tag{43}$$

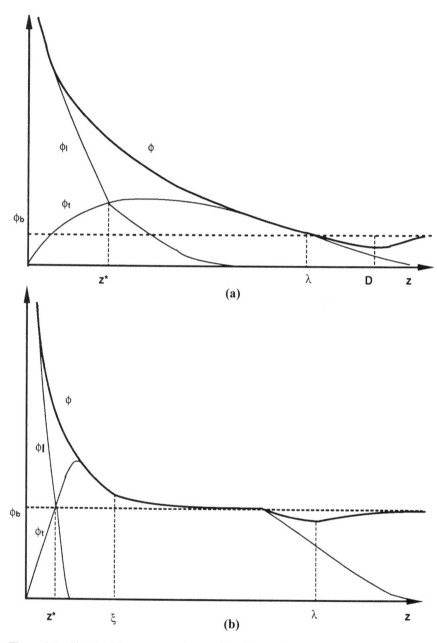

Figure 2.5 Sketch of the concentration profiles for an adsorbed polymer layer in equilibrium (a) with a dilute solution, (b) a semidilute solution in the semidilute tails regime.

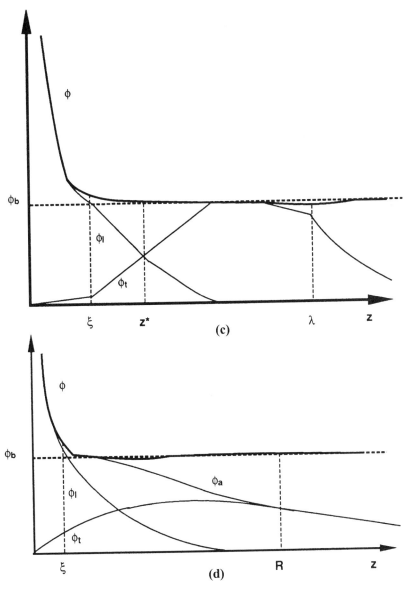

Figure 2.5   (c) in the semidilute loops regime, and (d) a concentrated solution.

The last term is the free chain concentration whereas the first term is the adsorbed chain concentration in the ground state dominance approximation where $Z^a_{ground} \propto \exp(N/\lambda^2 - z/\lambda)$. The total concentration reaches a minimal value at a finite distance from the wall where a depletion hole appears. The depletion is most pronounced at the overlap concentration $\phi^* \simeq 1/N$ where it is located at a distance close to λ from the surface. Upon dilution, the depletion hole becomes shallower and shifts outward at a distance $D = 2N/\lambda$.

At very large distances ($z > D$), the adsorbed chain contribution Z^a to the partition function becomes dominated by the free states variations and decays as a Gaussian $\propto \exp(-z^2/4N)$. The previous expression (Eq. (43)) for the total concentration is thus actually valid up to a distance D.

Inside the adsorbed layer ($z \ll R$), the order parameter approach is accurate for the free chains. As the free chain concentration here is much smaller than the bulk concentration (i.e., $\psi_f \ll 1$), the order parameter Equations (20) and (23) are the same as for the adsorbed chains, except for the boundary conditions. Therefore, $r_f = \varphi$ and $\psi_f = C\psi\int_0^z \psi^{-2}\,dz'$ where the boundary conditions have been taken into account. The hairpin contribution to the free chain concentration, given by the $\phi_f = \phi_b\psi_f^2$, dominates at the very edge of the layer down to the distance $z_1 \simeq \lambda^{4/3}/R^{1/3}$. The free chain concentration here is $\phi_f \simeq \phi_b(1/R^2\lambda^8)z^{10}$; the integration constant C is obtained by matching to the free chain concentration given by Eq. (42) at $z \simeq \lambda$. Further inside the layer, end penetration dominates and $\phi_f = (2/N)\phi_b\psi_f r_f$. In the tails sublayer ($l < z < z_1$), we obtain $\phi_f \simeq \phi_b(1/R^3\lambda^4)z^7$, whereas in the loops sublayer ($z < l$), $\phi_f \simeq \phi_b(l^3/R^3\lambda^4)z^4 \log(l/z)$.

2.3.4   ADSORBANCE

For infinite chains, at vanishing concentration, the concentration profile is $\phi_\infty(z) = 2/(z + b)^2$; this gives an adsorbance $\Gamma_\infty = 2/b$ in agreement with the general result of Eq. (30). The adsorbance is dominated by the short loops closest to the wall. For finite chains in the dilute regime, the concentration closest to the wall is unchanged up to corrections of order $1/N$, given by Eq. (28). The leading finite size corrections to the adsorbance come from the two crossover regions $z \simeq l$ and $z \simeq \lambda$. The crossover from the loop to the tail regime contributes a positive $N^{-1/3}$ correction due to the swelling of the profile by the tails. The cutoff region $z \simeq \lambda$ contributes a negative $N^{-1/2}$ correction. The next corrections, not retained here, would be of order $N^{-2/3}$. In the asymptotic limit $N \to \infty$, the leading correction is positive and Γ_∞ is approached from above. From the detailed solution of the dimensionless order parameter equations, we obtain

$$\Gamma = \frac{2}{b} + 7.07(Nb)^{-1/3}\left[\log\frac{N}{b^2}\right]^{-1/3} - 7.24N^{-1/2}\left[\log\frac{1}{\phi_b b^2}\right]^{1/2}. \tag{44}$$

In the most favorable case, $\phi_b \simeq 1/N$, the maximum of Γ is reached at $N = 4.3 \cdot 10^6 b^2$. The anomalous decrease of the adsorbance with molecular weight is thus only predicted for very high molecular weights.

2.4 Polymer Adsorption from Semidilute and Concentrated Solutions

Above the overlap concentration $\phi^* \simeq 1/N$, the correlation length ξ of the concentration fluctuations in the bulk polymer solution is shorter than the radius of gyration of the chains and defines a new relevant length scale. This length scale governs the decay of the concentration profile close to the wall. In the mean field theory, $\xi = 1/\sqrt{2\phi_b}$.

2.4.1 CLASSICAL ONE-ORDER PARAMETER THEORY

The standard mean field theory for a semidilute solution near a wall implicitly neglects tails. It is constructed with the single order parameter $\psi = \sqrt{\phi}$, which is given by

$$-\psi'' + \psi^3 - \phi_b\psi = 0. \tag{45}$$

When the polymer solution is adsorbing on the wall, the order parameter boundary conditions are

$$\left.\frac{\psi'}{\psi}\right|_0 = -\frac{1}{b}, \qquad \psi|_\infty = \sqrt{\phi_b}. \tag{46}$$

The concentration profile $\phi = \psi^2$ reads

$$\phi = \phi_b \coth^2(z/2\xi + \beta), \qquad \sinh(2\beta) = 2b. \tag{47}$$

If the polymer solution is repelled by a hard wall (depletion), the boundary condition is $\psi(0) = 0$ and the concentration is

$$\phi = \phi_b \tanh^2(z/2\xi). \tag{48}$$

The adsorption profile Eq. (47) remains valid in the loop-dominated region closest to the wall. The depletion profile Eq. (48) gives the correct large distance asymptotic behavior for the free chains close to the overlap concentration where ξ is larger than the layer thickness λ.

2.4.2 ADSORPTION FROM A SEMIDILUTE SOLUTION

In a very narrow regime ($1/N < \phi_b < \epsilon$) close to the overlap concentration, the bulk correlation length remains larger than the layer thickness λ. The situation is very similar to the dilute case. The depletion of the free chains outside the layer occurs over a distance ξ smaller than R. The free chain concentration profile crosses over smoothly with the depletion profile near a hard wall: $\phi_f = \phi_b \tanh^2(z/2\xi)$ for $z \gg \lambda$. Hairpin penetration dominates if $z \geq z_1 = \lambda^{4/3}R^{-1/3}(\xi/R)^{1/3}$ and $\phi_f = \phi_b z^{10}/\lambda^8\xi^2$. Deeper in the layer end penetration dominates: in the tails regime $\phi_f = \phi_b z^7/\lambda^4 R^2 \xi$, whereas in the loops regime $\phi_f = \phi_b z^4 l^3/\lambda^4 R^2 \xi$.

At higher concentrations, the correlation length ξ moves inside the adsorbed layer. As long as $\phi_b < \phi_1 \simeq (Nb)^{-2/3}$, ξ remains larger than l and the loops regime is qualitatively unaffected; we call this the semidilute tails regime. At concentrations higher than ϕ_1, ξ moves inside the loops sublayer; this is the semidilute loops regime [31]. In concentrated solutions ($\phi > \phi_2$) the adsorption becomes weak ($\epsilon N < 1$) and most of the adsorbed polymer is stored in large loops and tails of size R [31, 32]. This happens for $\phi_b R > 2/b$ when $\phi_2 \simeq (Nb^2)^{-1/2}$.

The Semidilute Tails Regime

In the semidilute tails regime, both the crossover length l and the concentration profiles in the loops sublayer ($z < l$) and in the dilute tails sublayer ($l < z < \xi$) have the same asymptotic values as in the dilute regime.

There is, however, a new region ($\xi < z < \lambda$) dominated by semidilute tails where the concentration has almost relaxed to its bulk value. At lowest order, $\phi_t = \phi_b$ and the order parameter φ is inversely proportional to the order parameter ψ, $\varphi = l^3\phi_b/\psi$. The order parameters φ and ψ are then given by $\varphi = z^2/4$, $\psi = 4l^3\phi_b z^{-2}$ wherefrom the concentration profiles and the molecular potential for $\xi < z < \lambda$:

$$\phi_l = 8l^6\xi^{-2}z^{-4}\phi_b, \qquad \phi_t = \phi_b + U, \qquad U = 6z^{-2}. \qquad (49)$$

The loop contribution to the concentration decreases as z^{-4} whereas the total concentration is higher than the bulk concentration by $6/z^2$.

As the bulk correlation length ξ is smaller than the layer thickness λ, the free chains contribute a finite fraction of the concentration at $z = \lambda$. The free monomer concentration inside the layer has to be matched to the bulk value ϕ_b at

$z \simeq \lambda$. In the new regime $\xi < z < \lambda$, we obtain $\phi_f \simeq (\phi_b/R^2\lambda^3)z^5(1 + R^2z/\lambda^3)$. Closer to the adsorbing surface ($z < \xi$), the free chain concentration is dominated by end penetration and the z dependence is the same as in the dilute regime, but the amplitudes are changed to obtain the correct behavior when $z = \xi$; in the dilute tails region ($l < z < \xi$) $\phi_f = (\phi_b/R^2\lambda^3\xi^2)z^7$, whereas in the loops regime ($z < l$) $\phi_f = (\phi_b l^3/R^2\lambda^3\xi^2)z^4 \log(l/z)$. End penetration is thus enhanced by a factor $\lambda R/\xi^2$ as compared to the dilute regime.

The depletion hole, located at the edge of the layer, is very shallow in this regime: $-\delta\phi/\phi_b \simeq \epsilon$. The various concentration profiles in the semidilute tails regime are sketched in Figure 2.5(b).

The Semidilute Loops Regime

At higher concentrations $\phi_b > \phi_1$, a new regime appears ($\xi < z < l$) where the loops strongly overlap as in a semidilute solution. At distances from the wall larger than ξ, the overall concentration is close to its bulk value ϕ_b. Introducing the order parameter $\eta = \psi + B\varphi$ rather than φ the first integral, Eq. (27) of the order parameter equations reads

$$\psi'\eta' + B\psi - \frac{1}{2}U^2 - \epsilon U = 0. \tag{50}$$

Inside the layer, ϵ is negligible with respect to U. In the range $z > \xi$, we may further neglect U^2 itself at the lowest order of approximation. We introduce the reduced variables, $\tilde{\psi} = \psi/\sqrt{\phi_b}, \tilde{\eta} = \eta/\sqrt{\phi_b}$ and the length scale $l = \phi_b^{1/4}B^{-1/2}$ so that Eq. (50) reduces to $\tilde{\psi}'\tilde{\eta}' + \tilde{\psi}/l^2 = 0$. Now inserting $\tilde{\psi}\tilde{\eta} = 1$ and solving, we obtain

$$\psi = \frac{\sqrt{\phi_b}}{(1 + z/2l)^2}, \qquad \eta = \sqrt{\phi_b}(1 + z/2l)^2, \qquad \varphi = l^2((1 + z/2l)^2 - (1 + z/2l)^{-2}),$$

$$U = \frac{3}{2l^2}\frac{1}{(1 + z/2l)^2} \tag{51}$$

where the integration constant is chosen so that $\varphi(0) = 0$. The loop and tail concentrations cross at a distance of order l. Equation (51) leads to the estimate $z \simeq 0.38l$ for the location of the crossing point. In the semidilute loops regime ($\xi < z < l$), we obtain the asymptotic laws

$$\psi = \sqrt{\phi_b}(1 - z/l), \quad \varphi = 2lz, \quad \phi_t = 2\phi_b z/l, \quad \phi_l = \phi_b, \quad U = \frac{3}{2l^2}(1 - z/l). \tag{52}$$

A systematic expansion for $z < l$ can be found in ref. [6]. In the semidilute tails regime ($\lambda > z > l$), the asymptotic laws are similar to those for $\phi_b < \phi_1$:

$$\psi = 4\sqrt{\phi_b}\left(\frac{l}{z}\right)^2, \quad \varphi = \frac{z^2}{4}, \quad \phi_l = 16\phi_b\left(\frac{l}{z}\right)^4, \quad \phi_t = \phi_b, \quad U = \frac{6}{z^2} \quad (53)$$

Using the expression for ψ given by Eq. (51), we obtain l from the constant $B = 2\Gamma/(N\int_0^\infty \psi(z)dz)$:

$$l = \phi_b \frac{Nb}{2}. \quad (54)$$

The thickness of the layer $\lambda = \epsilon^{-1/2}$ obtained from the equilibrium between bulk chains and adsorbed chains Eq. (24) reads

$$\epsilon = \frac{4}{N}\log\frac{R}{l} = \frac{4}{N}\log\frac{\phi_2}{\phi_b} \quad \text{and} \quad \lambda = \epsilon^{-1/2}. \quad (55)$$

The overall approach breaks down at the crossover to weak adsorption $\phi_b = \phi_2$.

The tail concentration in the dilute loops region $z < \xi$ is expected to be similar to that in the dilute concentration regime. In the order parameter equation for φ, corrections to the potential with respect to $2/z^2$ (which are of order ϕ_b) turn out to be important. It is consistent to keep these corrections and to neglect the 1 on the right-hand side of Eq. (17). The order parameters ψ and φ are solutions of the same equation with different boundary conditions. As a consequence, $\varphi = $ const $\psi\int_0^z dz'\ \psi^{-2}$ and $\phi_t = C\psi^2\int_0^z dz'\ \psi^{-2}$. One can actually directly check that the latter expression for ϕ_t remains valid in the semidilute loops region in Eq. (52). Throughout the loop-dominated regime $z < l$, the loop concentration is accurately described by the standard ground state dominance approximation theory and $\psi = \sqrt{\phi_b}\ \coth(z/2\xi)$, where β is taken to be zero for $z > b$. From this we obtain the tail monomer concentration

$$\phi_t = \frac{2\phi_b z}{l}\ f_t\left(z/2\xi\right)$$

with:

$$f_t(x) = \frac{1}{\tanh(x)}\left[\frac{1}{\tanh(x)} - \frac{1}{x}\right] \quad (56)$$

where the integration constant C entering ϕ_t has been chosen to match the semidilute loops regime ($l > z > \xi$) Eq. (52). The asymptotic expression for ϕ_t in the dilute loops regime $\phi_t = (2/3)\phi_b z/l$ essentially differs from that in the dilute concentration range by a factor log (l/z).

Free chain penetration into the outer part of the layer ($z > l$) dominated by semidilute tails occurs as in the semidilute tails regime ($\phi^* < \phi_b < \phi_1$), and the free monomer concentration calculated there is still valid. In the loops region, r_f and ψ_f are both proportional to φ. In the semidilute loops regime ($\xi < z < l$), we therefore have $\phi_f \propto z^2$ whereas in the dilute loops regime $\phi_f \propto z^4$ holds. Matching the free monomer concentrations at $z = \lambda$, we obtain $\phi_f = \phi_b(z^2 l^3/R^2\lambda^3)$ in the semidilute loops region ($\xi < z < l$) and $\phi_f = \phi_b(z^4 l^3/R^2\xi^2\lambda^3)$ in the dilute loops regime ($b < z < \xi$).

The free chains are depleted outside the adsorbed layer, and the relative depth of the depletion hole $-\delta\phi/\phi_b \simeq \epsilon$ is very weak in this regime. The various concentration profiles in the semidilute loops regime are sketched in Figure 2.5(c).

2.4.3 ADSORPTION FROM A CONCENTRATED SOLUTION

When the adsorption proceeds from a concentrated solution ($\phi > \phi_2$), the bound state is not dominant all over the layer ($\epsilon N < 1$). The concentration is close to its bulk value over most of the layer (which has a size of the order of the radius of gyration R) except in the very vicinity of the wall ($z < \xi$). From now on, for convenience we will define a rescaled order parameter $\overline{\psi} = \psi/\psi(\infty)$. In the ground state dominance approximation the monomer volume fraction $\phi(z) = \text{const } \psi_0^2(z) = \phi_b\overline{\psi}^2(z)$ is given by Eq. (47), and the order parameter is

$$\overline{\psi}(z) = \coth\left(\frac{z}{2\xi} + \beta\right). \tag{57}$$

Note that $\overline{\psi}(z)$ is very close to 1 in the region $z \gg \xi$; therefore the molecular potential U vanishes in this region.

The chain partition function $Z_N(z)$ in the ground state dominance approximation is directly equal to the order parameter: $Z_N^{gs}(z) = \overline{\psi}(z)$. (Here we again take into account the fact that $E_0 = 0$; hence Z_N^{gs} does not depend on N.) Beyond the ground state dominance approximation, all the high-order eigenfunctions of Eq. (9) contribute to Z_N:

$$Z_N(z) = \psi(z) + \sum_{q>0} K_q \psi_q(z) e^{-q^2 N}$$

where the index $q > 0$ runs over all the higher-order eigenfunctions corresponding to $E_q = q^2$ (note that $E_q > 0$ since $E_0 = 0$), and $K_q = \int_0^\infty \psi_q(z)dz$. The sum is exponentially cut off at high q's, so that only the eigenfunctions with $q \lesssim 1/\sqrt{N}$ contributes.

The eigenfunctions show the following asymptotic behavior in the region $z \gg \xi$ where $U \simeq 0$: $\psi_q(z) \propto \cos(qz + \text{const})$. On the other hand, in the region $z \sim \xi$

the wave-vector dependence of the eigenfunctions is weak (for $q \lesssim 1/\sqrt{N}$) if $\xi \ll R \equiv \sqrt{N}$: $\psi_q(z) \propto \psi_0(z)[1 + O(q^2\xi^2)]$ in this region. Implying a smooth crossover between the regions $z \sim \xi$ and $z \gg \xi$ and taking into account the normalization condition, we get

$$\psi_q(z) \simeq \sqrt{2}\,\cos(qz)\psi_0(z)$$

for $0 < q \lesssim 1/\sqrt{N}$. However, this expression is only approximate and leads to eigenfunctions ψ_q that are not exactly orthogonal to ψ_0. If $q > 0$, the amplitude K_q can be represented as

$$K_q = \int_0^\infty \psi_q(z)[1 - \overline{\psi}(z)]dz \simeq \sqrt{2}\psi_0(\infty)\eta_0$$

where

$$\eta_0 \equiv \int_0^\infty \overline{\psi}(z)[1 - \overline{\psi}(z)]dz. \tag{58}$$

Here we take into account the fact that $\overline{\psi}(z)$ and $\psi_q(z)$ form an orthonormal set of functions, and that $\eta(z) \equiv \overline{\psi}(z)[1 - \overline{\psi}(z)]$ vanishes in the region $z \gg \xi$, whereas $\cos(qz) \simeq 1$ for $z \lesssim \xi$. Thus, we obtain

$$Z_n(z) = \overline{\psi}(z) + 2\psi_0^2(\infty)\eta_0\overline{\psi}(z)\sum_q \cos(qz)e^{-q^2n}$$

$$= \overline{\psi}(z)[1 + \eta_0 f_Z(z, n)] \tag{59}$$

where $f_Z(z, n) = 2\psi_0^2(\infty) \sum_q \cos(qz)e^{-q^2n}$. In the relevant region $n \lesssim N$, $z \lesssim R = \sqrt{N}$, we get $f_Z \sim 1/R$. Equation (59) is valid if the correction to the ground state is small; i.e., if $\eta_0 f_Z \ll 1$. This condition can be rewritten as

$$|\eta_0| \ll R. \tag{60}$$

While it is possible to calculate explicitly the function $f_Z(z, n)$, it is more important to note the following general points. The correction to the ground state partition function is split into two factors; one of them, η_0, depends on the potential U but does not depend on the chain length n. The other factor, f_Z, is a universal function of z and n that does not depend on the potential U. It is straightforward to show that such a splitting is general—it is also valid for the correction to the total monomer density, which can be written as

$$\phi(z) = \phi_{gs}(z)[1 + \eta_0 f_\phi(z, N)] \tag{61}$$

where $\phi_{gs}(z)$ is defined in Eq. (47). Equation (61) remains valid when we take into account the fact that the potential U must be obtained in a self-consistent way, i.e., that it is determined itself by the monomer density profile, so that a correction to $\phi(z)$ induces a correction to $U(z)$.

The function $f_\phi(z, N)$ can be obtained in the following way. Let us consider the case of a weakly adsorbing wall, $b \to \infty$ (or physically $b \gg R$), when the function $\phi(z)$ can be calculated using a perturbation approach. Then we calculate $\phi_{gs}(z)$ and η_0 using Eqs. (47), (57), and (58) and finally find the universal function $f_\phi(z, N)$ using Eq. (61).

To proceed with the perturbative calculations, we first make use of the mirror-image principle: transform the half-space system with the boundary condition (2) to the full-space system (original solution plus its mirror image). The boundary condition is now replaced by the surface potential

$$U_s(z) = -\frac{2}{b} \delta(z).$$ (62)

The mirror-image principle then ensures that all the statistical properties of the full-space system coincide with those of the original half-space system as long as we do not need to distinguish between the original system and its image. The limit $b \to \infty$ means that the external potential is weak, so that the induced monomer density profile is given by linear response: $\phi(z) = \phi_b + \delta\phi(z)$,

$$\delta\phi(z) = -\int \kappa(z')U_s(z - z')dz'$$

where κ is the susceptibility of the concentrated polymer solution, which can be calculated using the random phase approximation. The result is [6, 33]

$$\kappa_q = \frac{\kappa_0(q)}{1 + \kappa_0(q)}$$

where $\kappa_q = \int \kappa(z)e^{iqz} \, dz$, and $\kappa_0(q) = \phi_b N g_D(Nq^2)$ is the susceptibility of the corresponding ideal system where $g_D(u) = (2/u^2)(u - 1 + e^{-u})$ is the Debye function. After some transformation, in the limit $\phi_b N \gg 1$ we get

$$\delta\phi(z) \simeq \frac{1}{b}\frac{1}{\xi}e^{-z/\xi} - \frac{1}{b\phi_b N}\int\frac{dq}{2\pi}e^{iqz}\frac{u(1 - e^{-u})}{u - 1 + e^{-u}}$$ (63)

where $u \equiv Nq^2$. The ground state dominance result is recovered in the limit $N \to \infty$:

$$\phi_{gs}(z) = \phi_b + \frac{1}{b}\frac{1}{\xi}e^{-z/\xi}, \quad \overline{\psi}(z) \equiv \sqrt{\phi_{gs}(z)/\phi_b} = 1 + \frac{\xi}{b}e^{-z/\xi}$$

(the same result is obtained using Eq. (57)). Thus,

$$\eta_0 = \int_0^\infty \overline{\psi}(1 - \psi)dz \simeq \frac{-1}{2\phi_b b}.$$

Finally, using Eq. (61) we obtain in the region $z \gg \xi$

$$f_\phi(z, N) = \frac{2}{N^{3/2}\phi_b} G(z/\sqrt{N}),$$
(64)

where

$$G(x) = \int_{-\infty}^{\infty} \frac{dq}{2\pi} \frac{1 - (1 + q^2)e^{-q^2}}{q^2 - 1 + e^{-q^2}} e^{iqx}.$$

We neglect here a $\delta(z)$ term that appears in $f_\phi(z, N)$. This term actually represents a small correction to $\phi_{gs}(z)$ localized in the region $z \sim \xi$ and is thus not important in the region $z \gg \xi$ that we consider.

The fact that the amplitude of the long-range inhomogeneity (with the scale $z \sim R$) induced by the adsorbing wall is proportional to η_0 can be physically interpreted in the following way [34, 26]. The total number of chains can be calculated in two ways with the ground state dominance approximation. One way is to first integrate the monomer density:

$$\mathcal{N}_m = \frac{1}{Nv} \int_0^{\infty} \psi_{gs}(z)dz = \frac{\phi_b}{Nv} \int_0^{\infty} \overline{\psi}^2(z)dz.$$

Another way is to calculate half of the total number of chain ends (since each pair of ends corresponds to one chain of N monomers): $\mathcal{N}_e = (1/2v)\int \phi_e(z)dz$, where $\phi_e(z) = (2\phi_b/N)Z_N^{gs}(z)$. Taking into account that $Z_N^{gs}(z) = \overline{\psi}(z)$, we get $\mathcal{N}_m = (\phi_b/Nv)\int \overline{\psi}(z)dz$. These two results do not coincide and the difference

$$\mathcal{N}_e - \mathcal{N}_m = \frac{\phi_b}{Nv} \int \overline{\psi}(1 - \overline{\psi})dz$$
(65)

is a measure of the intrinsic error of the ground state dominance approximation. It is thus natural to expect that the correction to the ground state distribution functions is proportional to the difference, that is, proportional to η_0. Note that $\mathcal{N}_e - \mathcal{N}_m$ is negative (since $\overline{\psi} > 1$), and a negative correction to $\phi_{gs}(z)$ is expected.

We now consider the case of strong adsorption. Using Eqs. (57), (58), (61), and (64), we get

$$\eta_0 \simeq -\frac{2}{b\phi_b}$$
(66)

and

$$\phi(z) - \phi_b = -\frac{4}{b\phi_b N^{3/2}} G(z/R)$$
(67)

in the region $\xi \ll z$. Note that the perturbation is negative as expected. For $z \lesssim R$, the factor $G(z/R)$ can be approximated as $G(z/R) \simeq 0.62 - 0.5z/R$; this factor nearly vanishes for $z > 3.5R$. It is interesting to note that the long-range concentration correction is increasing as the bulk concentration is decreased. Equation (67) is valid if $|\eta_0| \ll R$, that is, if $\phi_b \gg 1/\sqrt{Nb} \equiv \phi_2$. Note that at $\phi_b \sim \phi_2$ the typical amplitude of the concentration perturbation given by Eq. (67) is $\delta\phi \sim 1/N$; i.e., it smoothly crosses over with the prediction for the semidilute loops regime. The various concentration profiles in the concentrated regime are sketched in Figure 2.5(d).

2.5 Interaction Between Adsorbed Polymer Layers: Concentrated Solution

In the two following sections we discuss the interactions between two parallel surfaces immersed in a polymer solution. In Section 2.6 we consider in detail the case of a dilute or a semidilute solution. In this section, we start with the case of a concentrated solution that corresponds to a weak adsorption as explained above and that requires a specific treatment.

We consider two parallel adsorbing surfaces immersed in a concentrated polymer solution, $\phi_b > \phi_2$. The free energy of the system depends on the distance $2h$ between the surfaces; there is therefore an interaction force between the plates mediated by the polymer solution. We can distinguish between two cases: (1) irreversible adsorption when the amount of polymer in the gap is fixed; (2) reversible adsorption when the polymer in the gap is in equilibrium with the bulk solution ϕ_b. In this section only reversible adsorption is considered.

The interaction between reversibly adsorbing plates was shown by de Gennes [8] to be always attractive; the force per unit area is

$$f_{gs} = -\frac{1}{2v}(\phi(h) - \phi_b)^2 \tag{68}$$

where $\phi(h)$ is the monomer concentration at the midplane between the plates. Let us assume that $h \gg \xi$. For the case of one adsorbing plate, we get from Eqs. (47) and (57) $\phi(h) - \phi_b \simeq 4\phi_b e^{-h/\xi}$. In the presence of two plates, the concentration perturbation is twice larger: $\phi(h) - \phi_b \simeq 8\phi_b e^{-h/\xi}$. Therefore,

$$f_{gs} = -\frac{32}{v}\phi_b^2 e^{-2h/\xi}. \tag{69}$$

The attractive interaction force is exponentially weak in the region $h \gg \xi$.

As discussed in the previous section, the corrections to the ground state dominance approximation due to the finite molecular weight (end corrections) give rise to a concentration perturbation, Eq. (67), that is *not* exponentially small in the region $h \gg \xi$. It is thus natural to expect a similar noticeable interaction force between two plates in this region. Since the simple equation (68) is valid only within the ground state dominance approximation, we have to reconsider the two-plates problem to find the interaction force. Our next task is to find the end corrections to the free energy of an inhomogeneous polymer solution.

2.5.1   FREE ENERGY

The free energy of a concentrated polymer solution can be written as

$$F = \frac{1}{v}(F_{\text{conf}} + F_{\text{int}} + F_s) \tag{70}$$

where the factor $1/v$ is introduced in order to simplify the equations below. $F_s = \int U_s(z)\phi(z)dz$ corresponds to monomer–surface interactions, $F_{\text{int}} = (1/2)\int \phi^2 \, dz$ accounts for the monomer–monomer excluded volume interactions, and F_{conf} is the conformational free energy accounting for the reduction in conformational entropy of the polymer chains in an inhomogeneous state. (By definition, F_{conf} includes also the ideal gas term.) In the ground state dominance approximation, the conformational free energy is [33, 35, 36, 29, 30]

$$F_{\text{conf}}^{gs} = \frac{1}{4} \int \frac{(\nabla \phi)^2}{\phi} \, dz + \int \frac{\phi}{N} \ln \frac{\phi_b}{eN} \, dz \tag{71}$$

where $\nabla \phi = d\phi/dz$ for a one-dimensional heterogeneity. The second term on the right-hand side of Eq. (71) corresponds to the ideal gas entropy of the polymer chains (note that the addition of this term is equivalent to a simple renormalization of the monomer chemical potential). Equation (71) is obviously exact for a homogeneous solution with $\phi = \phi_b$. In the general case, we write

$$F_{\text{conf}} = F_{\text{conf}}^{gs} + F_{\text{conf}}^e \tag{72}$$

where F_{conf}^e is the end correction, which vanishes for $\phi \equiv \phi_b$ and thus can be expanded as a series of $\delta\phi(z) = \phi(z) - \phi_b$:

$$F_{\text{conf}}^e = A\delta\phi_0 + \frac{1}{2} \int \frac{dq}{2\pi} B(q)\delta\phi_q \delta\phi_{-q} + \dots \tag{73}$$

where $\delta\phi_q \equiv \int \delta\phi(z)e^{iqz}dz$. Higher-order terms are omitted here.

The considerations of Section 2.4.3 suggest that the proper order parameter is not $\delta\phi(z)$ but rather $\eta(z) = \overline{\psi}(z)[1 - \overline{\psi}(z)] = \sqrt{\phi(z)/\phi_b} - \phi(z)/\phi_b$. Formally expanding F^e_{conf} as a perturbation series in $\eta(z)$, we write

$$F^e_{conf} \simeq A'\eta_0 + \frac{1}{2}\int B'(q)\eta_q\eta_{-q}\frac{dq}{2\pi} \tag{74}$$

where $\eta_q = \int \eta(z)e^{iqz}dz$. Higher-order terms in this series can be omitted if $|\eta(z)| \ll 1$ everywhere; this condition ensures that $|\delta\phi(z)/\phi_b| \ll 1$, so that higher-order terms can be also neglected in Eq. (73). However, the region of validity of the η-perturbation expansion, Eq. (74), is actually much broader than for Eq. (73). Arguments similar to those presented in Section 2.4.3 show that the η-perturbation scheme is also valid if the concentration fluctuation is large ($\delta\phi \sim \phi_b$) but is localized in a thin region provided that $|\eta_0| \ll R$ (see condition (60)).

We are now in a position to calculate the kernel $B'(q)$ and A'. First we expand both terms in Eq. (74) as a series of $\delta\phi$ up to the second order, and then we compare the result with Eq. (73). Thus, we get

$$A = -\frac{A'}{2\phi_b}; \qquad B(q) = \frac{B'(q)}{4\phi_b^2} - \frac{A'}{4\phi_b^2}. \tag{75}$$

The parameters A and $B(q)$ can be found by comparing F_{conf} defined by Eqs. (71–73) with the well-known random phase approximation for the conformational free energy [33]:

$$F_{conf} = F_0 + \mu_0\delta\phi_0 + \frac{1}{2N\phi_b}\int \frac{1}{g_D(q^2N)}\delta\phi_q\delta\phi_{-q}\frac{dq}{2\pi} \tag{76}$$

where $F_0 = (1/N)\int \phi_b \ln(\phi_b/Ne)dz$ is the free energy of the homogeneous system with $\phi(z) \equiv \phi_b$, and $\mu_0 = (1/N)\ln(\phi_b/N)$ is the corresponding monomer chemical potential. Equating both the linear and quadratic terms in the two expressions for F_{conf}, we obtain

$$A = \frac{1}{N}; \qquad B(q) = \frac{1}{N\phi_b}\left[\frac{1}{g_D(Nq^2)} - \frac{Nq^2}{2}\right] \tag{77}$$

Using now Eqs. (74), (75), and (77), we finally get the end contribution to the conformational free energy:

$$F^e_{conf} \simeq \frac{2\phi_b}{N}\left[-\eta_0 + \frac{1}{2}\int \frac{dq}{2\pi}g(q^2N)\eta_q\eta_{-q}\right] \tag{78}$$

where

$$g(u) = \frac{1 - (u+1)e^{-u}}{u - 1 + e^{-u}}.$$

It is natural to expect that the expansion (78) is rapidly converging (i.e., the omitted higher-order terms are negligible) if the second-order term is much smaller than the first-order term. If the inhomogeneity is basically localized in a thin region ($\xi \ll \sqrt{N}$), then $\eta_q \sim \eta_0$ for $q \sim 1/\sqrt{N}$. Therefore we estimate the integral in Eq. (78) as $\eta_0^2/\sqrt{N}$. The condition of validity of Eq. (78) is $|\eta_0| \ll \sqrt{N}$, in agreement with our previous assumption.

Equation (78) was rigorously derived in ref. [26]. It was also shown that the general condition of its validity is $|\tilde{\eta}(z)| \ll 1$, where $\tilde{\eta}(z)$ is the order parameter smoothed over a distance of order $R = \sqrt{N}$.

Substituting Eqs. (71) and (78) in Eq. (72), we get the conformational free energy. The total free energy of the polymer solution is then obtained from Eq. (70). The free energy can be used to study the adsorption on a single plate already considered in Section 2.4.3. Using the surface potential of Eq. (62) (note that here we again use the mirror-image trick to map a semi-infinite system to an infinite one as in Section 2.4.3, we minimize the free energy F over the monomer volume fraction $\phi(z)$ and look for the asymptotic behavior in the region $z \gg \xi$. We get exactly the same result as before: Eq. (67). It is interesting to note that for a moderate adsorption strength ($b \sim \xi$) the square-gradient free energy (first term in Eq. (71)) is of order $1/\xi$, whereas the two end corrections given by Eq. (78) are of order ξ/N and $\xi^2/N^{3/2}$, respectively. The corrections are much smaller than the main term since $\phi_b \gg \phi_2$.

2.5.2   INTERACTION BETWEEN ADSORBING PLATES

The free energy expressions derived in the previous section are valid for an infinite system when the concentration ϕ is close to ϕ_b almost everywhere. Therefore these equations cannot be directly applied to a polymer solution in a slab between two adsorbing plates. However, the layer can be mapped to an infinite system by mirror-image transformations, with two mirrors located at the adsorbing surfaces. Thus we get an infinite periodic system; the effect of adsorbing surfaces is now accounted for by the surface potential U_s obtained as a periodic continuation of Eq. (62):

$$U_s = -\frac{2}{b} \sum_{n=-\infty}^{\infty} \delta(z - 2nh)$$

where $2h$ is the period or the distance between the plates. The free energy per period, the monomer concentration profiles, and the other statistical properties of the infinite periodic system exactly coincide with the properties of the original system.

According to general thermodynamic rules, the force between the plates is given by the derivative of the relevant thermodynamic potential Ω with respect to the distance $2h$:

$$f = -\frac{\partial \Omega}{\partial (2h)}. \tag{79}$$

Since the system is in equilibrium with a bulk solution with a monomer chemical potential $\mu_b = \phi_b + (1/N) \ln (\phi_b/N)$ and is also kept under external bulk pressure

$$\Pi_b = \frac{1}{v}\left(\frac{\phi_b}{N} + \frac{1}{2}\phi_b^2 \right),$$

the relevant thermodynamic potential is the grand canonical free energy:

$$\Omega(\mu_0, h) = F - \mu_b \frac{1}{v} \int \phi(z)dz + 2h\Pi_0. \tag{80}$$

The free energy F is defined by Eqs. (70), (72), (71), and (78), where we should insert a periodic function $\phi(z)$ and calculate the energy per period. Taking into account that the system is symmetric with respect to the midplane $z = h$, we rewrite Eq. (80) as

$$\Omega = \frac{2}{v}[\Omega_0 + \Omega_1] \tag{81}$$

where

$$\Omega_0 = \frac{1}{2}\int_0^h (\phi - \phi_b)^2 \, dz - \frac{1}{b}\,\phi(0) + \frac{1}{4}\int_0^h \frac{(d\phi/dz)^2}{\phi} \, dz \tag{82}$$

is the free energy in the ground state approximation and

$$\Omega_1 = -\frac{2\phi_b}{N}\,\eta_0 + \frac{\phi_b}{Nh}\sum_q g(Nq^2)\eta_q^2 \tag{83}$$

is the end correction (see Eq. (78)). Here $\eta_q \equiv \int_0^h \eta(z)\cos(qz)dz$, and $q = (2\pi/2h)n$, $n = 0, \pm 1, \pm 2, \ldots$. Note that the term $(1/N)\int \phi(z)\ln(\phi_b/eN)dz$ (the second term in Eq. (71)) does not contribute to Ω; this term is proportional to the total number of chains and has nothing to do with the interactions between the plates.

We can now consider Ω_1 as a small perturbation to the ground state free energy and substitute the η_q obtained in the ground state dominance approximation (i.e., η_q obtained by minimization of Ω_0). In the limit where $h \gg \xi$, the concentration profile close to one plate is nearly unperturbed by the presence of another plate. Thus we can use the one-plate result, Eq. (57), for the function $\eta(z) = \overline{\psi}(z)[1 - \overline{\psi}(z)]$, which is small everywhere apart from a thin layer of size ξ near each plate. The relevant wave vectors $q \sim 1/\sqrt{N}$ are small compared with $1/\xi$ and so we write $\eta_q \simeq \eta_0 \simeq -(2/\phi_b b)$. Equation (83) thus reduces to

$$\Omega_1 \simeq \frac{N}{b} + \frac{8}{NRb^2} \frac{1}{\phi_b} u_{\text{int}}(2h/R)$$

where

$$u_{\text{int}}(x) = \frac{1}{x} \sum_{n=-\infty}^{\infty} g\left(\frac{4\pi^2 n^2}{x^2}\right). \qquad (84)$$

The interaction force is calculated by derivation of the free energy and is written as $f = f_{gs} + f_1$, where f_{gs} is the ground state dominance result, and

$$f_1 = -\frac{1}{v} \frac{\partial \Omega_1}{\partial h} = -\frac{1}{v} \frac{16}{N^2 b^2 \phi_b} u'_{\text{int}}\left(\frac{2h}{R}\right)$$

is the contribution to the force due to the chain end (finite molecular weight) effects. Here $u'_{\text{int}}(x) = (d/dx)\, u_{\text{int}}(x)$. In particular, in the region $\xi \ll h \ll R$ Eq. (84) can be simplified as $u_{\text{int}}(x) \simeq 1/x$, so that the total force is

$$f \simeq -\frac{32}{v} \phi_b^2 e^{-2h/\xi} + \frac{1}{v} \frac{4}{N\phi_b b^2 h^2}. \qquad (85)$$

In the region $h \gg \xi$ the first term is exponentially small, and the force is dominated by the second term, which is due to end effects. The force is repulsive (in contrast to the attraction force predicted by the ground state dominance approximation) and is decreasing with the separation between the plates as $1/h^2$.

These results can be qualitatively explained as follows. First we note that adsorbing plates are less attractive for chain ends than for other monomers. In fact, the monomer concentration ϕ near a plate is increasing as $\overline{\psi}^2$ whereas the end concentration ϕ_e is proportional to $\overline{\psi}$, so that $\phi/\phi_e \propto \overline{\psi}$, where $\overline{\psi} \gg 1$ if $z \ll \xi$, $b \ll \xi$. Therefore, the monomer surface excess, $\Gamma = \int(\phi - \phi_b)dz \simeq 2/b$, is not balanced by a comparable excess of end monomers, so that $(2/N)\Gamma$ additional end monomers are distributed somewhere in the region far from the surface. (Note that exactly two ends correspond to each N monomers, hence the factor $2/N$.) Assuming that these additional ends are distributed nearly homogeneously throughout the layer, we get the excess concentration of ends, $\delta\phi_e \simeq (1/h)(2/N)\Gamma$. The corresponding excess free energy of an ideal gas of chain ends is

$$\delta F_e = \frac{1}{v} \int \delta \left(\phi_e \ln \frac{\phi_e}{e} \right) dz$$

$$\simeq \frac{1}{v} \int \left\{ \delta \phi_e \ln \phi_{e0} + \frac{1}{2} \frac{\delta \phi_e^2}{\phi_{e0}} \right\} dz$$

$$= \frac{1}{v} \frac{2}{N} \Gamma \ln \phi_{e0} + \frac{1}{2v} \left(\frac{2}{N} \Gamma \right)^2 \frac{1}{h \phi_{e0}}$$

where $\phi_{e0} = (2/N)\phi_b$ is the average concentration of ends. Therefrom, we get the effective force between the plates induced by the additional ends expelled from the adsorbed layers,

$$f_e \simeq -\frac{\partial (\delta F_e)}{\partial h} = \frac{4}{h^2 \phi_b v N b^2}$$

in agreement with Eq. (85).

2.6 Interaction Between Adsorbed Polymer Layers: Semidilute and Dilute Regimes

We now consider a less concentrated bulk solution in the dilute or semidilute range $\phi_b \ll \phi_2 = 1/\sqrt{N}b$. As for the adsorption on a single surface, we use here an approach based on two-order parameters. We first discuss the free energy of a solution confined in a gap of thickness $2h$ and then the interaction between the surfaces mediated by the polymer solution in the cases of both reversible and irreversible adsorption.

2.6.1 *FREE ENERGY OF A CONFINED POLYMER SOLUTION*

The free energy of a polymer solution in contact with surfaces can be split into three parts as for a concentrated solution $F = (1/v)(F_{conf} + F_{int} + F_s)$. The last two terms are easily calculated. We must now study in detail the conformational free energy F_{conf}. In the following discussion, we take advantage of the symmetry of the system with respect to the midplane $z = h$ and consider only one-half of the initial system, $0 < z < h$.

To get the conformational free energy F_{conf}, we first consider a solution of ideal chains near an adsorbing wall where the chains are also submitted to a fixed external potential $U(z)$, which induces a given monomer concentration profile $\phi(z)$. The free energy of this model system is

$$F = \frac{1}{v}(F_{\text{conf}} + F_s + F_U) \tag{86}$$

where

$$F_s = \int U_s(z)\phi(z) = -\frac{1}{b}\phi(0) \tag{87}$$

is the effective monomer–surface interaction free energy (note that $U_s(z) = 1/b\delta(z)$, which is different from Eq. (62): in that equation both the original surface potential U_s and its mirror image are taken into account, which results in doubling the potential), and

$$F_U = \int U(z)\phi(z)dz \tag{88}$$

is the external field contribution. The total partition function is written as a function of the total number of chains $\mathcal{N} = (1/Nv)\int \phi(z)dz$ and of the single chain partition function Z_N:

$$Z_{\text{tot}} = \frac{Z_N^{\mathcal{N}}}{\mathcal{N}!}. \tag{89}$$

The single chain partition function is obtained by summation of the partition function with a fixed endpoint $Z_N(z)$. The expansion over the eigenstates of Eq. (10) leads to

$$Z_N = \sum_i K_i^2 e^{-E_i N}. \tag{90}$$

The total free energy of the solution is $F = -\ln Z_{\text{tot}}$.

We focus on the short separation limit $h \ll R = \sqrt{N}$. In this regime, the separation between the ground state energy level E_0 and the next level E_1 is $E_1 - E_0 \sim 1/h^2 \gg 1/N$, so that the sum (90) is dominated by the ground state term $i = 0$. All the other terms are exponentially small:

$$Z_N \simeq K_0^2 e^{-E_0 N}. \tag{91}$$

Using Eqs. (86–89) and (91), the eigenstates given by Eq. (9), and the corresponding boundary conditions,

$$\frac{1}{\psi_i}\frac{\partial \psi_i}{\partial z} = \begin{cases} -1/b, & z = 0 \\ 0, & z = h \end{cases} \tag{92}$$

(the second condition is due to the symmetry of the slab), we obtain

$$F_{conf} = \int \eta' \psi' \, dz + v\mathcal{N} \ln \frac{v\mathcal{N} \int \psi^2 \, dz}{e(\int \psi dz)^2}. \tag{93}$$

The order parameter $\psi(z)$ is proportional to the ground state eigenfunction $\psi(z) = \Lambda\psi_0(z)$ (Λ is an arbitrary constant). It satisfies a Schrödinger-like equation similar to Eq. (9):

$$-\psi'' + (U - E_0)\psi = 0 \tag{94}$$

The new order parameter

$$\eta(z) = \frac{\Gamma}{\Lambda}\left(\psi_0(z) + \frac{2r(z)}{NK_0}\right)$$

is defined so that $\phi(z) = \psi(z)\eta(z)$. It satisfies the equation

$$-\eta'' + (U - E_0)\eta = 2v\mathcal{N}\left[\frac{1}{\int \psi \, dz} - \frac{\psi}{\int \psi^2 \, dz}\right] \tag{95}$$

which is similar to Eq. (13) for the rest function $r(z)$.

In the limit $N \rightarrow \infty$, the second term in Eq. (93) as well as the right-hand side of Eq. (95) vanish. Therefore the new order parameter η is proportional to ψ in this limit. When $N \rightarrow \infty$ we thus recover the ground state dominance approximation

$$F_{conf} = \frac{1}{4}\int \frac{(\phi')^2}{\phi} \, dz,$$

as expected.

We now return to the ideal chain solution in an external field $U(z)$ and minimize the total free energy given by Eq. (86), $F = F[\psi, \eta]$, with respect to the two independent fields $\psi(z)$ and $\eta(z)$. (The monomer concentration is $\phi = \psi\eta$; the total number of monomers, $\mathcal{N}$, is kept constant.) This leads to Eq. (94) for $\psi(z)$ and Eq. (95) for $\eta(z)$ with the proper boundary conditions. Hence, $F[\psi, \eta]$ is a proper thermodynamic potential for this problem.

The grand canonical potential of the system, Ω, is defined by Eq. (80). The minimization of Ω again leads to the order parameter equations (94) and (95). When $U(z) = \phi(z) - \phi_b$, the order parameter equations have a first integral

$$f(z) = \frac{1}{v}\left\{\psi' \eta' - \frac{1}{2}(\phi - \phi_b)^2 + \phi_e + E_0\phi - v\mathcal{N}\psi_0^2 - \frac{\phi_b}{N}\right\}. \tag{96}$$

Here $\phi_e(z) = 2v\mathcal{N}\psi(z)/\int \psi(z)dz$ is volume fraction of chain ends. We can directly check that $f(z) = $ const.

2.6.2 *INTERACTION BETWEEN REVERSIBLY ADSORBED LAYERS*

For the real polymer solution in a gap between two plates, the total free energy is given by Eq. (70) with the same conformational free energy (93) as for an ideal solution in a fixed external potential with the same concentration profile. The grand canonical potential is then obtained by Legendre transformation as in Eq. (80). The interaction force between the plates is found by looking for a small change of the grand potential Ω caused by a small change of the order parameters ψ and η when the distance between plates is changed from $2h$ to $2(h + \delta h)$. The result is

$$f = \frac{1}{v}\left\{-\frac{1}{2}(\phi(h) - \phi_b)^2 + \phi_e(h) + E_0\phi(h) - v\mathcal{N}\psi_0^2(h) - \frac{\phi_b}{N}\right\}. \quad (97)$$

The force coincides with the first integral defined in Eq. (96).

To calculate the force, we need to solve the order parameter equations (94) and (95) with boundary conditions similar to Eq. (92) and to substitute the result in Eq. (97). We describe in the following the main results for the dilute and the semidilute regimes, $\phi \ll \phi_2 = 1/b\sqrt{N}$. We also assume that the separation between the plates is small enough, $h \ll \lambda$, where $\lambda = 1/\sqrt{\epsilon}$ ($\epsilon = -E_0$ is always positive) is the cutoff length calculated in Sections 2.3 and 2.4. This length is smaller than the coil size $R = \sqrt{N}$ by a logarithmic factor, so that the condition $h \ll \lambda$ is essentially equivalent to $h \ll R$. It is possible to directly check that the last three terms in the curly brackets in Eq. (97) are subdominant with these approximations. Thus, we write

$$f \simeq \frac{1}{v}\left\{-\frac{1}{2}(\phi(h) - \phi_b)^2 + \phi_e(h)\right\}. \quad (98)$$

The first term, which is equivalent to the ground state dominance approximation, Eq. (68), is negative and gives an attractive contribution to the force. On the contrary, the second term, which represents the end correction, is always positive and gives a repulsive contribution to the force. The chain end contribution can be interpreted as an ideal gas pressure of chain ends at the midplane.

We start with a dilute solution, $\phi_b \ll 1/N$. In the region dominated by loops, $h \ll l$, where $l = (Nb \ln (N/b^2))^{1/3}$ is the crossover length calculated in Section 2.3, the ground state dominance approximation is accurate (the first term in curly brackets in Eq. (98) dominates). Thus, we recover the well-known results [9]:

$$f = \begin{cases} -\dfrac{1}{2vb^2h^2}, & h \ll b \\[2ex] -\dfrac{5.91}{vh^4}, & b \ll h \ll l \end{cases}. \quad (99)$$

The force is attractive. To explain qualitatively the interaction in these two regimes, we note that ϕ_b can be neglected with respect to $\phi(h)$. The relevant concentration is $\phi(h) \sim 1/bh$ when $h \ll b$; the monomer concentration is nearly uniform, so that $\phi(h)$ can be obtained by balancing the surface energy $-\phi/b$ and the excluded volume energy $(1/2)\phi^2 h$. When $h > b$, the relevant concentration is $\phi(h) \sim 1/h^2$. In this regime, $\phi \propto \psi^2$, and ψ is approximately the order parameter for the adsorption on a single plate given by Eq. (57).

In the tails-dominated regime, $l \ll h \ll \lambda$, the end correction to the force becomes essential, giving rise to a repulsive interaction:

$$f = \frac{420}{vh^4}, \quad l \ll h \ll \lambda. \tag{100}$$

By adjusting the factor Λ, the order parameter equations in this regime reduce to $-\psi'' + \eta\psi^2 = 0$, $-\eta'' + \psi\eta^2 = 1$ and the order parameters have the following scaling behaviors: $\eta \propto h^2$, $\psi \propto 1/h^4$. Substituting these results in Eq. (98), we obtain the same scaling behavior for both terms, $1/h^4$. The actual prefactor 420 has been calculated by numerical solution of the order parameter equations. The force versus separation profile is sketched in Figure 2.6.

In a semidilute solution, $1/N \ll \phi_b \ll \phi_1 = 1/(bN)^{2/3}$, the force is still given by Eqs. (99) if $h \ll l$ and Eq. (100) if $l \ll h \ll \xi$, where $\xi = 1/\sqrt{2\phi_b}$ is the bulk correlation length. In the region $h \gg \xi$, the monomer concentration (in the central part of the system) is nearly constant and is equal to the bulk concentration: $\phi = \psi\eta \simeq \phi_b$. The order parameter equations here can be simplified as $-\psi'' + U\psi = 0$, $-\eta'' + U\eta = 1$. (Note that only two unknown functions, say, U and ψ, are independent since $\eta \simeq \phi_b/\psi$.) A scaling analysis of the equations shows that $\eta \propto h^2$, $\psi \propto 1/h^2$, and $U = \phi - \phi_b \propto 1/h^2$. Hence, the end term in Eq. (98) scales as $1/h^2$, and the ground state term scales as $1/h^4$ and thus can be neglected. The exact result in this regime is [25]:

$$f \simeq \frac{\pi^2 \phi_b}{vh^2}, \quad \xi \ll h \ll \lambda. \tag{101}$$

For bulk concentrations higher than ϕ_1 (in the regime $\phi_1 \ll \phi \ll \phi_2$), the bulk correlation length is shorter than the crossover length: $\xi \ll l$, where $l \sim N\phi_b b$ (see Section 2.4). Here for small separations, $h \ll \xi$, the force is still given by Eqs. (99). In the region $\xi \ll h \ll l$, the end corrections to the monomer density are small so that the first term in Eq. (98) nearly coincides with the force, Eq. (69), obtained from the ground state dominance approximation. However, the end-induced force (the second term in Eq. (98)) cannot be neglected here. The

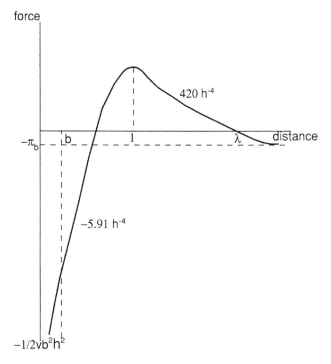

Figure 2.6 Equilibrium force between adsorbed polymer layers as a function of the separation between surfaces in the dilute regime.

total number of monomers between the plates is nearly constant in this regime, and the adsorbance is

$$\Gamma = \int_0^h \phi(z)dz \simeq \frac{2}{b}.$$

Hence, the total number of chain ends is

$$\mathcal{N}_e = \frac{2}{Nv} \int \phi(z)dz \simeq \frac{4}{Nvb}.$$

The chain ends are nearly uniformly distributed across the layer (the distribution is given by the function $\psi(z)$, Eq. (57), which is roughly constant in the region $\xi \ll z \ll h$), and the volume fraction of ends is

$$\phi_e(h) \simeq \frac{v\mathcal{N}_e}{h} \simeq \frac{4}{Nbh}.$$

Thus, the total interaction force is

$$f = f_{gs} + \phi_e(h)/v \simeq \frac{4}{Nbv} \frac{l}{h} - \frac{32}{v} \phi_b^2 e^{-2h/\xi}, \quad \xi \ll h \ll l. \quad (102)$$

Note that the force is repulsive if $h > \xi \ln (\phi_b/\phi_1)$ (the end contribution dominates in this region). For larger separations, $l \ll h \ll \lambda$, the interaction decays as $1/h^2$ according to Eq. (101).

In summary, the equilibrium interaction force is always attractive at short separations, and it is repulsive if the separation is larger than either l or ξ (in this last case the force is dominated by the ideal gas pressure of chain ends).

2.6.3 INTERACTION BETWEEN IRREVERSIBLY ADSORBED LAYERS

So far we have assumed that the layer between the adsorbing plates is in equilibrium with the bulk solution, ϕ_b, so that as h is decreased polymer chains can be expelled from the layer into the bulk. In practice this process might be very slow and occur over a timescale much larger than the experimental time scale. At experimental timescales, the total amount of polymer between the plates is therefore nearly fixed. This case of irreversible adsorption (note that we still allow for a complete conformational relaxation of polymer chains) is now considered. The interaction force $f(h)$ is determined by the total adsorbance (per plate), $\Gamma = \int_0^h \phi(z)dz$, rather than by ϕ_b. It is natural to expect a positive force (repulsion) for high Γ and a negative force for low Γ. The same formalism as described in the previous section can be applied to finding f. However, one must keep in mind that E_0 is determined not by the equilibrium with the bulk, but by the fixed total amount of polymer Γ. Accordingly, we formally set $\phi_b = 0$ and keep the term $E_0\phi(h)$ in Eq. (97). Omitting the last two terms in this equation we rewrite the force as

$$f = \frac{1}{v}\left\{-\frac{1}{2}\phi^2(h) + E_0\phi(h) + \phi_e(h)\right\}. \quad (103)$$

Let us assume that the adsorbance Γ was initially determined by an equilibrium of each plate with an extremely dilute solution, $\Gamma = \Gamma^*$. The ground state dominance approach [8, 9] predicts an exactly vanishing force in this situation, $f_0(h) \equiv 0$. When the end effects are taken into account, we expect two corrections to this vanishing force. One is related to the molecular weight dependence of Γ^*. As shown in Section 2.3, with a single adsorbing plate, the monomer concentration in the tail region $z > l$ is higher than that predicted by the classical mean field theory, leading to a higher saturation adsorbance, $\Gamma^* = \Gamma_0 + \delta\Gamma$, where $\Gamma_0 = 2/b$ is the ground state dominance result, and $\delta\Gamma \simeq 7.07(Nb \ln (N/b^2))^{-1/3}$.

The second correction is the chain end pressure term in Eq. (103). In the loop region $h \ll l$, this term is negligible so that the force can be calculated using the classical mean field approach. The result is [25]

$$f = \begin{cases} \delta\Gamma/(vbh^2), & h \ll b \\ \pi^2 \delta\Gamma/(2vh^3), & b \ll h \ll l \end{cases}$$

Note that the force is positive (repulsive interaction). This result is related to the fact that $\delta\Gamma > 0$: The adsorbed layer becomes thicker because of the end effects.

In the region $z \gg 1$, all three terms in Eq. (103) are comparable. The order parameters approximately satisfy the equations $-\psi'' + (\psi\eta - E_0)\psi = 0$, $-\eta'' + (\psi\eta - E_0)\eta = 1$, which imply the scaling behavior $\eta \propto h^2$, $\psi \propto 1/h^4$, $E_0 \propto 1/h^2$. The force scales then as $f \propto 1/h^4$. The prefactor is obtained from a numerical solution of the order parameter equations

$$f \simeq \frac{1670}{vh^4}, \quad l \ll h \ll R.$$

The "irreversible" force is repulsive and higher than the force obtained in Eq. (100) when the adsorbed polymer is in equilibrium with a dilute solution.

2.7   Concluding Remarks

In this chapter we have summarized our recent work emphasizing the role of the chain end points in the adsorption of a polymer solution on a surface. We have limited our presentation to a mean field theory in the limit of large molecular weights.

The main theoretical idea is to go beyond the classical mean field approximation that makes a so-called ground state dominance approximation and implicitly assumes that the chains are infinite, thus ignoring the end points. As stated in the introduction, the same mean field theory (though on a lattice and not in a continuous space) has been extensively studied numerically by the Dutch group of Scheutjens and Fleer. Our work thus gives the asymptotic version of the Scheutjens-Fleer theory in the limit of large molecular weights. The numerical study of the mean field equations provides detailed information about adsorbed layers in the case when the molecular weights are not too high. This has proven very useful in many experimental situations. The asymptotic theory gives simple asymptotic laws and an insight into the physical mechanisms involved. The two approaches are thus complementary.

A detailed comparison of the concentration profiles obtained from the two order parameter theory and from the numerical solutions can be found in ref. [37]. Although the various length scales are not always well separated in the numerical work as required for the asymptotic laws to be valid, the agreement between the two approaches is extremely good. The scaling laws for the two characteristic length scales, the crossover length, l, and the layer thickness, λ, are well satisfied. The numerical concentration profiles are also well described by the order parameter approach. A few new results have been obtained from the analytical approach. The most striking one is the variation of the adsorbance with the molecular weight of the adsorbed chains. We find an adsorbance that decreases with molecular weight at large molecular weights. This is in contradiction with both the numerical solution of the mean field equations and the experimental results. However, this asymptotic behavior is reached only at extremely large molecular weights, and for intermediate values we predict an adsorbance increasing with molecular weight. The order parameter approach can also be readily extended to include various solvency conditions, various polymer architectures [39], or polydispersity [40].

The essential limitation of the mean field theory is the fact that it ignores the correlations due to the excluded volume interactions between monomers. Starting from the statistics of the loop and tail sizes and using good solvent statistics for the chain conformation, a scaling theory of polymer adsorption that takes into account explicitly the effect of the chain ends has been constructed in refs. [27] and [6]. This theory provides a description of adsorbed polymer layers in qualitative agreement with the mean field theory. Close to the adsorbing surface, the layer is dominated by loops; farther away, it is dominated by tails. The crossover length between these two regions scales as $l \propto N^{1/2}$, and up to a logarithmic correction, the size of the layer is proportional to the chain radius of gyration. In both the tail and loop regions, the concentration decays as a power law $c(z) \propto c^{-4/3}$ in agreement with de Gennes self-similar construction. This scaling theory, however, does not allow for a detailed calculation of the concentration profiles [38].

One of the most important applications of polymer adsorption is colloidal stabilization. One then needs to understand the interaction between adsorbed polymer layers. As first pointed out by de Gennes, the interaction strongly depends on the reversibility of the adsorption. We have only considered here the case of full thermodynamic equilibrium with a reservoir and the case of a fixed adsorbed amount of polymer when the local chain conformation can reequilibrate freely.

The main result of the mean field theory is that in addition to the classical contributions to the force, there is an extra repulsive contribution that can be interpreted as an ideal gas pressure of the chain ends. We have found a force that is always repulsive when the adsorbed polymer amount is constant and nonmonotonic (attractive at short distances and repulsive at large distances) at thermodynamic equilibrium. To go beyond the mean field arguments, a scaling theory can be constructed; it provides the scaling variation of the force f with the distance h between interacting surfaces ($f \propto h^{-3}$) but it does not allow the study of the sign of the force. This is a major limitation, at least for the equilibrium case. A more complete theory is needed to decide whether the mean field approach provides the correct qualitative picture.

All these results point toward the importance of the chain ends in adsorbed polymer layers. Indirect evidence for this has been provided by the numerous quantitative comparisons between the results of the Scheutjens-Fleer theory and experiments. Although we do not know of any experiment that directly probes the chain end points, one could for example, label the end points in a scattering experiment or measure directly the bridging interactions between two surfaces [41]. Unveiling direct evidence of the role of chain end points remains a challenge for future experiments.

Acknowledgments

A. N. S. acknowledges the support of the NATO's Scientific Affairs Division in the framework of the Science for Stability Program.

References

1. D. Napper, *Polymer Stabilization of Colloidal Dispersions,* Academic Press, London, 1983.
2. A. Silberberg, *J. Chem. Phys.* **48:** 2835 (1968).
3. C. Hoeve, *J. Polym. Sci.* **C34:** 1 (1971).
4. R. Roe, *J. Chem. Phys.* **60:** 4192 (1974).
5. J. Scheutjens and G. Fleer, in *The Effect of Polymers on Dispersion Properties,* Th. F. Tadros Ed., Academic Press, New York, 1982.
6. A. N. Semenov, J. Bonet Avalos, A. Johner, and J.-F. Joanny, *Macromolecules* **29:** 2179 (1996).
7. P.-G. de Gennes, *Macromolecules* **14:** 1637 (1981).
8. P.-G. de Gennes, *Macromolecules* **15:** 492 (1982).

9. P.-G. de Gennes, *Adv. Colloid. Interface Sci.* **27**: 189 (1987).

10. L. Auvray and J. Cotton, *Macromolecules* **20**: 202 (1987).

11. L. Lee, O. Guiselin, B. Farnoux, and A. Lapp, *Macromolecules* **24**: 2518 (1991).

12. O. Guiselin, L. Lee, B. Farnoux, and A. Lapp, *J. Chem. Phys.* **95**: 4632 (1991).

13. M. Kawagushi and A. Takahashi, *Macromolecules* **16**: 1465 (1983).

14. P. Déjardin and R. Varoqui, *J. Chem. Phys.* **75**: 4115 (1981).

15. J. Klein, in *Liquids at Interfaces,* J. Charvolin, J. F. Joanny and J. Zinn-Justin Eds., North Holland, Amsterdam, 1989.

16. I. Schmitt and K. Binder, *J. Phys. (France)* **46**: 1631 (1985).

17. I. Carmesin and I. Noolandi, *Macromolecules* **22**: 1683 (1989).

18. K. Ingersent, J. Klein, and P. Pincus, *Macromolecules* **19**: 1374 (1986).

19. A. Johner and J.-F. Joanny, *J. Phys. II* **1**: 181 (1991).

20. G. Rossi and P. Pincus, *Europhys. Lett.* **5**: 641 (1988).

21. G. Fleer, M. Cohen-Stuart, J. Scheutjens, T. Cosgrove, and B. Vincent, *Polymers at Interfaces,* Chapman and Hall, London, 1993.

22. C. van der Linden, Ph.D. thesis (Agricultural University Wageningen, 1995).

23. D. Shaefer, J.-F. Joanny, and P. Pincus, *Macromolecules* **13**: 1280 (1980).

24. J. Bonet-Avalos, J.-F. Joanny, A. Johner, and A. Semenov, *Europhys. Lett.* **35**: 97 (1996).

25. A. N. Semenov, J.-F. Joanny, A. Johner, and J. Bonet Avalos, *Macromolecules in press:* (1997).

26. A. N. Semenov, *J. Phys. II* **6**: 1759 (1996).

27. A. N. Semenov and J.-F. Joanny, *Europhys. Lett.* **29**: 279 (1995).

28. S. Edwards, *Proc. Phys. Soc.* **85**: 613 (1965).

29. I. Lifshitz, *Zh. Eksp. Teor. Fiz.* **55**: 2408 (1968).

30. I. Lifshitz, *Sov. Phys. JETP* **28**: 1280 (1969).

31. C. Marques and J.-F. Joanny, *J. Phys. (Paris)* **49**: 1103 (1988).

32. P. Auroy, L. Auvray, and L. Lèger, *Macromolecules* **24**: 5158 (1991).

33. P.-G. de Gennes, *Scaling Concepts in Polymer Physics,* Cornell Univ. Press, Ithaca, NY, 1978.

34. G. D. T. Wu, G. Fredrickson, J. Carton, A. Ajdari, and L. Leibler, *J. Polym. Sci. B–Polym. Phys. Ed.* **33**: 2373 (1995).

35. I. M. Lifshits, A. Grosberg, and A. Khokhlov, *Rev. Mod. Phys.* **50**: 683 (1978).

36. I. Lifshits, A. Grosberg, and A. Khokhlov, *Sov. Phys. Uspekhi* **22**: 123 (1979).

37. A. Johner, J. Bonet Avalos, C. van der Linden, A. N. Semenov, and J.-F. Joanny, *Macromolecules* **29**: 3629 (1996).

38. M. Aubouy, O. Guiselin, E. Raphaël, *Macromolecules* **29**: 7261 (1996).

39. J. Joanny and A. Johner, *J. Phys. II* **6**: 511 (1996).

40. J. Baschnagel, A. Johner, and J.-F. Joanny, *Phys. Rev. E* **55**: xx (1997).

41. A. Johner and J.-F. Joanny, *Macromolecules Theory Simul.,* **6**: 479 (1997).

Chapter 3 | Replica Field Theory Methods in Physics of Polymer Networks

S. V. Panyukov[†] and Y. Rabin

Department of Physics, Bar-Ilan University
Ramat-Gan, Israel

[†]*Permanent address: Lebedov Physics Institute, Russian Academy of Sciences*
Moscow, Russia

Abstract

We present a comprehensive study of the statistical mechanics of randomly cross-linked polymer gels made of phantom chains with excluded volume, starting from the microscopic Edwards model and ending with analytical expressions for density correlation functions. Using replica field theory, we derive and solve the mean field equations and show that the ground state is described by a solution with spontaneously broken symmetry with respect to translation in replica space. The mean field solution contains statistical information about the fluctuations of monomers and of cross-links about their average positions in the network. We construct the free energy functional of the density field and of a Gaussian random field that represents the quenched inhomogeneous structure of the gel. This free energy is then used to obtain all the statistical information about the density fluctuations in a deformed network, in terms of the average number of cross-links and the thermodynamic parameters (temperature, density, and quality of solvent) of the initial and the deformed states. We calculate the density correlation functions that describe quenched and thermal density fluctuations. Analytical expressions for the correlators are obtained both in the long-wavelength and the short-wavelength limits. We use the mean field solution to obtain statistical information about the effect of macroscopic deformation on the stretching of individual network chains and discuss the deviations from affine response to deformation. Finally, we discuss recent extensions and generalizations of the ideas presented in this work.

3.1 Introduction

Polymer gels are fascinating materials that differ in many respects from ordinary solids. Although they possess all the normal characteristics of solids such as stability of shape, resistance to shear, etc., they can absorb solvent and swell to dimensions much larger than their dry size and exhibit linear elastic response even

83

THEORETICAL AND MATHEMATICAL
MODELS IN POLYMER RESEARCH

Copyright © 1998 by Academic Press.
All rights of reproduction in any form reserved.
ISBN 0-12-304140-6/$25.00.

at very large deformations. The fact that a gel is a solid permeated by solvent means that it can be considered as a combination of a solid and a liquid, and that its state of equilibrium is determined by the interplay between the two components. When the network is deformed, the osmotic pressure of the liquid component adjusts itself to balance the local elastic stresses. Conversely, if the liquid component changes its characteristics (by change of temperature, solvent, etc.), the network stresses adjust to the new osmotic pressure, resulting in a new equilibrium. This interplay determines the response of the gel to all external perturbations under which the network maintains its integrity (i.e., does not break).

Polymer gels can be synthesized by cross-linking a polymer solution or a melt, or by polymerizing a mixture of monomers and cross-linking molecules. Since all the above cross-linking processes are random, the resulting gels are disordered solids, the structure of which is determined by the state of the solution from which they were formed. The memory of the initial state is frozen into the network during its formation and affects all its subsequent history long after the process of cross-linking is terminated. Thus, if one attempts to model the response of the gel to some external deformation, one has to know not only the state of the gel immediately prior to the deformation but also the conditions under which it was prepared.

At first sight, the existence of memory effects suggests that to understand the behavior of polymer networks, one needs to have complete information about their complicated frozen structure; i.e., one has to specify a vast number (of the order of the Avogadro number) of parameters. If this were the case, it would undermine any attempt to obtain a *probabilistic* description of the physics of polymer gels by the usual methods of statistical physics. This seemingly intractable problem can be overcome by noticing, as was done by Edwards and his coworkers [1, 2], that if the gel is formed by *instantaneous* cross-linking of a polymer solution, the probability of observing a particular network structure is identical to the probability of observing a state of a polymer liquid in which some monomers (a fraction of which will be cross-linked immediately afterward) are in contact with each other. The latter probability distribution can be characterized by only a small number of parameters that define the conditions of preparation, such as temperature, solvent quality, degree of cross-linking, and density in the initial state.

Using the above approach, the problem can be reformulated in statistical terms in which the answers to all experimentally relevant questions about the behavior of polymer gels are given in terms of averages of the physical quantities we are interested in. In this way the problem reduces to finding the probability distribution associated with the physical quantity of interest and averaging with respect to this distribution. The possibility of obtaining such a reduced description is related to the fact that, in spite of their complexity and frozen randomness, gels have a unique state of microscopic equilibrium [3]. Although there is an obvious loss of ergodicity associated with the process of cross-linking (of the same kind that accompanies the crystallization of a liquid), gels differ from glasses in the sense that once they are formed by irreversible cross-linking, their equilibrium state is uniquely determined by (and only by) the parameters that characterize this state (temperature, quality of solvent, etc.) and does not depend on their "history" *after* preparation (as long as the integrity of the network is maintained). For example, if a gel is synthesized at temperature $T^{(0)}$ and subsequently studied at a temperature T, its state will depend only on T and not on the history of heating process (heating to $T' > T$ and subsequently cooling to T results in the same final state of equilibrium as heating directly from $T^{(0)}$ to T). In glasses, the final state will depend, in general, on the history of its preparation (after the synthesis of the glass).

This work is based on the Edwards model of *instantaneously cross-linked networks of Gaussian chains with excluded volume* [1]. Following Edwards, no additional constraints are introduced to describe the fact that real chains cannot cross each other and, therefore, this model does not account for the contribution of permanent topological entanglements to the elasticity of polymer networks. (Such permanent entanglements in irreversibly cross-linked networks differ from temporary entanglements in polymer solutions [4].) In choosing the above model of polymer networks we are guided by considerations of simplicity. Although more complicated models (including entanglements and non–Gaussian elasticity) are more realistic, our aim is to start with the simplest well-defined microscopic theory and to present a *strict* mathematical analysis of the problem that will serve as a point of reference for future generalizations. We will show that this model of the simplest polymer solid can be solved on a level of mathematical rigor similar to that of the most sophisticated theories of polymer solutions and, in the process, obtain important insights about the physics of polymer gels. We would like to stress that although this work uses state-of-the-art machinery of theoretical physics, the mathematical concepts involved are fairly standard and simple. Although the material covered in this chapter has strong overlap with our earlier work, ref. [3], we attempt here to focus on the mathematical aspects of the theory of polymer networks and to present all the derivations in a detailed and coherent fashion, without sacrificing mathematical rigor for the sake of simplicity. It is our hope that this work will be useful to polymer theorists (researchers and graduate students alike) and to experimentalists with a strong background in theoretical physics.

In Section 3.2 we introduce the Edwards model [1]. We show that the model predicts the existence of the *cross-link saturation threshold,* which defines the maximal density of cross-links that can be achieved by instantaneous cross-linking of a polymer solution. When this threshold is approached, the length scale associated with the quenched heterogeneity of network structure diverges and static inhomogeneities appear on the largest length scales in the gel.

We use the self-averaging property of the total free energy to express it as an average (with respect to the probability of synthesis of a given network structure) of the logarithm of the partition function of the deformed gel. To avoid the inconvenient averaging of the logarithm, we introduce the standard *replica trick* (define one replica of the initial state and m replicas of the final state) and express the true thermodynamic free energy as the derivative of the replica free energy with respect to $m,$ in the limit $m \to 0$. The constraints introduced by the cross-links are replaced by an effective attractive potential through the introduction of the grand canonical representation, which is then extended to all the monomers (i.e., both the number of monomers and the number of cross-links are allowed to fluctuate, so that only their average numbers are determined by the monomer chemical potential and by the fugacity of cross-links). This illustrates the dramatic computational simplification produced by the transformation to the abstract replica space (defined as the space of the coordinates in all the $1 + m$ replicas): *Frozen inhomogeneities of network structure (in real space) can be treated as thermal fluctuations in replica space* and the seemingly intractable calculation of frozen disorder can be performed using the usual methods of equilibrium statistical mechanics. While excluded volume interactions act independently in each of the replicas, the interactions that represent the cross-links act identically in all the replicas (this is an expression of the solid character of the gel) and, therefore, introduce a coupling between the replicas. Finally, the thermodynamic free energy is expressed through the grand canonical partition function of the replica system.

Although the interactions (both excluded volume ones and those associated with cross-links) are nonlocal along the chain contour, they are local both in the three-dimensional physical space and in the $3(1 + m)$-dimensional replica space. This fact is used in Section 3.3 where we transform to collective coordinates (field theory) and rewrite the interactions in terms of replica space densities (for the cross-links) and densities in each of the replicas (for the excluded volume). We then use a generalization of de Gennes' $n = 0$ method [4] to eliminate the elastic entropy term in the replica partition function by introducing a field theoretical representation of the entropy in terms of a n-component vector field $\vec{\varphi}$ (the limit $n = 0$ is taken at the end of the calculation). We then relate this field to the density in replica space. The details of the transformation to the field representation for Gaussian chains are given in the appendices. We represent the replica partition function as a functional integral of Boltzmann weights defined by a replica generalization of a $\vec{\varphi}^4$-type field Hamiltonian. We also discuss the various continuous and discrete symmetries of this Hamiltonian (rotations in n vector space, permutations of the replicas of the final state, rotations in each of the replicas, and translations in replica space).

In Section 3.4 we proceed to look for a mean field solution that minimizes the field Hamiltonian and, therefore, gives the steepest descent estimate of the replica partition function. We derive the field equations, the solutions of which correspond to the extrema of the Hamiltonian. Guided by the expectation that the ground state solution must have the maximal possible symmetry, we first consider the constant (in replica state) solution that has the full symmetry of the underlying Hamiltonian. However, this solution corresponds to the saddle point rather than to a minimum of the Hamiltonian and must be rejected.

The analogy with crystalline solids, which can be thought of as solutions with spontaneously broken translational symmetry (in real space) that minimize a translationally invariant (i.e., which has the symmetry of a liquid) Hamiltonian [5], suggests that we look for a solution with *spontaneously broken translational symmetry in replica space* that obeys the physical condition that it gives rise to a constant mean density in the three-dimensional space associated with each of the replicas [6]. The above condition dictates the dependence of the mean field solution on the replica space coordinates, and we find that the solution must be invariant under simultaneous translation (by a constant) along the principal axes of deformation in each of the replicas of the final state. Introducing a partition of the replica space into a three-dimensional longitudinal (along the principal axes of deformation in each of the final state replicas) and a $3m$-dimensional transverse subspace, we show that the mean field solution depends only on the coordinates of the transverse subspace and is invariant under rotations in this subspace. This allows the replacement of the $3(1 + m)$-dimensional nonlinear partial differential equation by a simple nonlinear differential equation from which the mean field solution is calculated numerically. We find that the solution is localized around a three-dimensional surface (the longitudinal subspace) in replica space, defined by the *affine* relation between the coordinates in the initial and in each of the final replicas, with a characteristic width of the order of the mesh size of the network. The fact that the solution is *replica symmetric* (i.e., invariant under permutations of the replicas of the final state) means that *the position of any given monomer is nearly identical (up to thermal fluctuations on the scale of a mesh size) in all of the replicas of the final state and that the average position of each monomer changes affinely with the deformation of the network.* We proceed to analyze the physical content of the mean field expression for the density of monomers in replica space, which is analogous to the Edwards-Anderson order parameter [7] familiar from the theory of spin glasses, and find that it contains *statistical* information about the frozen structure of the network. It is shown that this order parameter defines the probability distribution of deviations of network monomers from their mean (i.e., affinely displaced) positions, from which the average localization length that determines the length scale of thermal fluctuations of monomers, is calculated. Contrary to the Flory assumption [8], we find that *both the monomers and the cross-links fluctuate over length scales of the order of the mesh size.*

Using this inhomogeneous solution, we perform the steepest descent calculation of the replica partition function and obtain the mean field thermodynamic free energy of the gel. Our free energy coincides with the Deam and Edwards variational estimate [1], which is qualitatively similar (apart from a numerical coefficient and logarithmic corrections) to that of classical theories of network elasticity due to Flory and Rehner [8, 9] and James and Guth [10].

In Section 3.5 we proceed to examine the stability of the different (homogeneous and inhomogeneous) mean field solutions to find out whether they correspond to true minima of the replica Hamiltonian. To this end we calculate all the eigenvalues and eigenfunctions of the operator, which gives the energy of fluctuations about these solutions, using the expressions for the replica space correlation functions (first derived in ref. [3]). We demonstrate that the fluctuation energy evaluated on the homogeneous solution has some negative eigenvalues and therefore corresponds to a saddle point rather than to a true minimum of the Hamiltonian. On the other hand, all the eigenvalues (corresponding to rotations in the space of the n-vector model and to shear and density modes in replica space) associated with our inhomogeneous solution are positive. This proves that *the inhomogeneous solution minimizes the Hamiltonian and is stable with respect to arbitrary small fluctuations in replica space, including those which break the symmetry with respect to permutations of the replicas of the final state.*

Section 3.6 begins with the observation that to make contact with scattering experiments that probe the static inhomogeneities and the thermal density fluctuations of the monomer density in swollen and stretched networks, we have to eliminate (i.e., integrate over) the contributions of the shear modes and of the density modes in the state of preparation. To make the calculation feasible we assume that the deviations from the average density are small and can be treated within the *random phase approximation* (RPA) [4], which corresponds to keeping only quadratic terms in these deviations. We show that while this assumption always holds for frozen density inhomogeneities for gels prepared away from the cross-link saturation threshold, it breaks down for thermal density fluctuations in gels in good solvents where such fluctuations are strong. (These strong fluctuations on length scales larger than the "blob" size can be accounted for by an appropriate renormalization of the RPA parameters; see ref. [3].) The elimination of the "irrelevant" shear and density modes is done by introducing auxiliary fields that couple between the replicas of the final state. Diagonalization of these couplings allows us to calculate the nonaveraged free energy functional of the Fourier components of the monomer density (ρ_q) and of a random field (n_q) that represents the frozen structure of the network, and to derive the Gaussian distribution function $P(n_q)$ (the probability to observe a given amplitude of the random field n_q in the gel). We find the *equilibrium density distribution* ρ_q^{eq} that minimizes the free energy functional and show that it corresponds to the *static inhomogeneous density profile of the gel, which is uniquely defined by the structure of the network and by the thermodynamic conditions in the final deformed state.* The existence of the time-independent inhomogeneous density profile leads to the appearance of static speckle patterns in the intensity of light scattered from gels. We show that the field n_q can be interpreted as the inhomogeneous equilibrium density profile of the elastic reference state (i.e., that of a stretched network, without excluded volume interactions). Using the free energy functional (quadratic in ρ_q and n_q) and the distribution function $P[n]$, we compute all the statistical information about the static density inhomogeneities and thermal density fluctuations in a deformed gel. We relate our theoretical predictions to scattering experiments that measure static density correlations (averaged over both space and time) by showing that averaging over the ensemble of all possible network structures (consistent with given thermodynamic conditions in the state of preparation) is equivalent to averaging over the volume of a single polymer gel and that averaging over the ensemble of gels with a given network structure (thermal averaging) is equivalent to time averaging over the configurations of a single gel.

All the information that enables us to calculate the experimentally observable density correlation functions is contained in two functions, g_q and ν_q, where the former is the correlator of thermal density fluctuations in the elastic reference state and the latter is the structure averaged correlator, which measures the spatial correlations of the inhomogeneous equilibrium density profile in this state. (Explicit analytical expressions for these functions, in both the short-wavelength and the long-wavelength limits, are given in Section 3.7.) An exact (within the RPA) expression for the total structure factor, valid in the entire range of scattering wave vectors, is obtained for gels in the state of preparation. Asymptotic expressions (in both the long wavelength and the short wavelength limits) for the experimentally observable density correlators in the final deformed state are also given. It is shown that all the structural information about the gel is contained in two parameters: the density of cross-links in the state of preparation and the *heterogeneity parameter*, which measures the distance from the cross-link saturation threshold.

In Section 3.8 we return to our inhomogeneous mean field solution for the replica space density and show that it contains information on the deformation of labeled chains of given contour length under given deformation of the entire network. We present the results of calculations [11, 12]) that show that the average deformation of such chains depends both on their length and on the local environment in which they are embedded and that *only chains larger than the local mesh size are stretched affinely with the macroscopic deformation.* Strong deviations from affinity are obtained for shorter network chains (those much shorter than the local mesh size react to deformation only through disinterpenetration). These intriguing results also follow from the observation that the rms distance between the ends of a network chain is the sum of a contribution of the average distance between the ends (which deforms affinely with the network) and a fluctuation contribution (which is not affected by the deformation) and that the latter is important only on length scales of the order of the mesh size (the characteristic length scale of thermal fluctuations). Under uniaxial extension, the average mesh size deforms affinely along the direction of stretching and its transverse dimensions are not affected by the deformation.

In Section 3.9 we summarize the main results obtained in this work. We argue that gels do not belong to any of the known classes of solids. Unlike amorphous materials (e.g., glasses), once they are formed they have a single well-defined state of microscopic equilibrium. Unlike crystalline solids, they possess no long-range order and their "atoms" (i.e., monomers and cross-links) fluctuate over distances that exceed the average distance between neighboring "atoms." This new class of materials can be called *soft disordered equilibrium solids.* We end this work by discussing the limitations of our theory and reviewing its recent extensions and generalizations.

3.2   The Model

Consider the following situation: a chemically cross-linked polymer network is immersed in a good solvent and subjected to mechanical deformation. As long as the deformation does not affect the chemical structure of the gel (i.e., as long as the network does not break), the response will be determined by both the external conditions (deformation, solvent quality, temperature, etc.), which can be varied at will, and by the fixed *network structure*. The structure of the network is uniquely defined by specifying which monomers are joined at each cross-link point and is fixed once and for all at the time of preparation of the gel. It can depend on the method of cross-linking (irradiation, chemical reaction, etc.) and on the physical conditions (solvent quality, temperature, etc.) to which the system was subjected during synthesis. In the following, we will refer to the state of preparation (following cross-linking) as the *initial* state. The *final* state of the swollen and deformed network depends indirectly on the conditions of network preparation, since they determine the frozen structure of the network.

We start with the Edwards model of a *randomly cross-linked network of Gaussian chains with excluded volume* [1]. This microscopic model does not contain explicit topological constraints that would account for the presence of permanent entanglements in real networks. While such entanglement effects can be treated by the ad hoc introduction of an "entanglement tube" into the present model, they cannot be described with the same degree of mathematical rigor as the simpler "phantom" chains. (The concept of an "entanglement tube" does not arise naturally in the microscopic model and has to be introduced by hand into our formulation [13].)

Following Deam and Edwards [1], we neglect dangling ends and assume that the network is formed by cross-linking very long chains, well above the gelation point (which corresponds to the minimal concentration of cross-links at which an infinite connected network is formed). The case of a network prepared by instantaneous cross-linking of a melt of chains is studied in ref. [14]. In the present work we assume that the network is prepared by cross-linking chains in a semidilute solution in a good solvent, in which the interaction between monomers can be described by an effective second virial coefficient, $w^{(0)}$. Since chain end effects can be neglected for sufficiently long precursor chains, the polymer solution can be replaced by a single chain of N_{tot} monomers, where N_{tot} is the total number of monomers in the original solution. (This replacement is allowed as long as we do not consider the conformations of individual chains.) The constraint of average monomer density $\rho^{(0)}$ in the pre-cross-linked polymer solution is satisfied by confining the chain to a volume $V^{(0)}$ such that $\rho^{(0)} = N_{tot}/V^{(0)}$.

3.2.1 PRE-CROSS-LINKED POLYMER "SOLUTION"

The conformation of a polymer in a solvent is defined by the spatial position $\mathbf{x}_i(s)$ of the sth monomer (s takes values from 1 to the number of monomers in a chain) of the ith chain. In a continuum description of the chain, s becomes a continuous contour parameter that varies between 0 and N_{tot}, and the polymer is modeled by the Edwards Hamiltonian [15],

$$\frac{\mathcal{H}^{(0)}[\mathbf{x}(s)]}{T^{(0)}} = \frac{1}{2a^2} \int_0^{N_{tot}} ds \left(\frac{d\mathbf{x}}{ds}\right)^2 + \frac{w^{(0)}}{2} \int_0^{N_{tot}} ds \int_0^{N_{tot}} ds' \, \delta[\mathbf{x}(s) - \mathbf{x}(s')] \quad (2.1)$$

where a is the monomer size and $T^{(0)}$ is the temperature in the state of preparation. (Here and in the following we take the Boltzmann constant to be unity.)

The statistical weight of a particular configuration of the chain $\{\mathbf{x}(s)\}$ is given by the canonical distribution function

$$\mathcal{P}_{liq}[\mathbf{x}(s)] = Z_{liq}^{-1} \exp(-\mathcal{H}^{(0)}[\mathbf{x}(s)]/T^{(0)}),$$

$$Z_{liq} = \int D\mathbf{x}(s) \exp(-\mathcal{H}^{(0)}[\mathbf{x}(s)]/T^{(0)}). \quad (2.2)$$

Here $\int D\mathbf{x}(s)$ implies functional integration over all the configurations of the chain and Z_{liq} is the partition function of the polymer. (The subscript liq refers to the liquidlike state of the polymer, prior to the introduction of cross-links.)

3.2.2   *INSTANTANEOUS CROSS-LINKING AND CROSS-LINK SATURATION THRESHOLD*

To elucidate the physics of the process of cross-linking in the above model, let us consider an instantaneous configuration of a *constrained* polymer in a good solvent (with average monomer density $\rho^{(0)}$) in which there are exactly K binary contacts between monomers. (These contacts form and disappear due to thermal fluctuations.) During a contact event, two monomers share a contact volume v. (In the mean field approximation, the definition of the contact volume coincides with the definition of the excluded volume parameter, $v = w^{(0)}$, which appears in the Edwards Hamiltonian, Eq. (2.1).) The partition function of the constrained polymer is given by

$$Z_{liq}(K) = \int D\mathbf{x}(s) \, \exp(-\mathcal{H}^{(0)}[\mathbf{x}(s)]/T^{(0)}) \int d\mathbf{S}_K \prod_{\{i,j\}} v\delta[\mathbf{x}(s_i) - \mathbf{x}(s_j)] \quad (2.3)$$

where the product is taken over all the K pairs of monomers that are in contact with each other. The integration over $\mathbf{S}_K$ goes over the contour positions of the $2K$ monomers that participate in the contacts

$$\int d\mathbf{S}_K \equiv \frac{1}{2^K K!} \prod_{\{i,j\}} \int_0^{N_{tot}} ds_i \int_0^{N_{tot}} ds_j \quad (2.4)$$

and accounts for the fact that such contacts can occur with equal probability at any location along the chain contour. The normalization factor $K!$ arises because all contact pairs are indistinguishable and the factor 2^K accounts for the indistinguishability of the two monomers that form each contact.

We can now define the probability that the polymer has exactly K contacts as

$$P(K) = Z_{\text{liq}}(K) \Big/ \sum_{K=0}^{N_{\text{tot}}/2} Z_{\text{liq}}(K). \tag{2.5}$$

The average number of contacts is given by

$$\overline{K} = \sum_{K=0}^{N_{\text{tot}}/2} KP(K) \tag{2.6}$$

and can be easily estimated from mean field arguments. Notice that since the excluded volume of a monomer is given by the second virial coefficient $w^{(0)}$, each monomer can be represented by an impenetrable sphere of volume $w^{(0)}$. The volume fraction occupied by such spheres is $w^{(0)}\rho^{(0)}$ and, therefore, the average number of binary contacts between the spheres is

$$\overline{K} \simeq \frac{1}{2} N_{\text{tot}} w^{(0)} \rho^{(0)} = \frac{1}{2} V^{(0)} w^{(0)} (\rho^{(0)})^2. \tag{2.7}$$

When a polymer is *instantaneously* cross-linked by irradiation or by other means, a fraction of all monomers at a distance of the order of $(w^{(0)})^{1/3}$ from each other (i.e., that form a contact) become cross-linked (see Figure 3.1). The number of such cross-linked monomers $(2N_c)$ depends on the intensity of irradiation and determines the average number of monomers between neighboring cross-links, $\overline{N} = N_{\text{tot}}/(2N_c)$. As long as the required density of cross-links $\rho^{(0)}/(2\overline{N})$ is smaller than the average density of monomer contacts $w^{(0)}(\rho^{(0)})^2/2$ in the polymer liquid (prior to irradiation), the former can be increased by increasing the intensity of irradiation. The saturation density of cross-links (or, equivalently, the minimal chain length between cross-links) is obtained by equating the two densities $(\rho^{(0)}/2\overline{N})^{\max} \simeq w^{(0)}(\rho^{(0)})^2/2$ and yields the *cross-link saturation threshold*

$$w^{(0)} \rho^{(0)} \overline{N}^{\min} = 1. \tag{2.8}$$

We conclude that the physically meaningful range of parameters describing the initial state corresponds to $w^{(0)}\rho^{(0)}N > 1$. This simple mean field estimate should be revised for semidilute gels in good solvents where strong thermal fluctuations on length scales smaller than the "blob" size should be taken into account (by scaling methods [4]). The effect of such fluctuations was considered in

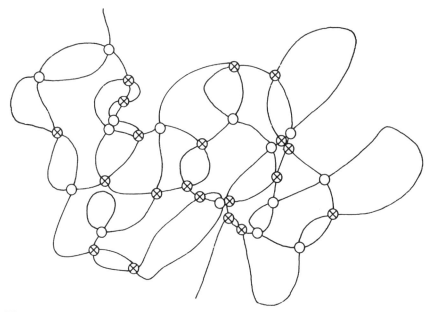

Figure 3.1   Schematic drawing of a network in the moment of cross-linking. Inter-monomer contacts (○) and cross-link points (⊗) are shown.

ref. [3], where we showed that the maximal attainable density of cross-links corresponds to a situation in which there is one cross-link per blob and that the cross-link saturation threshold condition becomes

$$(a^2 w^{(0)})^{3/5} \rho^{(0)} (\overline{N}^{\min})^{4/5} \simeq 1. \qquad (2.9)$$

Note that the crossover between the mean field and the scaling regimes takes place at $\rho^{(0)} \simeq w^{(0)}/a^6$, which coincides with the usual limit of applicability of mean field treatments of excluded volume effects [15].

What happens when the cross-link saturation threshold is approached (e.g., by increasing the intensity of irradiation)? We will show later in this work that the structure of the gel becomes increasingly inhomogeneous in the sense that the characteristic length scale of density fluctuations increases dramatically as the saturation threshold is approached. We would like to emphasize that although the existence of the saturation threshold has only been demonstrated here for instantaneous cross-linking of long chain polymers and therefore may be considered as an artifact of the present model, similar phenomena have been observed in computer simulations where other methods of cross-linking were used [16].

3.2.3   EDWARDS FORMULATION

Consider a network containing N_c cross-links that has been prepared by instantaneous cross-linking. Since at each cross-link point a chemical bond is formed between two monomers of the chain (the functionality of cross-links is 4), the resulting N_c cross-links are characterized by the set of monomers $\{i, j\}$ with corresponding positions $\{s_i, s_j\}$ on the chain contour. The set $\mathbf{S} \equiv \{s_i, s_j\}$ uniquely defines the structure (topology) of the network. The probability distribution that describes the gel under conditions of preparation is given by that of a polymer in a solvent, Eq. (2.2), supplemented by the constraint of a given configuration $\mathbf{S}$ of N_c monomer contacts:

$$\mathcal{P}^{(0)}[\mathbf{x}(s), \mathbf{S}] = [Z^{(0)}(\mathbf{S})]^{-1} \exp\left(-\mathcal{H}^{(0)}[\mathbf{x}(s)]/T^{(0)}\right) \prod_{\{i,j\}} \delta[\mathbf{x}(s_i) - \mathbf{x}(s_j)] \quad (2.10)$$

where the partition function of the gel in the state of preparation is defined by the normalization condition $\int D\mathbf{x}(s)\mathcal{P}^{(0)}[\mathbf{x}(s), \mathbf{S}] = 1$. (The integration goes over all the configurations in the volume $V^{(0)}$ occupied by the gel in the state of preparation.)

$$Z^{(0)}(\mathbf{S}) = \int D\mathbf{x}(s) \exp\left(-\mathcal{H}^{(0)}[\mathbf{x}(s)]/T^{(0)}\right) \prod_{\{i,j\}} \delta[\mathbf{x}(s_i) - \mathbf{x}(s_j)]. \quad (2.11)$$

The above distribution function gives the complete statistical mechanical description of the gel in the state of preparation, including its response to small perturbations (linear response). However, unlike usual solids, polymer networks display linear elastic response to stretching and swelling well into the large deformation regime, which cannot be described by $Z^{(0)}$. The partition function $Z(\mathbf{S})$ of an arbitrarily deformed gel differs from that of the undeformed one in the following respects: First, the swelling modifies the effective second virial coefficient; i.e., $w^{(0)}$ is replaced by w in the Edwards Hamiltonian and in the subsequent equations. Second, the deformation changes the volume ($V^{(0)}$) occupied by the network and the integration over the polymer coordinates extends over the new volume V. Third, one has to introduce the forces that act on the surface of the gel and produce the stretching. The effect of these forces will be represented by introducing the appropriate *deformation ratios* $\{\lambda_\alpha\}$ along the principal axes of deformation. Furthermore, in general, we have to allow for the possibility that the temperature in the final state, T, differs from that in the state of preparation, $T^{(0)}$ (see Figure 3.2).

Since the calculation of the partition function for a given realization of the network structure $\mathbf{S}$ is prohibitively difficult, we proceed to simplify the problem by the use of the *self-averaging* property of the free energy of a macroscopic sys-

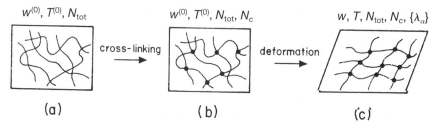

Figure 3.2   (a) Polymer solution prior to cross-linking (characterized by parameters $w^{(0)}$, $T^{(0)}$, N_{tot}). (b) Initial undeformed gel (parameters $w^{(0)}$, $T^{(0)}$, N_{tot}, N_c). (c) Final deformed gel (parameters w, T, N_{tot}, N_c, $\{\lambda_\alpha\}$).

tem. This property follows from the additivity of the free energy. Imagine that we divide the entire sample into a large number of small but still macroscopic domains, each of which has its own unique structure (Figure 3.3). In the limit of an infinitely large number of such domains, the probability $\mathcal{P}(\mathbf{S})$ of the appearance of a domain with a given structure $\mathbf{S}$ is determined by the process of cross-linking. Since, in the case of instantaneous cross-linking of a polymer in a solvent,

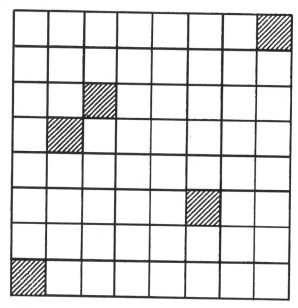

Figure 3.3   Partitioning of the gel into macroscopic regions characterized by different network structures. Domains with a particular structure $\mathbf{S}$ are shaded.

the initial state of the gel prior to deformation is a particular realization of the equilibrium state of this polymer and since the solution is ergodic, the probability is given by the Gibbs distribution function

$$\mathcal{P}(\mathbf{S}) = Z^{(0)}(\mathbf{S}) / \int d\mathbf{S}' Z^{(0)}(\mathbf{S}'). \tag{2.12}$$

The total free energy $\mathcal{F}(\mathbf{S})$ of a macroscopic network with a given structure $\mathbf{S}$ can be written as the sum of free energies of such domains. This sum can be replaced by the sum over all possible realizations of network structures. The contribution of each of these structures is weighted by the distribution function (2.12), yielding:

$$\mathcal{F}(N_{\text{tot}}, N_c) = -T \int d\mathbf{S}\, \mathcal{P}(\mathbf{S}) \ln Z(\mathbf{S}). \tag{2.13}$$

The integration over $\mathbf{S}$ goes over the contour positions of the $2N_c$ monomers that participate in the contacts

$$\int d\mathbf{S} \equiv \frac{1}{2^{N_c} N_c!} \prod_{\{i,j\}} \int_0^{N_{\text{tot}}} ds_i \int_0^{N_{\text{tot}}} ds_j. \tag{2.14}$$

In writing down this formula, we took the thermodynamic limit in which the free energy is independent of the particular choice of network structure and depends only on thermodynamic properties such as the total numbers N_{tot} and N_c of monomers and of cross-links, on volume, temperature, and quality of solvent.

Replica Formalism

The averaging in (2.13) can be performed using the *replica* method, which is based on the identity [1]

$$\ln Z = \lim_{m \to 0} \frac{Z^m - 1}{m}. \tag{2.15}$$

The trick consists of introducing the "replica" free energy $\mathcal{F}_m(N_{\text{tot}}, N_c)$ of a network with N_{tot} monomers and N_c cross-links

$$\exp[-\mathcal{F}_m(N_{\text{tot}}, N_c)/T] \equiv \int d\mathbf{S}\, Z^{(0)}(\mathbf{S}) Z^m(\mathbf{S}). \tag{2.16}$$

At this stage, $\mathcal{F}_m$ has no obvious physical meaning and is introduced only to avoid the cumbersome averaging of the logarithm in (2.13). It is easy to show that the physical free energy $\mathcal{F}(N_{\text{tot}}, N_c)$ (2.13) is related to the replica free energy by the expression

$$\mathcal{F}(N_{\text{tot}}, N_c) = \frac{d\mathcal{F}_m(N_{\text{tot}}, N_c)}{dm} \bigg|_{m=0}. \tag{2.17}$$

Since we are only interested in the $m \to 0$ limit, the function $\mathcal{F}_m$ can be expanded in a power series in m. Equation (2.17) implies that in the calculation of $\mathcal{F}_m$ one should retain terms only up to first order in m.

Some intuition about the physical meaning of the replica free energy can be gained by considering the limit $m \to 0$ in Eq. (2.16). This yields

$$\exp[-\mathcal{F}_0(N_{tot}, N_c)/T] = \int d\mathbf{S}\, Z^{(0)}(\mathbf{S}) \equiv Z_{liq}(N_c) \qquad (2.18)$$

and we conclude that in this limit the replica free energy reduces to the free energy $-T \ln Z_{liq}(N_c)$ of the constrained partition function of a polymer in a good solvent, with a given number (N_c) of binary contacts between monomers.

Going back to Eq. (2.16) we note that, for integer m, the product $Z^{(0)}(\mathbf{S})Z^m(\mathbf{S})$ can be interpreted as the partition function of a replica system that consists of $1 + m$ noninteracting systems (replicas). The zeroth replica represents the initial nondeformed gel (with partition function $Z^{(0)}(\mathbf{S})$ corresponding to a particular realization $\mathbf{S}$ of the network structure) and the other m identical replicas represent the final deformed gel (each with partition function $Z(\mathbf{S})$). The Hamiltonian and the partition function of the zeroth replica are given by (2.1) and (2.11), respectively, where for clarity of notation we label the monomer coordinates by the superscript (0). Similarly, the Hamiltonian of the kth replica $(1 \le k \le m)$ is

$$\frac{\mathcal{H}^{(k)}[\mathbf{x}^{(k)}(s)]}{T} = \frac{1}{2a^2} \int_0^{N_{tot}} ds \left(\frac{d\mathbf{x}^{(k)}}{ds}\right)^2 + \frac{w}{2} \int_0^{N_{tot}} ds \int_0^{N_{tot}} ds'\, \delta[\mathbf{x}^{(k)}(s) - \mathbf{x}^{(k)}(s')] \quad (2.19)$$

where we have used the fact that the quality of solvent and the temperature are identical in all the replicas $(k \ge 1)$ corresponding to the final state of the gel, i.e., $w^{(k)} = w$ and $T^{(k)} = T$, respectively. The partition function of the kth replica is

$$Z^{(k)}(\mathbf{S}) = \int D\mathbf{x}^{(k)}(s) \exp(-\mathcal{H}^{(k)}[\mathbf{x}^{(k)}(s)]/T) \prod_{\{i,j\}} \delta[\mathbf{x}^{(k)}(s_i) - \mathbf{x}^{(k)}(s_j)] \qquad (2.20)$$

where the integration goes over all chain configurations in the volume V occupied by the deformed gel.

Substituting expressions (2.11) and (2.20) into Eq. (2.16) yields

$$\exp[-\mathcal{F}_m(N_{tot}, N_c)/T] = \int d\mathbf{S} \prod_{k=0}^{m} \int D\mathbf{x}^{(k)}(s)$$

$$\times \exp(-\mathcal{H}^{(k)}[\mathbf{x}^{(k)}(s)]/T) \prod_{\{i,j\}} \delta[\mathbf{x}^{(k)}(s_i) - \mathbf{x}^{(k)}(s_j)]. \qquad (2.21)$$

Since the replica Hamiltonians in the above equation do not contain the cross-link contour coordinates $\mathbf{S}$, we can pull the integrals over these coordinates in front of the product of the δ-functions and perform N_c integrations over the cross-linked pairs of monomers $\{i, j\}$. Each of these integrations produces a factor

$$B[\{\mathbf{x}^{(k)}\}] = \int_0^{N_{\text{tot}}} ds_i \int_0^{N_{\text{tot}}} ds_j \prod_{k=0}^m \delta[\mathbf{x}^{(k)}(s_i) - \mathbf{x}^{(k)}(s_j)] \tag{2.22}$$

and, therefore, the integration over $\mathbf{S}$ introduces the factor $(B[\{\mathbf{x}^{(k)}\}])^{N_c}$ into the integrand in (2.21).

Grand Canonical Representation

Instead of working directly with the constraints introduced by substituting the δ-functions (Eq. (2.22)) into Eq. (2.21), we can replace them by effective interactions, using the following identity

$$B^{N_c} = \frac{N_c!}{2\pi i} \oint \frac{dz_c}{z_c^{N_c+1}} e^{z_c B}. \tag{2.23}$$

This relation is analogous to the usual thermodynamic transformation from the canonical to the grand canonical ensemble, which suggests that z_c can be interpreted as the *fugacity* that defines the average number of cross-links in the latter ensemble. Apart from mathematical convenience, the use of the grand canonical ensemble reflects the physical observation that only the average number of cross-links can be fixed by any physical or chemical method of gel preparation.

To complete the transformation to the grand canonical ensemble, we introduce the chemical potential μ of monomer units (it differs from the usual thermodynamic definition of the chemical potential by a factor of T) through the identity

$$\delta(N_{\text{tot}} - M) = \frac{1}{2\pi i} \oint d\mu \, e^{\mu(N_{\text{tot}} - M)}. \tag{2.24}$$

As the result, the replica partition function (2.16) takes the form

$$\exp[-\mathcal{F}_m(N_{\text{tot}}, N_c)/T] = \left[\frac{1}{2\pi i} \oint d\mu \, e^{\mu N_{\text{tot}}}\right]\left[\frac{1}{2\pi i} \oint \frac{dz_c}{z_c^{N_c+1}}\right] \Xi_m(\mu, z_c) \tag{2.25}$$

where Ξ_m is the grand canonical partition function:

$$\Xi_m(\mu, z_c)$$
$$= \int_0^\infty dM \, e^{-\mu M} \int D\hat{\mathbf{x}}(s) \exp\left\{-\frac{1}{2a^2} \int_0^M ds \left(\frac{d\hat{\mathbf{x}}}{ds}\right)^2 + \frac{z_c}{2} \int_0^M ds \int_0^M ds' \, \delta[\hat{\mathbf{x}}(s) - \hat{\mathbf{x}}(s')]\right.$$
$$\left. - \sum_{k=0}^m \frac{w^{(k)}}{2} \int_0^M ds \int_0^M ds' \, \delta[\mathbf{x}^{(k)}(s) - \mathbf{x}^{(k)}(s')]\right\}. \tag{2.26}$$

In writing (2.26) we introduced the $3(1 + m)$-dimensional vector $\hat{\mathbf{x}}$ in the space of the replicas (replica space), with components $x_\alpha^{(k)}$ ($\alpha = x, y, z; k = 0, \ldots, m$) and used the identities

$$\left(\frac{d\hat{\mathbf{x}}}{ds} \right)^2 = \sum_{k=0}^{m} \left(\frac{d\mathbf{x}^{(k)}}{ds} \right)^2 \tag{2.27}$$

and

$$\delta[\hat{\mathbf{x}}(s) - \hat{\mathbf{x}}(s')] = \prod_{k=0}^{m} \delta[\mathbf{x}^{(k)}(s) - \mathbf{x}^{(k)}(s')]. \tag{2.28}$$

Notice that the expression inside the curly brackets in the exponent in Eq. (2.26) can be interpreted as (minus) the effective Hamiltonian of a polymer chain in an abstract replica space. Network constraints due to the cross-links (which were present in the original physical space) are replaced in this replica space by an effective attractive interaction with strength proportional to z_c. Unlike the usual excluded volume interaction, which is expressed as the sum of δ-functions with coefficients $w^{(k)}$ and therefore is *diagonal* in the replicas, the effective attractive interaction due to the cross-links appears as a product of δ-functions (Eq. (2.28)), which *couples* the different replicas. This statement can be further clarified by the observation that a total Hamiltonian describes a set of noninteracting systems only if it can be represented as the sum of the Hamiltonians of the constituents.

On a more physical level, the difference between the effects of excluded volume and cross-links stems from the different ways in which they enter the replica formulation of the statistical mechanics of polymer networks. Thermal fluctuations take place *independently* in the different replicas and, therefore, different monomer pairs interact via excluded volume in different replicas. On the other hand, all the replicas have, by definition, the *same* network structure (they all correspond to the same realization of this structure) and thus, the same monomer pairs interact through attractive interactions (which account for the presence of cross-links), in all the replicas.

The cross-link-induced coupling between the replicas has a dramatic effect on the typical conformation of the polymer in the abstract replica space. While the attractions due to the cross-links can be stabilized by excluded volume repulsions *in each* of the replicas, there is nothing to balance the attractions *between different* replicas, with the consequence that (as will be shown in detail in the following) the true ground state of the replica system corresponds not to a state of uniform density in replica space but rather to a *collapsed* state of the polymer in this space.

In the thermodynamic limit, N_{tot}, $N_c \to \infty$, the integrals over μ and z_c in (2.25) can be evaluated by the method of steepest descent, with the result

$$\mathcal{F}_m(N_{tot}, N_c)/T = -\ln \Xi_m(\mu, z_c) - N_{tot}\mu + N_c \ln z_c \qquad (2.29)$$

where the fugacity z_c of cross-links and the chemical potential μ of monomers can be obtained by minimizing the right-hand side of (2.29):

$$N_{tot} = -\partial \ln \Xi_m(\mu, z_c)/\partial\mu, \qquad N_c = \partial \ln \Xi_m(\mu, z_c)/\partial \ln z_c. \qquad (2.30)$$

Substituting Eq. (2.29) into Eq. (2.17), we relate the physical free energy to the replica partition function of the grand canonical ensemble:

$$\mathcal{F}(N_{tot}, N_c) = -T \frac{\partial \ln \Xi_m(\mu, z_c)}{\partial m}\bigg|_{m=0} - T \frac{\partial \ln \Xi_m(\mu, z_c)}{\partial \mu}\bigg|_{m=0} \frac{\partial \mu}{\partial m}\bigg|_{m=0}$$

$$- T \frac{\partial \ln \Xi_m(\mu, z_c)}{\partial \ln z_c}\bigg|_{m=0} \frac{\partial \ln z_c}{\partial m}\bigg|_{m=0} - N_{tot}\frac{\partial \mu}{\partial m}\bigg|_{m=0} + N_c \frac{\partial \ln z_c}{\partial m}\bigg|_{m=0}$$

$$= -T \frac{\partial \ln \Xi_m(\mu, z_c)}{\partial m}\bigg|_{m=0} \qquad (2.31)$$

where the last equality is obtained using (2.30). The monomer chemical potential and the fugacity of cross-links, which parameterize the grand canonical partition function in Eq. (2.31), should be expressed in terms of the parameters N_{tot} and N_c using Eq. (2.30), in the limit $m = 0$.

To calculate the free energy, Eq. (2.31), one has to evaluate functional integrals over the set of trajectories $\{\hat{x}(s)\}$ in replica space, in Eq. (2.26). Direct calculation of such integrals is prohibitively difficult and, following the usual approach in polymer physics [15], we will transform the problem into a more tractable one by going over to collective coordinates (field theory).

3.3 Field Theory

Inspection of the grand canonical partition function reveals the source of the mathematical difficulties that arise in theories of interacting stringlike objects. Whereas the elastic term (Eqs. (2.26) and (2.27)) is local in this representation (i.e., depends only on a single coordinate along the chain, s), the interaction terms in Eq. (2.26) are nonlocal (depend on two coordinates, s and s'). An attempt to confront these difficulties head-on was made by Deam and Edwards [1], who used a variational method to calculate the thermodynamic free energy of an instantaneously cross-linked polymer gel. Such an approach is known to give good results for the ground state energy (we will see that the thermodynamic free energy obtained by the above authors coincides with our mean field estimation) but is difficult to apply to the calculation of density fluctuations for which one has to have complete information about the ground state. In this work we take a different path, which was also proposed by Edwards and Vilgis [18] (in the context of end-linked networks, without excluded volume) and which is based on the realization that although the interactions are nonlocal along the chain contour, due to the presence of the δ-functions they are local in real (and in replica) space. Therefore, it is advantageous to transform to a description in terms of fields over spatial coordinates (i.e., to collective coordinates), in which the local character of the interaction terms is made explicit [15]. We now proceed to construct the field theoretical representation of the grand canonical partition function. For this purpose, in Section 3.3.1 we first introduce the order parameter for an amorphous solid and discuss its properties. Then, in Section 3.3.2 we construct the density functional representation for the grand canonical partition function, Eq. (2.26). In Section 3.3.3 we introduce the field theoretical representation of the entropy of a polymer chain with given monomer density in replica space and use it in Section 3.3.4 to derive the field theory of polymer networks.

3.3.1   ORDER PARAMETER

We start with the definition of the microscopic (i.e., nonaveraged) monomer density in replica space

$$\rho(\hat{\mathbf{x}}) \equiv \int ds \; \delta[\hat{\mathbf{x}} - \hat{\mathbf{x}}(s)] = \int ds \; \prod_{k=0}^{m} \delta[\mathbf{x}^{(k)} - \mathbf{x}^{(k)}(s)]. \tag{3.1}$$

The thermal average of this expression, $\langle \rho(\hat{\mathbf{x}}) \rangle$, is an extension of the Edwards-Anderson order parameter in the theory of spin glasses [7] to the case of a system which "remembers" the state of its preparation. The memory of this state is fixed in the network structure, which is the same for initial (undeformed) and final (deformed) states of the network. To see why this order parameter can be considered as a measure of the correlations between the replicas of the *initial* and different *final* states of the network, let us examine its qualitative properties. The order parameter vanishes (in the thermodynamic limit) in the "liquid" phase where there is no correlation between the monomer positions in different replicas. This state has full translational and rotational symmetry characteristic of a liquid phase.

The *translational symmetry* of the system *is spontaneously broken* in the "solid" phase because of the localization of monomers near their average positions in the cross-linked gel. This situation corresponds to *spontaneously broken ergodicity*. In principle, there can be further ergodicity breaking associated with the possibility of existence of a large number of stable states separated by infinite (in the thermodynamic limit) barriers. The presence of such spin-glass-like phases can be detected by the inspection of the order parameter (3.1) since they are characterized by broken replica symmetry. As will be demonstrated in the following, our mathematical analysis shows that the spin-glass-type scenario does not take place in polymer gels. *The physical reason for the absence of multiple localized states is the presence of long-range elastic forces in the polymer network,* which make these strongly inhomogeneous states energetically unfavorable.

The order parameter $\langle \rho(\hat{\mathbf{x}}) \rangle$ has a nonvanishing value in the solid phase since the conformations of the network in the different replicas are strongly correlated due to the fixed structure of the network, which is identical in all the replicas [19]. The observation that there is no replica-symmetry-breaking in the theory of polymer networks is of great importance since it allows us to characterize each state of the network with the given structure **S** by a unique order parameter $\langle \rho(\hat{\mathbf{x}}) \rangle$ that is invariant under the *permutations* of the m replicas of the *final state*. All the possible states of such a network with spontaneously broken translational symmetry are related by global translations and rotations of the gel as a whole. These states can be characterized by the order parameter $\langle \rho(\hat{\mathbf{x}}') \rangle$, where the components $\mathbf{x}'^{(k)}$ ($k = 1, \ldots, m$) of the replica vector $\hat{\mathbf{x}}'$ are related to $\mathbf{x}^{(k)}$ by the corresponding translation and rotation transformations. In contrast, in a spin-glass-like phase there would be states not related by global translations and rotations of the system. Such a phase cannot be characterized by a unique order parameter $\langle \rho(\hat{\mathbf{x}}) \rangle$, and one has to introduce the distribution function of the order parameters.

In the undeformed final state, the order parameter is invariant under the permutation of all $m + 1$ replicas. This symmetry is broken for networks deformed (stretched, swollen, or compressed) with respect to the preparation state. To fix the corresponding "gauge" of the order parameter, we consider the case when the average monomer density is constant in real space, in both the initial and the final states of the network,

$$\rho^{(k)}(\mathbf{x}) = \text{constant.} \tag{3.2}$$

The density in the kth replica can be calculated by integrating the abstract density in replica space over the coordinates of all the other replicas

$$\rho^{(k)}(\mathbf{x}^{(k)}) \equiv \prod_{\{l \neq k\}} \int d\mathbf{x}^{(l)} \rho(\hat{\mathbf{x}}) = \int ds\,\delta[\mathbf{x}^{(k)} - \mathbf{x}^{(k)}(s)] \tag{3.3}$$

which can also be represented in the form

$$\rho^{(k)}(\mathbf{x}) = \int d\hat{\mathbf{x}} \, \rho(\hat{\mathbf{x}}) \delta[\mathbf{x} - \mathbf{x}^{(k)}]. \tag{3.4}$$

We now show that this form is consistent with Eq. (3.2) if $\rho(\hat{\mathbf{x}})$ depends on the coordinates of the replicas only through the linear combinations

$$\mathbf{y}^{(k)} \equiv \mathbf{x}^{(k)} - \lambda^{(k)} \star \mathbf{x}^{(0)}, \quad k = 1, \ldots, m \tag{3.5}$$

where $\lambda^{(k)}$ are arbitrary constants (the symmetry of the order parameter with respect to permutation of replicas with $k \geq 1$ implies that one should take $\lambda_\alpha^{(k)} = \lambda_\alpha$ for all these replicas) and where we define the $\star$ operation by

$$(\lambda \star \mathbf{x})_\alpha \equiv \lambda_\alpha x_\alpha \tag{3.6}$$

(no summation over α). A similar functional dependence of the order parameter on the replica coordinates (in the momentum representation) was introduced by Castillo and coworkers [20] for the case of an undeformed gel, $\{\lambda_\alpha = 1\}$.

Clearly, we can always replace $\rho(\hat{\mathbf{x}}) = \rho(\mathbf{x}^{(0)}, \{\mathbf{x}^{(k)}\})$ by a different function of the arguments $x_\alpha^{(0)}$ and $y_\alpha^{(k)}$ ($k \geq 1$); i.e., we can write $\rho(\hat{\mathbf{x}}) = f(\mathbf{x}^{(0)}, \{\mathbf{y}^{(k)}\})$. From Eq. (3.4), the density in the kth replica can be expressed as

$$\rho^{(0)}(\mathbf{x}^{(0)}) = \prod_{\{l \neq 0\}} \int d\mathbf{y}^{(l)} f(\mathbf{x}^{(0)}, \{\mathbf{y}^{(l)}\}), \quad \text{for } k = 0 \tag{3.7}$$

$$\rho^{(k)}(\mathbf{x}^{(k)}) = \int d\mathbf{x}^{(0)} f^{(k)}(\mathbf{x}^{(0)}, \mathbf{x}^{(k)} - \lambda \star \mathbf{x}^{(0)}), \quad \text{for } k \neq 0 \tag{3.8}$$

where

$$f^{(k)}(\mathbf{x}^{(0)}, \mathbf{y}^{(k)}) = \prod_{\{l \neq k\}} \int d\mathbf{y}^{(l)} f(\mathbf{x}^{(0)}, \{\mathbf{y}^{(l)}\}). \tag{3.9}$$

In Eqs. (3.7) and (3.9), we shifted the integrations by $d\mathbf{x}^{(l)} \to d\mathbf{y}^{(l)}$. Our proof follows immediately from inspection of the above integrals, which shows that for $\rho_{mf}^{(k)}$ to be independent of $\mathbf{x}^{(k)}$ (for all k) as required, f must depend on $\mathbf{x}^{(k)}$ *only* through the linear combination $\mathbf{x}^{(k)} - \lambda^{(k)} \star \mathbf{x}^{(0)}$ (for example, if f was a function of both $\mathbf{x}^{(0)}$ and $\{\mathbf{y}^{(l)}\}$, upon performing the integration in Eq. (3.7), we would obtain a function of $\mathbf{x}^{(0)}$ and not a constant). Furthermore, in order for Eqs. (3.7) and (3.9) to be meaningful, the integrals that appear in these equations must be convergent; i.e., the function $f(\{\mathbf{y}^{(k)}\})$ must be a sufficiently rapidly decreasing function of its arguments.

To relate the parameters λ_α to the characteristics of the deformed network, we use the fact that the function $f^{(k)}$ depends only on $\mathbf{y}^{(k)}$ and, therefore, we can replace $\int d\mathbf{x}^{(0)}$ by $\int d\mathbf{y}^{(k)}/(\lambda_x \lambda_y \lambda_z)$. Inserting the expression for $f^{(k)}$ (Eq. (3.9)) and comparing with (3.7), we obtain a relation between the mean density of the undeformed gel and that of the deformed one (i.e., any one of the identical replicas of the deformed state):

$$\rho^{(k)} = \rho^{(0)}/(\lambda_x \lambda_y \lambda_z), \quad k \neq 0. \tag{3.10}$$

Notice that under an arbitrary stretching or compression, the linear dimensions of the system, L_{0x}, L_{0y}, L_{0z}, change to L_x, L_y, L_z. The volume changes from $V_0 = L_{0x}L_{0y}L_{0z}$ to $V = L_x L_y L_z$ and, thus, the relation between the final density ρ and the density prior to the deformation, ρ_0, is given by $\rho = \rho_0(L_{0x}L_{0y}L_{0z}/L_x L_y L_z)$. Comparison with Eq. (3.10) shows that λ_x, λ_y, and λ_z must be identified with the extension ratios along the principal axes of the deformation of the network,

$$\lambda_\alpha = L_\alpha/L_{0\alpha}, \quad \text{for } \alpha = x, y, z. \tag{3.11}$$

In general, only a fraction of monomers (*gel*) are localized in the vicinity of their average positions and the other monomers (*sol*) are delocalized and can span all the volume occupied by the system. In the thermodynamic limit only the gel fraction contributes to the order parameter $\langle \rho(\hat{\mathbf{x}}) \rangle$. This situation is considered in detail in ref. [14]; here we restrict our attention to gels prepared far beyond the gel-formation threshold, when there are no sol molecules in the system.

3.3.2 *DENSITY FUNCTIONAL*

We now return to the expression for the grand canonical partition function, Eq. (2.26), and derive its density functional representation. Contour integration over a nonlocal (in the contour coordinates s) δ-function can be replaced by a spatial integral over a local (in space) function of the density, as follows. We introduce the identity

$$\int ds \int ds' \, \delta[\hat{\mathbf{x}}(s) - \hat{\mathbf{x}}(s')] \equiv \int ds \int ds' \int d\hat{\mathbf{x}} \, \delta[\hat{\mathbf{x}} - \hat{\mathbf{x}}(s)] \, \delta[\hat{\mathbf{x}} - \hat{\mathbf{x}}(s')]$$
$$= \int d\hat{\mathbf{x}} \, \rho^2(\hat{\mathbf{x}}) \tag{3.12}$$

where the second equality is obtained by changing the order of the integrals and using the definition of $\rho(\hat{\mathbf{x}})$, Eq. (3.1). Similarly, we derive an analogous relation for the kth replica:

$$\int ds \int ds' \, \delta[\mathbf{x}^{(k)}(s) - \mathbf{x}^{(k)}(s')] \equiv \int d\mathbf{x}^{(k)} [\rho^{(k)}(\mathbf{x}^{(k)})]^2. \tag{3.13}$$

The next step is to replace the integration over the monomer coordinates $\{\hat{\mathbf{x}}(s)\}$ by the integration over the collective coordinates $\{\rho(\hat{\mathbf{x}})\}$. This is done by inserting the representation of unity

$$1 \equiv \int D\rho(\hat{\mathbf{x}}) \delta\left(\rho(\hat{\mathbf{x}}) - \int_0^M ds \, \delta[\hat{\mathbf{x}} - \hat{\mathbf{x}}(s)] \right) \tag{3.14}$$

in front of the exponential in (2.26) and moving the integration over $\rho(\hat{\mathbf{x}})$ to the leftmost end of the expression on the right-hand side of this equation. Using identities (3.12) and (3.13), the grand canonical partition function can be represented as a functional integral over the replica density field:

$$\Xi_m(\mu, z_c) = \int D\rho(\hat{\mathbf{x}})\exp\left\{S(\mu, [\rho(\hat{\mathbf{x}})]) + \frac{z_c}{2}\int d\hat{\mathbf{x}}\, \rho^2(\hat{\mathbf{x}})\right.$$

$$\left. - \sum_{k=0}^{m} \frac{w^{(k)}}{2}\int d\mathbf{x}^{(k)}[\rho^{(k)}(\mathbf{x}^{(k)})]^2\right\}. \tag{3.15}$$

Here, the term proportional to z_c accounts for the contribution of cross-links, the terms proportional to $w^{(k)}$ represent the excluded volume interactions, and $S(\mu, [\rho(\hat{\mathbf{x}})])$ is the replica analog of the elastic entropy of the polymer chain, with the given monomer density $\rho(\hat{\mathbf{x}})$ in replica space:

$$\exp\{S(\mu, [\rho(\hat{\mathbf{x}})])\} \equiv \int_0^\infty dM\, e^{-\mu M}\int D\hat{\mathbf{x}}(s)$$

$$\times \delta\left(\rho(\hat{\mathbf{x}}) - \int_0^M ds\, \delta[\hat{\mathbf{x}} - \hat{\mathbf{x}}(s)]\right)\exp\left\{-\frac{1}{2a^2}\int_0^M ds\left(\frac{d\hat{\mathbf{x}}}{ds}\right)^2\right\}. \tag{3.16}$$

3.3.3   ELASTIC ENTROPY

We now derive a field theoretical representation of $\exp\{S\}$. Introducing the exponential representation of the δ-function

$$\delta\left(\rho(\hat{\mathbf{x}}) - \int_0^M ds\,\delta[\hat{\mathbf{x}} - \hat{\mathbf{x}}(s)]\right) \equiv \int Dh(\hat{\mathbf{x}})\exp\left[i\int d\hat{\mathbf{x}}\, h(\hat{\mathbf{x}})\rho(\hat{\mathbf{x}}) - i\int_0^M ds\, h(\hat{\mathbf{x}}(s))\right] \tag{3.17}$$

and moving the integration over $h(\hat{\mathbf{x}})$ to the leftmost end of the term on the right-hand side of Eq. (3.16), gives

$$\exp\{S(\mu, [\rho(\hat{\mathbf{x}})])\} = \int Dh(\hat{\mathbf{x}})\exp\left[i\int d\hat{\mathbf{x}}\, h(\hat{\mathbf{x}})\rho(\hat{\mathbf{x}})\right]$$

$$\times \int d\hat{\mathbf{x}}_1\int d\hat{\mathbf{x}}_2 \hat{G}\{\hat{\mathbf{x}}_1, \hat{\mathbf{x}}_2, [ih(\hat{\mathbf{x}})]\} \tag{3.18}$$

where

$$\hat{G}\{\hat{\mathbf{x}}_1, \hat{\mathbf{x}}_2, [ih(\hat{\mathbf{x}})]\} = \int_0^\infty dM\, e^{-\mu M}\int_{\hat{\mathbf{x}}_1}^{\hat{\mathbf{x}}_2} D\hat{\mathbf{x}}(s)$$

$$\times \exp\left\{-\int_0^M ds\left[\frac{1}{2a^2}\left(\frac{d\hat{\mathbf{x}}}{ds}\right)^2 + ih(\hat{\mathbf{x}}(s))\right]\right\} \tag{3.19}$$

can be interpreted as the grand canonical partition function of an ideal Gaussian chain, with ends fixed at points $\hat{\mathbf{x}}_1$ and $\hat{\mathbf{x}}_2$, in an external field $ih(\hat{\mathbf{x}})$ (in a $3(1+m)$-dimensional replica space). Comparison of Eqs. (3.16) and (3.18) shows that the constraint of fixed density distribution is eliminated at the expense of introducing an additional stochastic field h. To avoid handling the constraints associated with the connectivity of the chain, we transform the Gaussian chain problem into a field theory [21] (Appendix 3.A). Using the replica space generalization ($\mathbf{x} \rightarrow \hat{\mathbf{x}}$) of the usual trick of relating the polymer problem to the $n = 0$ limit of the n-vector model (Appendix 3.A), $\hat{G}$ is represented as a functional integral over an n-component vector field $\vec{\varphi}(\hat{\mathbf{x}})$, with components $\varphi_i(\hat{\mathbf{x}})$ ($i = 1, \ldots, n$). Analytic continuation to the limit $n \rightarrow 0$ yields

$$\hat{G}\{\hat{\mathbf{x}}_1, \hat{\mathbf{x}}_2, [ih(\hat{\mathbf{x}})]\} = \int D\vec{\varphi}\, \varphi_1(\hat{\mathbf{x}}_1)\varphi_1(\hat{\mathbf{x}}_2)\exp\{-H_0[ih(\hat{\mathbf{x}}), \vec{\varphi}(\hat{\mathbf{x}})]\} \quad (3.20)$$

where the effective (dimensionless) Hamiltonian H_0 has the form

$$H_0[ih(\hat{\mathbf{x}}), \vec{\varphi}(\hat{\mathbf{x}})] = \int d\hat{\mathbf{x}}\left[\frac{1}{2}(\mu + ih(\hat{\mathbf{x}}))\,\vec{\varphi}^2(\hat{\mathbf{x}}) + \frac{a^2}{2}(\hat{\nabla}\vec{\varphi}(\hat{\mathbf{x}}))^2\right]. \quad (3.21)$$

Here $\hat{\nabla}$ is the $3(1+m)$-dimensional gradient operator with respect to replica space coordinates. Notice that we now have three different spaces and, to avoid confusion, we use different designations for vectors embedded in them: $\mathbf{x}$ is the usual three-dimensional vector with components x_α ($\alpha = x, y, z$), the vector $\hat{\mathbf{x}}$ is defined in $3(1+m)$-dimensional replica space and has components $x_\alpha^{(k)}$ ($\alpha = x, y, z; k = 0, \ldots, m$), and the n-dimensional vector $\vec{\varphi}$ has components φ_i ($i = 1, \ldots, n$).

We now substitute Eqs. (3.20) and (3.21) into (3.18) and integrate over the field h, using the identity

$$\int Dh(\hat{\mathbf{x}})\exp\left[i\int d\hat{\mathbf{x}}\, h(\hat{\mathbf{x}})(\rho(\hat{\mathbf{x}}) - \vec{\varphi}^2(\hat{\mathbf{x}})/2)\right] \equiv \delta(\rho(\hat{\mathbf{x}}) - \vec{\varphi}^2(\hat{\mathbf{x}})/2). \quad (3.22)$$

The resulting field theoretical representation of the elastic entropy S, is

$$\exp\{S(\mu, [\rho(\hat{\mathbf{x}})])\} = \int D\vec{\varphi}(\hat{\mathbf{x}})\left[\int d\hat{\mathbf{x}}\, \varphi_1(\hat{\mathbf{x}})\right]^2 \delta(\rho(\hat{\mathbf{x}}) - \vec{\varphi}^2(\hat{\mathbf{x}})/2)$$
$$\times \exp\left\{-\int d\hat{\mathbf{x}}\left[\frac{1}{2}\mu\vec{\varphi}^2(\hat{\mathbf{x}}) + \frac{a^2}{2}(\hat{\nabla}\vec{\varphi}(\hat{\mathbf{x}}))^2\right]\right\}. \quad (3.23)$$

Note that, as a by-product, we obtain the important relation between the vector field $\vec{\varphi}$ and the monomer density in the replica space:

$$\rho(\hat{\mathbf{x}}) = \vec{\varphi}^2(\hat{\mathbf{x}})/2. \quad (3.24)$$

This formula is an exact relation between the two fluctuating fields $\vec{\varphi}$ and ρ. It is the generalization of a well-known relation in the $n = 0$, φ^4 formulation of the excluded volume problem [4] (which relates the *average* of the square of the abstract field φ to the physically observable mean density $\langle\rho\rangle$).

3.3.4 FIELD HAMILTONIAN

Substituting Eq. (3.23) into Eq. (3.15) and carrying out the trivial (due to the δ-function) integration over the field $\rho(\hat{\mathbf{x}})$, we obtain an explicit representation for the grand canonical partition function of a Gaussian phantom network, in terms of the field $\vec{\varphi}(\hat{\mathbf{x}})$:

$$\Xi_m(\mu, z_c) = \int D\vec{\varphi}(\hat{\mathbf{x}}) \left[\int d\hat{\mathbf{x}}\ \varphi_1(\hat{\mathbf{x}})\right]^2 \exp\{-H[\vec{\varphi}(\hat{\mathbf{x}})]\}. \qquad (3.25)$$

In evaluating this expression, we have to perform the functional integration over the field $\vec{\varphi}$ and then make the analytic continuation from integer n to $n = 0$ (where n is the number of components of this vector field). The Hamiltonian H is given by

$$H[\vec{\varphi}] = \int d\hat{\mathbf{x}} \left[\frac{1}{2}\mu\vec{\varphi}^2(\hat{\mathbf{x}}) + \frac{a^2}{2}(\hat{\nabla}\vec{\varphi}(\hat{\mathbf{x}}))^2 - \frac{z_c}{8}(\vec{\varphi}^2(\hat{\mathbf{x}}))^2\right]$$

$$+ \sum_{k=0}^{m}\frac{w^{(k)}}{8}\int d\mathbf{x}^{(k)}\left[\prod_{l\neq k}\int d\mathbf{x}^{(l)}\vec{\varphi}^2(\hat{\mathbf{x}})\right]^2. \qquad (3.26)$$

This effective Hamiltonian is a straightforward extension of the φ^4 zero-component field theory of a polymer chain with excluded volume to the $3(1 + m)$-dimensional replica space. It has a number of discrete and continuous symmetries:

1. Arbitrary rotations in the abstract space of the n-vector model.
2. Permutation of the replicas of the final state.
3. Arbitrary rotations in the space of *each* of the replicas. Due to the presence of the excluded volume terms, the Hamiltonian is not invariant under arbitrary rotations in replica space that would, in general, mix the different replicas. (The densities in each of the replicas, $\rho^{(k)}(\mathbf{x}^{(k)})$, that enter the excluded volume interaction term in the Hamiltonian are not invariant under rotations in replica space $\{\hat{\mathbf{x}}\}$ that mix the different replicas.)
4. Translation by an arbitrary constant vector in replica space.

The existence of the symmetry under translation in replica space suggests (wrongly!) that our field theory describes a polymer "solution" in replica space, with cross-links replaced by effective attractions between monomers. Consider, for example, the single-replica, $m = 0$, version of our model. In this case, $\prod_{l \neq 0} \int d\mathbf{x}^{(l)}$ is replaced by unity and the corresponding Hamiltonian (3.26) becomes

$$
H[\vec{\varphi}]\big|_{m=0} = \int d\mathbf{x}^{(0)} \left[\frac{1}{2} \mu \vec{\varphi}^2(\mathbf{x}^{(0)}) + \frac{a^2}{2} (\nabla \vec{\varphi}(\mathbf{x}^{(0)}))^2 \right.
$$
$$
\left. + \frac{(w^{(0)} - z_c)}{8} (\vec{\varphi}^2(\mathbf{x}^{(0)}))^2 \right]
\tag{3.27}
$$

which is identical to the Hamiltonian of the $n \to 0$ model of a polymer chain without cross-links but with an excluded volume parameter $w^{(0)} - z_c$. Substitution of this Hamiltonian to Eq. (3.25) gives the grand canonical partition function of a constrained polymer with a second virial coefficient $w^{(0)} - z_c$, confined to a volume $V^{(0)}$. The reduction of the excluded volume parameter compared to its bare value $w^{(0)}$ reflects the fact that some of the intermonomer contacts in any configuration of this polymer (N_c of them, on the average) represent the cross-links and do not contribute to the excluded volume interaction energy.

Although the above conclusion is perfectly valid for the single replica case, it misses the fact that our model contains not only the replica of the initial state but also m replicas of the final state and that the calculation has to be performed in the $3(1 + m)$-dimensional replica space *before* taking the limit $m \to 0$. Inspection of Eq. (3.15) shows that whereas excluded volume repulsions act only within the individual replicas (only the sum $\sum_{k=0}^{m} w^{(k)}[\rho^{(k)}(\mathbf{x})]^2$ appears in the exponent in Eq. (3.15)), cross-link-induced attractions ($z_c[\rho(\hat{\mathbf{x}})]^2$) introduce a coupling between all the replicas. We show in the next section that this coupling leads to the localization of network monomers near their average positions and reflects the fact that our model describes a *solid*.

3.4 Mean Field Solution

3.4.1 THE MEAN FIELD SOLUTION

In the preceding section we constructed a formally exact field theoretical representation for the partition of a randomly cross-linked network. We now proceed to calculate the functional integral (3.23). Due to the presence of the φ^4 terms, this integral is not Gaussian and cannot be calculated exactly. Instead, we resort to a mean field estimate by the method of *steepest descent,* which is equivalent to finding the solution $\vec{\varphi}_{mf}$ that minimizes the effective Hamiltonian (3.26). The condition that $\vec{\varphi}_{mf}$ corresponds to an extremum of H is

$$\left(\mu + \sum_{k=0}^{m} w^{(k)} \rho_{mf}^{(k)}(\mathbf{x}^{(k)}) - a^2 \hat{\nabla}^2 - \frac{z_c}{2} \varphi_{mf}^2(\hat{\mathbf{x}}) \right) \vec{\varphi}_{mf}(\hat{\mathbf{x}}) = 0 \tag{4.1}$$

where, from Eqs. (3.1)–(3.4), the mean field density of monomer units in the kth replica is

$$\rho_{mf}^{(k)}(\mathbf{x}^{(k)}) \equiv \prod_{l \neq k} \int d\mathbf{x}^{(l)} \rho_{mf}(\hat{\mathbf{x}}) \quad \text{where } \rho_{mf}(\hat{\mathbf{x}}) = \vec{\varphi}_{mf}^2(\hat{\mathbf{x}})/2. \tag{4.2}$$

The thermodynamic parameters μ and z_c that appear in (4.1) can be related to the physical parameters that characterize the gel in the state of preparation, i.e., the average monomer density $\rho^{(0)} = N_{tot}/V^{(0)}$ and the average number of monomers between cross-links that are nearest neighbors along the chain contour, $\overline{N} = N_{tot}/(2N_c)$. The parameters μ and z_c should be calculated from Eq. (2.30), with the corresponding derivatives evaluated at $m = 0$. In the mean field approximation, this equation can be replaced by

$$N_{tot} = \frac{\partial H[\vec{\varphi}_{mf}]}{\partial \mu} \bigg|_{m=0}, \qquad N_c = -\frac{\partial H[\vec{\varphi}_{mf}]}{\partial \ln z_c} \bigg|_{m=0} \tag{4.3}$$

where the Hamiltonian H (for $m = 0$) is defined in Eq. (3.27).

Since we are looking for a solution of Eq. (4.1) that describes the spatially homogeneous initial state of the gel (with density $\rho^{(0)}$), the mean field solution is obtained by setting the expression in the brackets in this equation to zero. We obtain

$$\rho^{(0)} = \varphi_{mf}^2/2 = \mu/(z_c - w^{(0)}). \tag{4.4}$$

Substituting this solution into Eq. (3.27) and using Eqs. (4.3) yields $N_{tot} = \rho^{(0)} V^{(0)}$ and $N_c = z_c V^{(0)}(\rho^{(0)})^2/2$. Using these relations and the definition of $\overline{N}$, we obtain the mean field expressions for μ and z_c in terms of $\rho^{(0)}$ and $\overline{N}$

$$\mu = 1/\overline{N} - w^{(0)}\rho^{(0)}, \qquad z_c = 1/(\rho^{(0)}\overline{N}). \tag{4.5}$$

We would like to emphasize that to calculate the free energy, Eq. (2.31), it is not enough to obtain the solution of the mean field Eq. (4.1) and that there are two further conditions that must be satisfied by this mean field solution.

1. We have to verify that the solution *minimizes* the effective Hamiltonian. Notice that since Eq. (4.1) was obtained from the condition $\delta H/\delta \vec{\varphi} = 0$, its solutions correspond to the *extrema,* but not necessarily to the mini-

ma of H. A solution of this equation minimizes H if the second derivative operator

$$K_{ij}(\hat{\mathbf{x}}, \hat{\mathbf{x}}') \equiv \frac{\delta^2 H}{\delta\varphi_i(\hat{\mathbf{x}})\delta\varphi_j(\hat{\mathbf{x}}')}\bigg|_{\vec{\varphi} = \vec{\varphi}_{mf}} \qquad (4.6)$$

whose eigenvalues $\{\Lambda\}$ give the "energies" of small (but otherwise arbitrary) fluctuations ($\{\delta\varphi_i\}$) about the mean field ground state, has only nonnegative eigenvalues. The spectrum of eigenvalues can be found from the secular equation

$$\sum_{j=1}^{n}\int d\hat{\mathbf{x}}' K_{ij}(\hat{\mathbf{x}}, \hat{\mathbf{x}}')\psi_j(\hat{\mathbf{x}}') = \Lambda\psi_i(\hat{\mathbf{x}}) \qquad (4.7)$$

where $\vec{\psi}$ are the eigenfunctions corresponding to these eigenvalues (the indices i and j enumerate the components of the n-vector field). A complete analysis of this eigenvalue problem will be presented in Section 3.5.

2. We have to show that our solution corresponds to the true ground state of the Hamiltonian since, in the thermodynamic limit, the steepest descent estimate of the replica partition function is dominated by the lowest minimum of H (ground state dominance). This problem has been considered in ref. [3]; here we will only discuss the results obtained in that work.

3.4.2 HOMOGENEOUS SOLUTION

We now proceed to look for a solution of Eq. (4.1) for arbitrary integer m. In principle, this complicated nonlinear equation may admit many solutions that would correspond to different extrema of H. One should find all the solutions, compare their "energy" (by substituting the corresponding solution in the definition of H, Eq. (3.26)), and then find the one that describes the true ground state of the Hamiltonian. A less tedious strategy is based on symmetry arguments, i.e., on the expectation that such a solution must have the full symmetry of H, provided, of course, that it minimizes the Hamiltonian. The trivial solution, $\vec{\varphi}_{mf}(\hat{\mathbf{x}}) = 0$, does not satisfy condition (4.4) and, therefore, we have to consider a solution with broken symmetry with respect to rotations in the space of the n-vector model. We conclude that the gel corresponds to the low-temperature phase of this model. Upon inspection of H, Eq. (3.26) shows that this solution is invariant under the displacement of the replica space coordinates by an arbitrary constant vector, $\hat{\mathbf{x}} \to \hat{\mathbf{x}} + \hat{\mathbf{u}}$, a condition that is trivially satisfied by the spatially homogeneous (in replica space) solution

$$\vec{\varphi}_{mf}(\hat{\mathbf{x}}) = \vec{n}\varphi_{mf} \tag{4.8}$$

where $\vec{n}$ is a constant unit vector in the space of the n-vector model. It is easy to check that Eq. (4.1) has indeed a constant (in replica space) solution that can be determined from this equation by setting the expression in the brackets to zero. In the limit $m \to 0$, we get

$$\varphi_{mf} = \sqrt{\frac{2}{z_c \overline{\overline{N}}}}. \tag{4.9}$$

However, as will be shown in Section 3.5.1, the analysis of the spectrum of eigenvalues of the second derivative operator K defined in Eq. (4.6) (evaluated on this homogeneous solution) shows that some of the eigenvalues are negative, and therefore the constant solution corresponds to a *saddle point* rather than to a minimum of H. This solution remains stable with respect to uncorrelated density fluctuations in each of the replicas, but it is unstable with respect to fluctuations that are correlated in different replicas (i.e., with respect to the formation of an amorphous solid state, with strongly correlated monomer positions in different replicas). We conclude that the homogeneous solution corresponding to the liquid state must be rejected and proceed to look for another solution of the mean field equation.

3.4.3   *INHOMOGENEOUS SOLUTION*

The fact that the homogeneous (in replica space) solution does not minimize H forces us to look for a solution that has a lower symmetry than the Hamiltonian, a situation commonly referred to as *spontaneous symmetry breaking*. Such a phenomenon arises in crystalline solids in which the energy is invariant under arbitrary translations but the ground state is invariant only under translations by multiples of lattice vectors, along the symmetry axes of the crystal lattice [5]. This suggests an interesting analogy between crystalline and amorphous solids: From the knowledge that spontaneous symmetry breaking of translational symmetry in *real space* gives rise to *crystalline solids,* we expect that the breaking of this symmetry in *replica space* leads to the general class of *disordered solids*. Note that, according to this view, the difference between a disordered solid and a liquid stems only from the breaking of translational symmetry in the former, and no additional symmetry breaking is necessary in general. Therefore, since the Hamiltonian is invariant under the permutation of the ($k \neq 0$) replicas, we will first look for replica symmetric solutions and examine whether they minimize H. As was discussed in Section 3.3.1, such solutions describe a solid, the monomers of which are localized in the sense that their average positions are uniquely determined by the thermodynamic conditions and that fluctuations about these average positions are limited to microscopic distances. After we find the solution that minimizes the Hamiltonian, we will argue that there are no other solutions that satisfy the condition of local equilibrium (the proof of this statement is given in ref. [3]); therefore, this solution corresponds to the unique ground state of the Hamiltonian.

Spontaneous Breaking of Translational Symmetry in Replica Space

We proceed to look for a mean field solution with spontaneously broken translational symmetry that is inhomogeneous in replica space. (The analysis below was first given in ref. [6].) It is shown in Section 3.3.1 that the order parameter $\rho(\hat{\mathbf{x}})$ (and, consequently, the field $\vec{\varphi}(\hat{\mathbf{x}})$) of a spatially homogeneous gel) depends only on the combination of replica coordinates $\mathbf{y}^{(k)} = \mathbf{x}^{(k)} - \lambda \star \mathbf{x}^{(0)}$ (see Eq. (3.5)). It is easy to check that any function (e.g., our solution) that depends only on the arguments $\mathbf{y}^{(k)}$ is invariant under the following displacement of replica space coordinates:

$$\hat{\mathbf{x}} \to \hat{\mathbf{x}} + \sum_\alpha u_\alpha \hat{\mathbf{e}}_\alpha \qquad (4.10)$$

where u_α are the components of an arbitrary constant three-dimensional vector and $\hat{\mathbf{e}}_\alpha$ is a unit vector in the $3(1 + m)$-dimensional replica space:

$$\hat{\mathbf{e}}_\alpha = \left(\frac{\mathbf{e}_\alpha}{(1 + m\lambda_\alpha^2)^{1/2}}, \frac{\lambda_\alpha \mathbf{e}_\alpha}{(1 + m\lambda_\alpha^2)^{1/2}}, \dots, \frac{\lambda_\alpha \mathbf{e}_\alpha}{(1 + m\lambda_\alpha^2)^{1/2}} \right). \qquad (4.11)$$

Here $\mathbf{e}_\alpha$ is a unit vector along the direction α ($\alpha = x, y, z$) in the usual three-dimensional space.

The condition of invariance under an arbitrary displacement along the axes $\hat{\mathbf{e}}_x$, $\hat{\mathbf{e}}_y$, and $\hat{\mathbf{e}}_z$ (Eq. (4.10)) singles out these three directions in replica space. An arbitrary vector $\hat{\mathbf{x}}$ can be decomposed as

$$\hat{\mathbf{x}} = \hat{\mathbf{x}}_L + \hat{\mathbf{x}}_T \qquad (4.12)$$

where $\hat{\mathbf{x}}_L \equiv \sum_\alpha (\hat{\mathbf{e}}_\alpha \cdot \hat{\mathbf{x}})\hat{\mathbf{e}}_\alpha$ and $\hat{\mathbf{x}}_T$ is orthogonal to $\hat{\mathbf{x}}_L$. Condition (4.10) means that our solution depends only on the transverse components $\hat{\mathbf{x}}_T$ of the vector $\hat{\mathbf{x}}$. Furthermore, in the basis defined by the vectors $\hat{\mathbf{e}}_x$, $\hat{\mathbf{e}}_y$, and $\hat{\mathbf{e}}_z$ and their orthogonal complements (which can be constructed by the Graham-Shmidt procedure), only the first three components ($\hat{\mathbf{e}}_\alpha \cdot \hat{\mathbf{x}} \equiv x_{L\alpha}$) of $\hat{\mathbf{x}}_L$ are nonvanishing. Thus, $\mathbf{x}_L$ can be thought of as a three-dimensional vector. In the same way, we can define a $3m$-dimensional vector $\mathbf{x}_T$.

Returning to Eq. (4.1), we note that since we are looking for a solution that depends only on $\mathbf{x}_T$, the Laplacian $\hat{\nabla}^2$ can be replaced by ∇_T^2, where the gradient is taken only with respect to the components of the transverse vector. Since we are looking for the ground state of the Hamiltonian and the spherically symmetric solution has lower energy than the ones that break rotational symmetry (in the $3m$-dimensional subspace defined by the transverse coordinates), we conclude that the solution can depend only on the magnitude of $\mathbf{x}_T$, i.e., on the scalar combination (see Figure 3.4)

$$\varsigma \equiv \frac{1}{2}(\mathbf{x}_T)^2 = \frac{1}{2}\left[\hat{\mathbf{x}}^2 - \sum_\alpha (\hat{\mathbf{e}}_\alpha \cdot \hat{\mathbf{x}})^2\right]$$

$$= \sum_\alpha \left[\sum_{k=1}^m (\mathbf{y}_\alpha^{(k)})^2 - \frac{\lambda_\alpha^2}{1 + m\lambda_\alpha^2}\left(\sum_{k=1}^m \mathbf{y}_\alpha^{(k)}\right)^2\right] \tag{4.13}$$

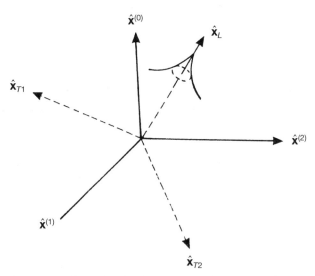

Figure 3.4   The $3(1 + 2)$-dimensional replica space (each axis in the figure represents a three-dimensional subspace) in the original and the rotated (longitudinal and transverse) coordinates. The cylindrical symmetry of the inhomogeneous mean field solution about the longitudinal subspace $\hat{\mathbf{x}}_L$ (depends only on ς) is illustrated by the cusp-shaped feature.

where the last equality is obtained, after some algebra, from the definition (3.5), $y_\alpha^{(k)} = x^{(k)} - \lambda \star x^{(0)}$. Therefore, the most general form of a solution with spontaneously broken symmetry under translation in replica space can be written as [6]

$$\vec{\varphi}_{mf}(\hat{x}) = \vec{n}\varphi_{mf}(s) \tag{4.14}$$

where the constant unit vector $\vec{n}$ was defined in Eq. (4.8).

Calculation of the Inhomogeneous Mean Field Solution

Inserting the mean field solution, Eq. (4.14), into the field Hamiltonian, Eq. (3.26), we notice that the solution enters the Hamiltonian only through the combinations $\vec{\varphi}_{mf}^2(\hat{x})$ and $[\vec{\nabla}\vec{\varphi}_{mf}(\hat{x})]^2$. Upon the substitution $(\vec{n})^2 = 1$, the resulting Hamiltonian becomes independent of the number of components n of the field and taking the limit $n \to 0$ does not affect the final results. (The nontrivial character of the n-vector field will play a role only when one considers fluctuation corrections due to excluded volume effects.)

The calculation of $\varphi_{mf}(s)$ can be further simplified by the observation that the effective Hamiltonian has to be known only up to the *first order* in m. This follows since the free energy is obtained by calculating

$$\frac{dH}{dm}\bigg|_{m=0} = \frac{\partial H[\varphi]}{\partial m}\bigg|_{m=0} + \int d\hat{x}\, \frac{\delta H[\varphi]}{\delta \varphi(\hat{x})}\bigg|_{m=0} \frac{\partial \varphi(\hat{x})}{\partial m}\bigg|_{m=0} = \frac{\partial H[\varphi]}{\partial m}\bigg|_{m=0} \tag{4.15}$$

where the second term in the first equation vanishes since we consider only solutions that minimize the Hamiltonian. According to Eq. (4.15), one should first compute the analytic continuation to the $m \to 0$ limit of the mean field solution φ_{mf} and then substitute it into H and take the derivative with respect to m. Alternatively, the $m \to 0$ limit of φ_{mf} can be directly calculated from Eq. (4.1) by keeping only the $k = 0$ term in the sum and dropping the term proportional to m in the spherical part of the Laplacian, $(\nabla_T^2)_{\mathrm{sph}} = 2s\partial^2/\partial s^2 + 3m\partial/\partial s$. This results in the equation for the function $\varphi_{mf}(s)$,

$$\left(\frac{1}{\overline{N}} - 2a^2 s\frac{\partial^2}{\partial s^2} - \frac{z_c}{2}\varphi_{mf}^2(s)\right)\varphi_{mf}(s) = 0 \tag{4.16}$$

where we used the equality $\mu + w^{(0)}\rho_{mf}^{(0)} = 1/\overline{N}$ (see Eq. (4.5)). Equation (4.16) can be reduced to a dimensionless form by introducing the dimensionless variable $t \equiv s/(2a^2\overline{N})$ and writing

$$\varphi_{mf}(s) = \sqrt{\frac{2}{z_c\overline{N}}}\chi(t). \tag{4.17}$$

Substituting the expression (4.17) into Eq. (4.16), we find that the dimension-less function χ obeys the equation [6]

$$t\chi''(t) = \chi(t) - \chi^3(t) \tag{4.18}$$

with the boundary conditions $\chi(0) = 1$ and $\chi(t \to \infty) \to 0$. (These choices are dictated by the form of the equation, assuming that χ'' is finite at the origin and that χ is finite at infinity, respectively.) The asymptotic $(t \to \infty)$ behavior of this function can be easily found since in this limit the function goes to zero and one can neglect the nonlinear term in Eq. (4.18). Direct integration of the resulting linear equation yields

$$\chi(t) \sim t^{1/4} \exp(-2t^{1/2}), \quad t \gg 1. \tag{4.19}$$

For arbitrary t the function $\chi(t)$ is computed numerically (see Figure 3.5).

A self-consistent scheme to calculate the mean field density in replica space (which is the Edwards-Anderson order parameter of this model) was proposed in ref. [20], and numerical results were obtained for gels prepared near the gelation threshold. (A more general result, valid for arbitrary conversion ratios, was obtained in ref. [14].) The above authors have also calculated the analog of the spin glass nonlinear susceptibility (two-point replica space density correlation function) and showed that this function diverges at the gelation threshold [22].

To gain some understanding about the information content of our solution, we notice that the mean field density in replica space, $\rho_{mf}(\hat{\mathbf{x}}) = \varphi_{mf}^2(\varsigma)/2$, can be expressed in terms of the Laplace transform $\Pi(\sigma)$ of $\rho_{mf}(\hat{\mathbf{x}})/\rho_{mf}(0)$:

$$\frac{\rho_{mf}(\hat{\mathbf{x}})}{\rho_{mf}(0)} = \int_0^\infty d\sigma \Pi(\sigma) \exp(-\sigma\varsigma). \tag{4.20}$$

Using the definition $\varsigma = \hat{\mathbf{x}}_T^2/2$ (Eq. (4.13)), we can express the exponential as

$$\exp(-\sigma\varsigma) = \exp\left[-\frac{\sigma}{2} \sum_\alpha \sum_{k=0}^m (x_\alpha^{(k)})^2 + \frac{\sigma}{2} \sum_\alpha \left(\sum_{k=0}^m x_\alpha^{(k)} e_\alpha^{(k)} \right)^2 \right]$$

$$= \int \frac{d\mathbf{r}}{(2\pi)^{3/2}} \sigma^{3/2} \exp\left[-\frac{\sigma}{2} \sum_\alpha \sum_{k=0}^m (x_\alpha^{(k)} - r_\alpha e_\alpha^{(k)})^2 \right] \tag{4.21}$$

where the second equality can be easily checked by performing the Gaussian integration and using the normalization condition, $\sum_{k=0}^m (e_\alpha^{(k)})^2 = 1$. Since σ has the dimensions of inverse length squared, it is convenient to introduce a length R and define $\sigma \equiv 1/R^2$. With the above substitutions, Eq. (4.20) is recast into the form

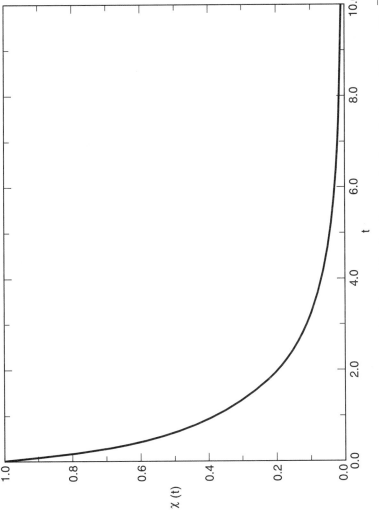

Figure 3.5 The normalized mean field solution $\chi(t)$ is plotted versus the dimensionless variable $t = s/(2a^2\overline{N})$.

117

$$\frac{\rho_{mf}(\hat{\mathbf{x}})}{\rho_{mf}(0)} = \int d\mathbf{r} \int_0^\infty dR P(R) \prod_\alpha \prod_{k=0}^m W(x_\alpha^{(k)} - r_\alpha e_\alpha^{(k)}|R) \qquad (4.22)$$

where

$$P(R) \equiv \frac{2}{R^3} \Pi\left(\frac{1}{R^2}\right) \qquad (4.23)$$

and

$$W(\Delta x|R) \equiv \frac{1}{(2\pi R^2)^{1/2}} \exp\left[-\frac{(\Delta x)^2}{2R^2}\right]. \qquad (4.24)$$

Both functions are normalized to unity, $\int dR P(R) = \int d\sigma \Pi(\sigma) = 1$, and $\int d(\Delta x) W(\Delta x|R) = 1$, and admit a simple physical interpretation. According to the definition of the function $\rho(\hat{\mathbf{x}})$ (Eq. (3.1)), the left-hand side of Eq. (4.22) is simply the mean field expression for the probability to find a given monomer at points $\{\mathbf{x}^{(k)}\}$ in each of the replicas (i.e., $\mathbf{x}^{(0)}$ in the zeroth replica, $\mathbf{x}^{(1)}$ in the first replica, etc.). The function $W(x_\alpha^{(k)} - r_\alpha e_\alpha^{(k)}|R)$ can be interpreted as the conditional probability to observe a fluctuation of the monomer position $x_\alpha^{(k)}$ about its mean position $r_\alpha e_\alpha^{(k)}$ in the kth replica, given that the rms deviation from this mean position (*localization* length) is R. The function $P(R)$ is then the probability of finding a monomer with a localization length R. This function is expressed through the Laplace transform of the mean field density and has the same characteristic scale of variation, $a\overline{N}^{1/2}$ (independent of the deformation λ_α), as the mean field solution $\varphi_{mf}(\varsigma)$. It can be calculated numerically and is shown in Figure 3.6. The function $\Pi(\sigma)$, which determines the distribution of localization lengths $P(R)$ (see Eq. (4.23)), was calculated numerically for undeformed networks close to the gelation threshold, in Ref. [20].

The physical meaning of the above probabilities becomes clear in the limit $m \to 0$. In this case, Eq. (4.11) yields $e_\alpha^{(0)} = 1$ and $e_\alpha^{(k)} = \lambda_\alpha$ and the argument of the function W becomes $x_\alpha^{(0)} - r_\alpha$ and $x_\alpha^{(k)} - \lambda_\alpha r_\alpha$ (for $k > 0$). Thus, if the average position of a monomer in the initial state is $\langle x_\alpha^{(0)} \rangle = r_\alpha$, the corresponding average position in all the replicas of the deformed state is $\langle x_\alpha^{(k)} \rangle = \lambda_\alpha \langle x_\alpha^{(0)} \rangle$, which means that *the average position of every monomer changes affinely with the deformation of the network*. The mean deviation from affine behavior (under deformation) of a typical monomer is given by R and, therefore, the localization length R can be interpreted as the length scale for thermal fluctuations of a typical monomer. It must be considered as a random variable that fluctuates in the space of the network (the probability of observing such a fluctuation is $P(R)$), both due to the frozen inhomogeneity of the structure of the network and due to the fact

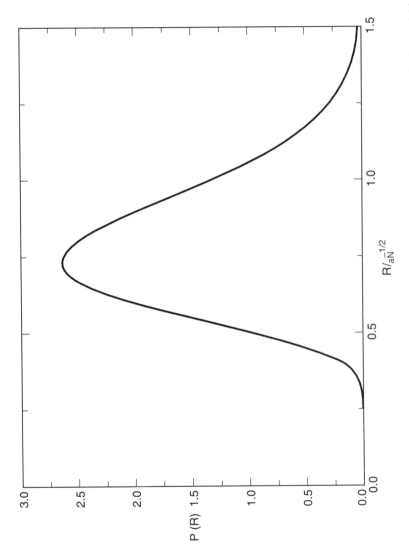

Figure 3.6 The probability $P(R)$ of finding a monomer with localization length R is plotted versus the dimensionless variable $R/(a\sqrt{N})$.

that the thermal fluctuations of a given monomer depend on its position along the chain contour.

The localization of the monomers in the final state of the network (in the sense that their average position is uniquely defined by their initial state and by the macroscopic deformation of the gel) arises as the consequence of spontaneous breaking of translational symmetry in replica space and reflects the *solid character of the network*.

We can show that the probability of having nonfluctuating monomers vanishes in our model ($P(0) = 0$), which means that all monomers, including crosslinks, fluctuate about their average positions. If the cross-links were strictly pinned down (as in Flory's model [8]), one would expect this probability to be of the order of the fraction of cross-links, $P(0) \sim \overline{N}^{-1}$.

Up to this point we have considered the behavior of a single monomer and shown that, on the average, its position changes affinely with the deformation of the network and that deviations from affine behavior arise only due to fluctuations. More information about the network can be obtained if we consider two-monomer quantities such as the rms distance between two end monomers of a network chain. The results of such a calculation [11] are discussed in Section 3.8.

3.4.4 MEAN FIELD FREE ENERGY

Anticipating that our inhomogeneous mean field solution, with spontaneously broken symmetry with respect to translation in replica space, corresponds to the true minimum of the Hamiltonian, we will use it to calculate the thermodynamic free energy of the network via the steepest descent estimate of the functional integral, Eq. (3.25). To this end, we first calculate the mean field Hamiltonian obtained by substituting the inhomogeneous mean field solution, Eq. (4.17), into Eq. (3.26). (In this calculation, we have to keep terms up to first order in m.)

$$H_{mf} = \int d\hat{\mathbf{x}} \, \tilde{H}(\varsigma) + \frac{w^{(0)}}{2} V^{(0)}(\rho^{(0)})^2 + m \frac{w}{2} V \rho^2. \qquad (4.25)$$

Here $\tilde{H}(\varsigma)$ is the Hamiltonian density corresponding to the term in the first square bracket in Eq. (3.26). In writing down the last two contributions in the above equation, we use the facts that (1) the densities in all the replicas (evaluated on the mean field solution) are constant and (2) the densities in all the replicas of the final state are equal, $\rho \equiv \rho^{(1)} = \rho^{(2)} = \cdots = \rho^{(m)}$.

We first calculate the replica space integrals $\int d\hat{\mathbf{x}} \, \tilde{H}(\varsigma) = \int d\mathbf{x}_L \int d\mathbf{x}_T \, \tilde{H}(\varsigma)$ in Eq. (4.25). Since the integrand depends only on the variable $\varsigma \equiv \frac{1}{2}(\hat{\mathbf{x}}_T)^2$, the $\mathbf{x}_T$ integration can be split into the product of a trivial angular factor $\int d\Omega$, which gives the surface area of a $3m$-dimensional unit sphere, $S_{3m} = 2\pi^{3m/2}/\Gamma(3m/2)$ (Γ is the gamma-function) and an integral of the form $\int d|\mathbf{x}_T||\mathbf{x}_T|^{3m-1}$. The last integral can be written as $\int_0^\infty d\varsigma(2\varsigma)^{3m/2-1}$ and combining both contributions and integrating by parts, we obtain

$$\int d\mathbf{x}_T \tilde{H}(\varsigma) = S_{3m} \int_0^\infty d\varsigma\,(2\varsigma)^{3m/2-1}\,\tilde{H}(\varsigma)$$

$$= -\int_0^\infty d\varsigma \frac{(2\pi\varsigma)^{3m/2}}{(3m/2)\cdot\Gamma(3m/2)} \frac{d\tilde{H}(\varsigma)}{d\varsigma}. \tag{4.26}$$

In obtaining the last equality we have used the relation $\tilde{H}(\infty) = 0$, which follows from the observation that the integrand (the Hamiltonian density, $\tilde{H}(\varsigma)$) is a polynomial in $\varphi_{mf}(\varsigma)$, and thus it decreases exponentially fast with ς, for $\varsigma \to \infty$ (see Eqs. (4.17) and (4.19)). We expand this to first order in m and use the relations $(3m/2) \cdot \Gamma(3m/2) = \Gamma(1 + 3m/2) \to 1 - 3m\gamma/2$ (γ is Euler's constant) and $(2\pi\varsigma)^{3m/2} \to 1 + (3m/2)\ln(2\pi\varsigma)$. This yields

$$\int d\mathbf{x}_T \tilde{H}(\varsigma) \overset{m\to 0}{\Rightarrow} \tilde{H}(0) - \frac{3m}{2}\int_0^\infty d\varsigma \ln(2\pi\varsigma e^\gamma)\frac{d\tilde{H}(\varsigma)}{d\varsigma}. \tag{4.27}$$

To calculate the above integral it is convenient to introduce the dimensionless variable t defined in Eq. (4.17). Writing $\ln \varsigma = \ln(a^2\overline{N}) + \ln t$ and neglecting the $\ln t$ term, we obtain $\tilde{H}(0)[1 + m \ln(a\sqrt{\overline{N}})^3]$.

The exponentially fast decrease (with ς) of the integrand in Eq. (4.26) means that dominant contribution to the integral comes from values of ς in the interval between 0 and $a^2\overline{N}$. We can use this fact to perform the remaining $d\mathbf{x}_L$ integration, which goes over an infinite region. (The integrand does not depend on the coordinates $\mathbf{x}_L$.)

Equation (4.13) shows that ς is a positively defined quadratic form of variables $y_\alpha^{(k)}$. Therefore the equality $\varsigma = 0$ can be satisfied only if $y_\alpha^{(k)} = 0$ (for all α and k). We conclude that the condition $\varsigma = 0$ defines a three-dimensional surface in the replica space, on which the affine relation between the coordinates of the initial and the final replicas, $\mathbf{x}^{(k)} = \lambda \star \mathbf{x}^{(0)}$, is satisfied. This surface will be called the *longitudinal* subspace in the following discussion. Its volume V_L is calculated in Eq. (4.29). Since the mean field solution decays on a length scale of the order of the average distance between cross-links, $a\overline{N}^{1/2}$ (see Eq. (4.19)), we conclude that the *average* positions of all the monomers in the replicas of the deformed state are uniquely defined by their positions in the initial undeformed state (they can be obtained from the latter by the affine transformation $\mathbf{x}^{(0)} \to \lambda \star \mathbf{x}^{(0)}$) and that deviations from affinity can only take place due to fluctuations about these average positions, on distances of the order of the mesh size.

Substituting the relation $x_\alpha^{(k)} = \lambda_\alpha x_\alpha^{(0)}$ into the definition of x_L (Eq. (4.12)),

$$x_{L\alpha} \equiv \sum_{k=0}^{m} e_\alpha^{(k)} x_\alpha^{(k)} = \frac{x_\alpha^{(0)} + \sum_{k=1}^{m} \lambda_\alpha x_\alpha^{(k)}}{(1 + m\lambda_\alpha^2)^{1/2}} = (1 + m\lambda_\alpha^2)^{1/2} x_\alpha^{(0)} \qquad (4.28)$$

and integrating over $\mathbf{x}^{(0)}$, we find that the integration over $x_{L\alpha}$ contributes the factor

$$V_L \equiv V^{(0)} \prod_\alpha (1 + m\lambda_\alpha^2)^{1/2} \qquad (4.29)$$

where $V^{(0)}$ is the initial volume of the gel. Finally, multiplying the results of the transverse and longitudinal integrations and expanding to first order in m yields

$$\int d\hat{\mathbf{x}}\, \tilde{H}(\varsigma) = \tilde{H}(0) V^{(0)} \left[1 + m \left(\ln(a\sqrt{\bar{N}})^3 + \frac{1}{2} \sum_\alpha \lambda_\alpha^2 \right) \right] \qquad (4.30)$$

where $\tilde{H}(0)$ is obtained from Eq. (4.12), $\tilde{H}(0) = \mu \rho_{mf}^{(0)} - z_c(\rho_{mf}^{(0)})^2$.

We proceed to calculate the contribution of the second term on the right-hand side of Eq. (4.25). The density in the zeroth replica is evaluated using the equality

$$N_{\text{tot}} = \int d\mathbf{x}^{(0)} \rho^{(0)} = \int d\hat{\mathbf{x}}\, \rho_{mf}(\varsigma) \qquad (4.31)$$

where the second equality is obtained from the definition (3.4). Notice that since no assumptions about the precise functional form of $\tilde{H}(\varsigma)$ were used in the derivation of Eq. (4.30), we can write (using $\rho_{mf}(0) = \rho_{mf}^{(0)}$)

$$N_{\text{tot}} = \rho_{mf}^{(0)} V^{(0)} \left[1 + m \left(\ln(a\sqrt{\bar{N}})^3 + \frac{1}{2} \sum_\alpha \lambda_\alpha^2 \right) \right]. \qquad (4.32)$$

This defines the density in the replica of the initial state, $\rho^{(0)} = N_{\text{tot}}/V^{(0)}$.

Collecting all the terms in (4.25) and using Eqs. (4.4) and (4.5), after some algebra we arrive at the expression

$$H_{mf} = \frac{1}{2} \mu (\rho_{mf}^{(0)})^2 V^{(0)} + m \left[N_c \left(\frac{1}{2} \sum_\alpha \lambda_\alpha^2 + \ln(a\sqrt{\bar{N}})^3 \right) + \frac{w}{2} V \rho^2 \right]. \qquad (4.33)$$

From this mean field Hamiltonian, we can calculate (using Eq. (3.25)) the steepest descent estimate for the grand canonical partition function (in the thermodynamic limit, $V \to \infty$): $\ln \Xi_m = -H_{mf}$. Substitution into Eq. (2.31) results in the following expression for the mean field free energy of the stretched polymer network

$$\frac{\mathscr{F}_{mf}\{\lambda_\alpha\}}{VT} = \nu\left[\frac{1}{2}\sum_\alpha \lambda_\alpha^2 + \ln(a\sqrt{\overline{N}})^3\right] + \frac{1}{2}w\rho^2 \qquad (4.34)$$

where $\rho = N_{tot}/V$ and $\nu \equiv \rho/(2\overline{N})$ is the density of cross-links in the final state of the gel. (Note that both ν and ρ depend only on the final volume of the gel and thus depend on $\{\lambda_\alpha\}$ only through the product $\lambda_x\lambda_y\lambda_z$.) This expression was previously derived by Deams and Edwards [1].

Equation (4.34) differs in two important ways from the free energy of the classical theories of polymer networks. First, the coefficient in front of the elastic entropy term is $\nu/2$ instead of the classical ν. Second, Flory's theory [8] contains a $\ln(\lambda_x\lambda_y\lambda_z)$ term that depends on the deformation of the network. The reasons for the discrepancies were discussed by Deams and Edwards [1]: The first is attributed to the neglect of cross-link fluctuations about their mean positions in both the Flory-Rehner [8] and the James-Guth [10] theories and the second to the uniform density assumption in the former (but not in the latter) model.

3.5   Stability of the Mean Field Solution

We found that the mean field equation admits two solutions (i.e., a homogeneous one and the one with spontaneously broken symmetry with respect to translation in replica space), both of which are symmetric with respect to permutation of the m replicas of the final deformed state of the gel. To check which of the mean field solutions corresponds to a minimum of the Hamiltonian, we need to consider their stability with respect to small but otherwise arbitrary fluctuations.

Another important issue is the question of replica symmetry breaking (RSB) [23]. Although there is a trivial lack of symmetry between the zeroth replica (of the initial undeformed gel) and all the other ones (those of the final deformed gel), the real issue is the presence or the absence of RSB between the m replicas of the final system. A related question is: Are gels regular solids characterized by a single ground state or do they belong to the class of spin glasses that have multiple "ground" states separated by infinite barriers? This issue was first considered by Goldbart and Goldenfeld [24], who favored the second option.

In ref. [24] it was conjectured that the ground state of the replica Hamiltonian of a network with excluded volume interactions is degenerate. Different "localized" ground states were identified with states characterized by topologically distinct entanglements of the network chains. The above identification of *excluded volume interactions* with *the effect of mutual impenetrability* originates from the expectation that the excluded volume interaction prevents chains from passing through each other. This argument relies implicitly on the kinetic picture of topological restrictions but disregards the question of initial conditions for the kinetic equations: the thermodynamic problem of separation of the phase space into regions with different entanglement of the chains is now transformed to that of the correct choice of initial positions of *all* the monomers in the network. However, since the above thermodynamic approach makes no selection of the allowed monomer positions, it is not clear how it could take such initial conditions into account. Moreover, note that the above identification violates the physical nature of topological restrictions: the breaking of replica symmetry is the result of the *frustration* phenomenon, when different equilibrium states have the same energy; the ground state of spin-glass-like systems is strongly *degenerate*. Such energy arguments can not be invoked for the separation of the phase space of the polymer system into topologically distinct regions since these regions have different phase volumes (and entropies, which are logarithms of such volumes). The partition of the phase space into topologically distinct regions is fixed once and for all in the process of network synthesis and is not affected by external conditions (temperature, quality of the solvent, and so on). On the other hand, the energy of topologically distinct localized states varies when these conditions are changed. A more careful examination of this intuitive statement reveals that the situation is more delicate. The Edwards Hamiltonian of polymer chains with excluded volume is, in fact, a free energy that can be derived (by coarse-graining) from a more microscopic level of description in which real chemical monomers and solvent molecules enter explicitly. Although it is customary to refer to the polymer segments described by the Edwards Hamiltonian as "monomers," strictly speaking they are statistical segments and are larger than the original monomers of the real polymers. The Edwards model is only applicable for the consideration of the long-wavelength behavior of polymers on length scales larger than the statistical segment size, and thus the theory is meaningful only if one introduces a short-wavelength cutoff in the coarse-grained Hamiltonian. The answer to the question of whether chains can pass through each other depends on how the short-wavelength limit is taken, an issue that is beyond the scope of the coarse-grained level of description. Different choices of dealing with the chain impenetrability constraint may or may not give rise to the appearance of entanglements in the theory of polymer solutions and melts. For example, when one attempts to simulate the excluded volume problem on a computer, one may use an algorithm in which the chain impenetrability constraint is explicitly taken into account. A different effective model is that of a chain of beads (with excluded volume), connected by phantom elastic springs. Both algorithms give the correct excluded volume exponents but only the former (e.g., the bond fluctuation method [25]) gives rise to entanglements.

The previous argument suggests that topological entanglements should be introduced into the Edwards model of gels as *constraints* that reduce the configurational space available to the polymer chains (similar to the way cross-links were introduced in the present work). In this way, topological entanglements are the analogs of hard walls in the well-known problem of gas molecules in a box; the walls act only as constraints, and do not appear in the Hamiltonian that describes the kinetic energy and the interaction between the particles in the gas. Different states of entanglement have, in general, different entropies (much in the same way that the entropy of the gas in a box depends on its size) and are, therefore, nondegenerate. Transitions between these states are strictly forbidden, independent of the size of the network. The situation is fundamentally different in spin glasses where local frustration leads to the appearance of many degenerate states, the barriers between which become infinite only in the thermodynamic limit.

The above discussion has two important ramifications: (1) The original Edwards model, on which this work (and that of ref. [24]) is based, does not describe *entangled gels*. (2) Even if one were to introduce entanglement constraints into the model, the resulting theory would belong in a different class of models than spin glasses and thus there is no reason to expect that it would have RSB and multiply degenerate ground states. Interestingly, in a more recent work ([26]) no RSB was found on a mean field level in the absence of excluded volume interactions, but it was conjectured that RSB would appear if these interactions were included. In the following, we will deal with the above question and investigate whether the ground state solution we found is stable against RSB (and other) fluctuations.

Consider small fluctuations $\delta \vec{\varphi}(\hat{x}) = \vec{\varphi}(\hat{x}) - \vec{\varphi}_{mf}(\hat{x})$ about the mean field solutions we found. To test the stability of these solutions, we have to calculate the spectrum of eigenvalues $\{\Lambda\}$ of the second derivative operator K (evaluated on the appropriate mean field solution), Eqs. (4.6) and (4.7), which gives the energy of these fluctuations and check whether all of them are positive (i.e., whether all small fluctuations increase the energy). In the harmonic approximation, the energy of these fluctuations, is given by quadratic corrections (in $\delta \vec{\varphi}$) to the mean field Hamiltonian:

$$\Delta H[\delta \vec{\varphi}] = H[\vec{\varphi}] - H_{mf} = \frac{1}{2} \sum_{ij} \int d\hat{x} \int d\hat{x}' K_{ij}(\hat{x}, \hat{x}') \delta \varphi_i(\hat{x}) \delta \varphi_j(\hat{x}'). \quad (5.1)$$

The operator $\overleftrightarrow{K}$, which gives the energy spectrum of fluctuations about a particular mean field solution $\vec{\varphi}_{mf}$, can be obtained by substituting $\vec{\varphi}(\hat{x}) = \vec{\varphi}_{mf}(\hat{x}) +$

$\delta\vec{\varphi}(\hat{\mathbf{x}})$ into the Hamiltonian (Eq. (3.26)) and expanding up to second order in $\delta\vec{\varphi}$ (terms linear in $\delta\vec{\varphi}$ vanish since mean field solutions correspond to extrema of the Hamiltonian). From the general form of the solution, $\vec{\varphi}_{mf}(\hat{\mathbf{x}}) = \vec{n}\,\varphi_{mf}(\hat{\mathbf{x}})$, we conclude that $\overleftrightarrow{K}$ can be decomposed using the projection operators $\overleftrightarrow{P}^{\parallel} \equiv \vec{n}\,\vec{n}$ and $\overleftrightarrow{P}^{\perp} \equiv \overleftrightarrow{1} - \vec{n}\,\vec{n}$, which project an arbitrary vector along directions parallel and perpendicular to $\vec{n}$, respectively. Thus,

$$\overleftrightarrow{K}(\hat{\mathbf{x}}, \hat{\mathbf{x}}') = K^{\parallel}(\hat{\mathbf{x}}, \hat{\mathbf{x}}')\overleftrightarrow{P}^{\parallel} + K^{\perp}(\hat{\mathbf{x}}, \hat{\mathbf{x}}')\overleftrightarrow{P}^{\perp} \qquad (5.2)$$

where

$$K^{\perp}(\hat{\mathbf{x}}, \hat{\mathbf{x}}') \equiv \delta(\hat{\mathbf{x}} - \hat{\mathbf{x}}')[1/\overline{N} - a^2\hat{\nabla}^2 - (z_c/2)\varphi_{mf}^2(\hat{\mathbf{x}})] \qquad (5.3)$$

and

$$K^{\parallel}(\hat{\mathbf{x}}, \hat{\mathbf{x}}') \equiv \delta(\hat{\mathbf{x}} - \hat{\mathbf{x}}')[1/\overline{N} - a^2\hat{\nabla}^2 - 3(z_c/2)\varphi_{mf}^2(\hat{\mathbf{x}})]$$
$$+ \varphi_{mf}(\hat{\mathbf{x}})\varphi_{mf}(\hat{\mathbf{x}}')\sum_{k=0}^{m} w^{(k)}\delta(\mathbf{x}^{(k)} - \mathbf{x}'^{(k)}). \qquad (5.4)$$

The operators $K^{\parallel}$ and $K^{\perp}$ have eigenfunctions $\psi^{\parallel}(\hat{\mathbf{x}})$ and $\psi^{\perp}(\hat{\mathbf{x}})$, with eigenvalues $\Lambda^{\parallel}$ and $\Lambda^{\perp}$, respectively, which are determined by the scalar (in the n-dimensional space) variants of Eq. (4.7),

$$\int d\mathbf{x}' K^{\parallel}(\hat{\mathbf{x}}, \hat{\mathbf{x}}')\psi^{\parallel}(\hat{\mathbf{x}}') = \Lambda^{\parallel}\psi^{\parallel}(\hat{\mathbf{x}}), \qquad \int d\hat{\mathbf{x}}' K^{\perp}(\hat{\mathbf{x}}, \hat{\mathbf{x}}')\psi^{\perp}(\hat{\mathbf{x}}') = \Lambda^{\perp}\psi^{\perp}(\hat{\mathbf{x}}). \quad (5.5)$$

An important simplification results from the observation that since we are interested in the limit $m \to 0$, all eigenvalues can be expanded around $m = 0$, i.e., $\Lambda = \Lambda_0 + m\Lambda_1 +$ (higher-order terms in m). To test the stability of the mean field solution, we only need to check whether the above eigenvalues are nonnegative in this limit and, therefore, as long as $\Lambda_0 \neq 0$, the higher order corrections to the spectrum need not be considered. However, if we want to calculate the eigenfunctions, we cannot set $m = 0$ since symmetry under permutations of the m replicas of the final state implies that these eigenfunctions are degenerate, and the degree of degeneracy must be kept a finite integer. The general rule to be followed in taking the limit $m \to 0$ is that one can safely take this limit in all analytical expressions, but one has to keep m finite in all other cases (e.g., in summations over the index k where $k = 1, \ldots, m$) until one arrives at analytic functions of m. This rule was already used to derive the mean field solution where it led to dramatic simplification of replica calculations.

Another important simplification results from the fact that, to consider the stability of the mean field solution, it is sufficient to obtain the eigenvalues corresponding to the lowest energy fluctuations. Since we expect that this energy is a monotonously decreasing function of the wavelength, in this section we will only study long-wavelength fluctuations (i.e., fluctuations on length scales much larger than the mesh size of the network, $a\overline{N}^{1/2}$).

An arbitrary fluctuation $\delta\varphi$ can be represented as a linear combination of the eigenmodes ψ. Since ψ is the solution of a linear equation, it is only defined up to a multiplicative constant. In the following, we will choose this constant to be unity so that for each fluctuation mode we have

$$\delta\varphi(\hat{\mathbf{x}}) = \psi(\hat{\mathbf{x}}). \tag{5.6}$$

3.5.1 HOMOGENEOUS SOLUTION

The first step is to calculate the spectrum of eigenvalues of the operator $K^{\parallel}$ (Eq. (5.4)) evaluated on the constant solution $\varphi_{mf}(\hat{\mathbf{x}}) = \text{const}$. We will show that some of the eigenvalues of this operator are negative and therefore will not study further the spectrum of the operator $K^{\perp}$ for the homogeneous mean field solution. The secular equation is obtained by substituting Eq. (5.4) into Eq. (5.5) and removing $\delta(\hat{\mathbf{x}} - \hat{\mathbf{x}}')$ by integrating over $\hat{\mathbf{x}}'$:

$$(-z_c\varphi_{mf}^2 - a^2\hat{\nabla}^2)\,\psi^{\parallel}(\hat{\mathbf{x}}) + \varphi_{mf}^2 \sum_{k=0}^{m} w^{(k)} \prod_{l \neq k} \int d\mathbf{x}^{(l)} \psi^{\parallel}(\hat{\mathbf{x}}) = \Lambda^{\parallel}\psi^{\parallel}(\hat{\mathbf{x}}). \tag{5.7}$$

As can be verified by direct substitution into the resulting equation, the solutions are plane waves, $\psi^{\parallel}(\hat{\mathbf{x}}) \sim \exp(i\hat{\mathbf{q}} \cdot \hat{\mathbf{x}})$. Using the identities $\int d\mathbf{x}\, \exp(i\mathbf{q} \cdot \mathbf{x}) = 0$ for $\mathbf{q} \neq 0$ and $\int d\mathbf{x}\, \exp(i\mathbf{q} \cdot \mathbf{x}) = V$ for $\mathbf{q} = 0$ (V is the volume of the final system), we have the following scenarios, depending on the direction of the wave vector $\hat{\mathbf{q}}$:

1. For wave vectors that lie in the ith ($\hat{\mathbf{q}}^{(i)} \equiv (0, \ldots, \mathbf{q}^{(i)}, \ldots, 0)$) sector of replica space, the only contribution comes from the ith term in the sum (since otherwise one of the integrations over the $l \neq k$ replicas will be over the coordinate $\mathbf{x}^{(i)}$ and the corresponding integral will vanish). In this case the factor $\exp(i\hat{\mathbf{q}} \cdot \hat{\mathbf{x}})$ can be taken outside the integrals and the integrations will result in the product of the volumes of all replicas except the ith one. We have to distinguish between two cases:

    (a) For wave vectors lying completely in the zeroth (i.e., $\hat{\mathbf{q}}^{(0)} \equiv (\mathbf{q}^{(0)}, 0, \ldots, 0)$) replica, the integration produces a factor of V^m. Thus, taking the limit $m \to 0$ and collecting the terms in Eq. (5.7), we find the eigenvalues

$$\Lambda^{\parallel}(\mathbf{q}^{(0)}) = (w^{(0)} - z_c)\varphi_{mf}^2 + a^2(\mathbf{q}^{(0)})^2 \tag{5.8}$$

(b) For wave vectors lying completely in the kth (i.e., $\hat{\mathbf{q}}^{(k)} \equiv (0, \ldots,$ $\mathbf{q}^{(k)}, \ldots, 0)$) sector of replica space ($k \neq 0$), the integration produces a factor of $V^{(0)}V^{m-1}$. In the limit $m \rightarrow 0$, we find the eigenvalues

$$\Lambda^{\|}(\mathbf{q}^{(k)}) = (wV^{(0)}/V - z_c)\varphi_{mf}^2 + a^2(\mathbf{q}^{(k)})^2 \tag{5.9}$$

2. For all other wave vectors, which are not restricted to these sectors (i.e., $\hat{\mathbf{q}} \equiv (\mathbf{q}^{(0)}, \ldots, \mathbf{q}^{(k)}, \ldots, \mathbf{q}^{(m)})$), the sum in Eq. (5.7) does not contribute to the eigenvalue equation and we obtain

$$\Lambda^{\|}(\hat{\mathbf{q}}) = -z_c \varphi_{mf}^2 + a^2 \hat{\mathbf{q}}^2. \tag{5.10}$$

Inspection of Eqs. (5.8) and (5.9) shows that, for large enough excluded volume parameters, the homogeneous solution is stable with respect to fluctuations of density in each of the replicas. The presence of the negative eigenvalues in Eq. (5.10) (for small enough values of q) shows that this solution corresponds to a saddle point rather than to a minimum of the Hamiltonian. We conclude that the solution that has the full translational invariance of H does not represent its ground state, and we proceed to examine the stability of solutions with spontaneously broken translational symmetry.

3.5.2   INHOMOGENEOUS SOLUTION

We turn to the calculation of the solutions of the secular equations, (5.5), which correspond to the inhomogeneous mean field solution $\varphi_{mf}(\hat{\mathbf{x}}) = \varphi_{mf}(s)$ (Eq. (4.17)). Note that the mean field solution does not depend on the three-dimensional vector $\mathbf{x}_L$, defined by the projection $x_{L\alpha}$ of the replica space vector $\hat{\mathbf{x}}$ on the *longitudinal* subspace spanned by the three vectors $\hat{\mathbf{e}}_\alpha$, $\alpha = x, y, z$ (Eqs. (4.11) and (4.12)). This fact can be used to simplify the calculations by performing a partial Fourier transform with respect to the coordinates $x_{L\alpha}$

$$f(\hat{\mathbf{x}}) = \int \frac{d\mathbf{q}_L}{(2\pi)^3} f_{\mathbf{q}_L}(\mathbf{x}_T) \exp(i\mathbf{q}_L \cdot \mathbf{x}_L) \tag{5.11}$$

where $\mathbf{x}_T$ is a $3m$-dimensional vector in the *transverse* subspace defined as the orthogonal complement to the longitudinal subspace. (Here $f(\hat{\mathbf{x}})$ is an arbitrary function of replica space coordinates.) The advantage of this representation becomes evident when we notice that when the Laplacian $\hat{\nabla}^2$ is applied to the function $f(\hat{\mathbf{x}})$, we obtain

$$\hat{\nabla}^2 f(\hat{\mathbf{x}}) = \int \frac{d\mathbf{q}_L}{(2\pi)^3} (\nabla_T^2 - \mathbf{q}_L^2) f_{\mathbf{q}_L}(\mathbf{x}_T) \exp(i\mathbf{q}_L \cdot \mathbf{x}_L) \tag{5.12}$$

where the Laplacian ∇_T^2 is taken only with respect to the $\mathbf{x}_T$ coordinates.

In the following, we will express all eigenfunctions $\psi(\hat{\mathbf{x}})$ in terms of their longitudinal Fourier components

$$\psi(\hat{\mathbf{x}}) = \psi_{\mathbf{q}_L}(\mathbf{x}_T) \exp(i\mathbf{q}_L \cdot \mathbf{x}_L) \tag{5.13}$$

and label the corresponding eigenvalues as $\Lambda(\mathbf{q}_L)$.

3.5.3   ROTATIONAL MODES

The eigenvalues and eigenfunctions of the operator $K^\perp$ can be obtained by substituting the mean field solution $\varphi_{mf}(\varsigma)$ into Eq. (5.5). Using the above-defined Fourier representation for the eigenfunctions $\vec{\psi}^\perp(\hat{\mathbf{x}})$, the corresponding secular equation becomes

$$[1/\overline{N} - a^2\nabla_T^2 + a^2\mathbf{q}_L^2 - \Lambda^\perp(\mathbf{q}_L) - (z_c/2)\varphi_{mf}^2(\varsigma)]\vec{\psi}_{\mathbf{q}_L}^\perp(\mathbf{x}_T) = 0. \tag{5.14}$$

The general solution of this equation, which satisfies the orthogonality condition $\vec{\psi}_{\mathbf{q}_L}^\perp(\mathbf{x}_T) \cdot \vec{n} = 0$, is of the form $\vec{\psi}_{\mathbf{q}_L}^\perp(\mathbf{x}_T) = \psi_{\mathbf{q}_L}^\perp(\mathbf{x}_T)\delta\vec{n}_{\mathbf{q}_L}$, where $\delta\vec{n}_{\mathbf{q}_L}$ is an arbitrary vector that satisfies the condition $\delta\vec{n}_{\mathbf{q}_L} \cdot \vec{n} = 0$. The function $\psi_{\mathbf{q}_L}^\perp(\mathbf{x}_T)$ is obtained from the scalar variant of Eq. (5.14). Since the vector $\delta\vec{n}_{\mathbf{q}_L}$ has $n - 1$ independent components, each of the eigenvalues $\Lambda^\perp(\mathbf{q}_L)$ is $n - 1$-fold degenerate.

Notice that the above equation has the form of a Shrödinger equation with a spherically symmetric potential. As is well-known from quantum mechanics, the ground state solution that corresponds to the minimal value of $\Lambda^\perp$ is spherically symmetric [27]. Although there are also nonspherically symmetric solutions, they have higher energy and do not have to be considered in the study of the stability of the inhomogeneous solution, $\varphi_{mf}(\varsigma)$.

The spherically symmetric solution can be found from the observation that Eq. (5.14) is identical in form to Eq. (4.1) and, therefore, the solution is simply

$$\vec{\psi}_{\mathbf{q}_L}^\perp(\mathbf{x}_T) = \varphi_{mf}(\varsigma)\delta\vec{n}_{\mathbf{q}_L} \tag{5.15}$$

where $\varsigma = \mathbf{x}_T^2/2$. The corresponding eigenvalues are obtained by substituting this solution back into Eq. (5.14) and using Eq. (4.1):

$$\Lambda^\perp(\mathbf{q}_L) = a^2\mathbf{q}_L^2. \tag{5.16}$$

These eigenfunctions and eigenvalues are associated with the rotations of the vector $\vec{n}$ in the abstract n-dimensional space. This can be demonstrated by showing how the field $\vec{\varphi}$ transforms under the infinitesimal rotation $\vec{n} \rightarrow \vec{n} + \delta\vec{n}$, with $\delta\vec{n}(\hat{\mathbf{x}}) \equiv \delta\vec{n}_{\mathbf{q}_L} \exp(i\mathbf{q}_L \cdot \mathbf{x}_L)$. Under this rotation

$$\delta^{\perp}\vec{\varphi}(\hat{\mathbf{x}}) = \vec{\varphi}_{mf}(\varsigma)|_{\vec{n} \rightarrow \vec{n} + \delta\vec{n}} - \vec{\varphi}_{mf}(\varsigma) = \varphi_{mf}(\varsigma)\delta\vec{n}(\hat{\mathbf{x}}). \qquad (5.17)$$

Since $\delta\vec{n}(\hat{\mathbf{x}})$ is orthogonal to $\vec{n}$, the deviation (5.17) corresponds to the transverse mode of the fluctuations. These solutions are gapless Goldstone modes; i.e., their eigenvalues are positive definite and vanish in the long-wavelength limit. The situation is equivalent to that of a ferromagnet (with $n \rightarrow 0$ spin components) where the Goldstone modes describe "soft" ($q \rightarrow 0$) rotations of the magnetization vector [28], although in our case these modes do not have a simple physical interpretation.

3.5.4 SHEAR AND DENSITY MODES

We proceed to calculate the spectrum of the operator $K^{\|}$. The secular equation (5.5) is obtained by substituting the mean field solution, expression (4.17), into Eq. (5.4), where for simplicity of notation, we drop the superscript $\|$ everywhere in the following:

$$[1/\bar{N} - a^2\hat{\nabla}^2 - 3(z_c/2)\varphi_{mf}^2(\varsigma)]\psi(\hat{\mathbf{x}}) + \varphi_{mf}(\varsigma)\sum_{k=0}^{m} w^{(k)}\delta\rho^{(k)}(\mathbf{x}^{(k)}) = \Lambda\psi(\hat{\mathbf{x}}) \qquad (5.18)$$

with $w^{(k)} = w$ for $k = 1, \ldots, m$. Here,

$$\delta\rho^{(k)}(\mathbf{x}^{(k)}) = \int d\hat{\mathbf{x}}'\, \delta(\mathbf{x}^{(k)} - \mathbf{x}'^{(k)})\varphi_{mf}(\varsigma')\psi(\hat{\mathbf{x}}') \qquad (5.19)$$

is the density fluctuation in the kth replica. (This identification follows from Eqs. (3.3), (3.24), and (5.6).)

Equations (5.18) and (5.19) admit two types of solutions that can be classified according to whether the density fluctuations $\delta\rho^{(k)}$ do or do not vanish identically in all the replicas (i.e., for all k).

Shear Modes

Consider the case

$$\delta\rho^{(k)}(\mathbf{x}^{(k)}) = 0, \quad \text{for all } k. \qquad (5.20)$$

Such fluctuations correspond to *pure shear* modes, i.e., to displacements in replica space that do not affect the density in each of the replicas. Substituting the partial Fourier transform, Eq. (5.13), into Eq. (5.18), we can recast the latter into the form

$$[1/\overline{N} - a^2\nabla_T^2 + a^2\mathbf{q}_L^2 - \Lambda(\mathbf{q}_L) - 3(z_c/2)\varphi_{mf}^2(\mathfrak{s})]\psi_{\mathbf{q}_L}(\mathbf{x}_T) = 0. \qquad (5.21)$$

It is convenient to represent this equation in a dimensionless form by introducing the $3m$-dimensional vector $\mathbf{r} \equiv \mathbf{x}_T/(aN^{1/2})$. Defining $\psi_{\mathbf{q}_L}(\mathbf{x}_T) \equiv \xi(\mathbf{r})$ and $t \equiv \mathbf{r}^2/2$ yields the dimensionless eigenvalue equation

$$(1 - \nabla_{\mathbf{r}}^2 - 3\chi^2(t))\xi(\mathbf{r}) = \varpi\xi(\mathbf{r}) \qquad (5.22)$$

where χ is defined in Eq. (4.17) and where the eigenvalues Λ and ϖ are related by

$$\Lambda(\mathbf{q}_L) = a^2\mathbf{q}_L^2 + \varpi/\overline{N}. \qquad (5.23)$$

Equation (5.22) has the standard form of a Shrödinger equation in a spherically symmetric potential. Its solution can be represented as a product of a radial (function of t only) and an angular (function of the direction of $\mathbf{r}$ only) part [27]. The latter is an eigenfunction of the angular momentum operator (in the $3m$-dimensional space) and is labeled by the angular momentum quantum number $\ell = 0, 1, 2, \ldots$ (the eigenvalue ϖ_ℓ is $(3m - 1)\ell + 1$ degenerate).

The lowest "energy" solution is spherically symmetric ($\ell = 0$) and is therefore a function of t only. Substitution into Eq. (5.19) shows that such a solution cannot satisfy the condition (5.20) and thus must be rejected. The $\ell = 1$ case corresponds to the dipole-type solution $\xi_1(\mathbf{r}) = \nabla_{\mathbf{r}}\chi(t)$, with the eigenvalue $\varpi_1 = 0$. This solution will be studied in detail below. Solutions that correspond to higher harmonics (with $\ell > 1$) have positive definite ϖ_ℓ, which increase monotonically with ℓ. The corresponding fluctuations have positive definite eigenvalues (Eq. (5.23)) and therefore do not affect the stability of our mean field solutions. They describe complicated distortions in replica space and will not be considered further in this section.

We now return to the case $\ell = 1$ and note that an equation of the same form as (5.21) can be obtained by applying the ∇_T operator to Eq. (4.1) and, therefore, $\nabla_T\varphi_{mf}(\mathfrak{s})$ is a solution of Eq. (5.21). The general solution of (5.18) is obtained by multiplying the transverse gradient by a constant vector $\overline{\mathbf{u}}_T$:

$$\psi_{\mathbf{q}_L}(\mathbf{x}_T) = (\overline{\mathbf{u}}_T \cdot \nabla_T)\varphi_{mf}(\mathfrak{s}). \qquad (5.24)$$

This solution has to satisfy the $1 + m$ conditions $\delta\rho^{(k)}(\mathbf{x}^{(k)}) = 0$ or, in terms of their Fourier transforms,

$$\rho_{\mathbf{q}(k)}^{(k)} \equiv \int d\hat{\mathbf{x}} \, \exp(-i\hat{\mathbf{q}}^{(k)} \cdot \hat{\mathbf{x}})\varphi_{mf}(\mathfrak{s})\psi(\hat{\mathbf{x}}) = 0 \qquad (5.25)$$

where, as before, we define $\hat{\mathbf{q}}^{(k)} \equiv (0, \dots, \mathbf{q}^{(k)}, \dots, 0)$. The integral can be calculated by separating the integration into longitudinal and transverse components, $d\hat{\mathbf{x}} \rightarrow d\mathbf{x}_L d\mathbf{x}_T$. To perform the integration over the longitudinal component, it is convenient to define the projections of these replica-space wave vectors on the parallel and the perpendicular directions to the subspace spanned by the three unit vectors $\hat{\mathbf{e}}_\alpha$ (Eq. (4.11)). The corresponding projections are given by $\hat{\mathbf{q}}_L^{(k)} = \Sigma_\alpha q_{L\alpha}^{(k)} \hat{\mathbf{e}}_\alpha$ and $\hat{\mathbf{q}}_T^{(k)} \equiv \hat{\mathbf{q}}^{(k)} - \hat{\mathbf{q}}_L^{(k)}$, where $q_{L\alpha}^{(k)} = \hat{\mathbf{q}}^{(k)} \cdot \hat{\mathbf{e}}_\alpha$ can be treated as the components of a three-dimensional vector $\mathbf{q}_L^{(k)}$. Similarly, the $3m$-dimensional vector $\mathbf{q}_T^{(k)}$ is defined by the projection of the replica vector $\hat{\mathbf{q}}^{(k)}$ on the transverse subspace. Substituting Eq. (4.11) for $\hat{\mathbf{e}}_\alpha$ yields (in the limit $m \rightarrow 0$)

$$\mathbf{q}_L^{(0)} = \mathbf{q}^{(0)}, \qquad (\mathbf{q}_T^{(0)})^2 = 0, \qquad \mathbf{q}_L^{(k)} = \lambda \star \mathbf{q}^{(k)}; \quad \text{for } k \neq 0 \qquad (5.26)$$

where the second equality is obtained from the first one by noticing that $(\hat{\mathbf{q}}_T^{(0)})^2 \equiv (\hat{\mathbf{q}}^{(0)})^2 - (\hat{\mathbf{q}}_L^{(0)})^2 = 0$. The rather strange peculiarity of vectors in replica space is that in the limit $m \rightarrow 0$, we may have $(\mathbf{q}_T^{(0)})^2 = 0$ but $\mathbf{q}_T^{(0)} \cdot \mathbf{q}_T^{(k)} \neq 0$. (As an example, consider the $3m$-component vector $\mathbf{a}_T = (1, 1, \dots, 1)$ whose elements are all different from zero but whose norm vanishes in the limit $m \rightarrow 0$.)

With the above definitions, the replica-space product in the exponent of Eq. (5.25) can be written as the sum of longitudinal and transverse contributions:

$$\hat{\mathbf{q}}^{(k)} \cdot \hat{\mathbf{x}} = \mathbf{q}_L^{(k)} \cdot \mathbf{x}_L + \mathbf{q}_T^{(k)} \cdot \mathbf{x}_T. \qquad (5.27)$$

Making the above replacement in Eq. (5.25) and using Eq. (5.13), the integration over $d\mathbf{x}_L$ gives a δ-function, $\delta(\mathbf{q}_L^{(k)} - \mathbf{q}_L)$. Substituting Eq. (5.12) into the remaining integral over $d\mathbf{x}_T$ and moving the constant $\bar{\mathbf{u}}_T$ outside the integral, condition (5.25) becomes

$$\bar{\mathbf{u}}_T \cdot \int d\mathbf{x}_T \exp(-i\mathbf{q}_T^{(k)} \cdot \mathbf{x}_T) \varphi_{mf}(\varsigma) \nabla_T \varphi_{mf}(\varsigma) = 0. \qquad (5.28)$$

Using $\varphi_{mf}(\varsigma) \nabla_T \varphi_{mf}(\varsigma) = \nabla_T \varphi_{mf}^2(\varsigma)/2$ and integrating by parts, we finally obtain (the surface term vanishes since $\varphi_{mf}(\varsigma \rightarrow \infty) \rightarrow 0$)

$$(\bar{\mathbf{u}}_T \cdot \mathbf{q}_T^{(k)}) \int d\mathbf{x}_T \exp(-i\mathbf{q}_T^{(k)} \cdot \mathbf{x}_T) \varphi_{mf}^2(\varsigma) = 0. \qquad (5.29)$$

Since the above integral is, in general, nonvanishing, the vector $\bar{\mathbf{u}}_T$ must obey the condition $\bar{\mathbf{u}}_T \cdot \mathbf{q}_T^{(k)} = 0$.

The eigenvalues are calculated by comparing Eq. (5.21) with the equation for $\nabla_T \varphi_{mf}(\varsigma)$ (obtained by applying the gradient to Eq. (4.1)). This gives

$$\Lambda_S(\mathbf{q}_L) = a^2 \mathbf{q}_L^2 \qquad (5.30)$$

i.e., shear modes are gapless Goldstone modes. We show below that the eigenfunctions (5.24) describe the infinitesimal displacement $\mathbf{x}_T \rightarrow \mathbf{x}_T + \mathbf{u}_T(\mathbf{x}_L)$ of the coordinate $\mathbf{x}_T$ (in the abstract transverse $3m$-dimensional subspace). Under the displacement $\mathbf{u}_T(\mathbf{x}_L) = \mathbf{u}_{T\mathbf{q}}\exp(i\mathbf{q} \cdot \mathbf{x}_L)$, the variation of the field $\varphi(\hat{\mathbf{x}})$ has the form

$$\delta_S\varphi(\hat{\mathbf{x}}) = \varphi_{mf}(\varsigma)|_{\mathbf{x}_T \to \mathbf{x}_T + \mathbf{u}_T} - \varphi_{mf}(\varsigma) = \mathbf{u}_T(\mathbf{x}_L) \cdot \nabla_T\varphi_{mf}(\varsigma) = \psi_{\mathbf{q}_L}(\mathbf{x}_T). \quad (5.31)$$

Under this displacement, the argument of the function $\varphi_{mf}(\varsigma)$ changes as $\varsigma \to \varsigma + \mathbf{x}_T \cdot \mathbf{u}_T$ and, expanding to first order in $\mathbf{u}_T$, we obtain the second equality in Eq. (5.31). The third equality follows from comparison with (5.24), upon identifying the arbitrary vector $\bar{\mathbf{u}}_T$ with the displacement $\mathbf{u}_T$. The additional condition $\mathbf{u}_T \cdot \mathbf{q}_T^{(k)} = 0$ (see Eq. (5.29)) means that the displacement $\mathbf{u}_T$ has to be orthogonal to the $m + 1$ vectors $\mathbf{q}_T^{(k)}$. This condition imposes $m + 1$ constraints on the $3m$ components of the vector $\mathbf{u}_T$, and we conclude that the shear modes are $2m - 1$ degenerate.

Density Modes, General Consideration

We now return to Eq. (5.19) and consider the general case in which at least one of the density fluctuations $\delta\rho^{(k)}(\mathbf{x}^{(k)})$ is not identically zero. It is convenient to work with functions over the usual three-dimensional space (i.e., $\mathbf{x}^{(k)}$) in each of the replicas, instead of the $3(1 + m)$-dimensional replica space. This is achieved by recasting Eq. (5.18) into an equation for $\delta\rho^{(k)}(\mathbf{x}^{(k)})$. This equation can be used to express ψ through $\delta\rho^{(k)}$,

$$\psi(\hat{\mathbf{x}}) = -\int d\hat{\mathbf{x}}' D(\Lambda; \hat{\mathbf{x}}, \hat{\mathbf{x}}')\varphi_{mf}(\varsigma')\sum_{k=0}^{m} w^{(k)}\delta\rho^{(k)}(\mathbf{x}'^{(k)}) \quad (5.32)$$

where D is defined by the equation

$$[1/\bar{N} - \Lambda - a^2\hat{\nabla}^2 - 3(z_c/2)\varphi_{mf}^2(\hat{\mathbf{x}})]D(\Lambda; \hat{\mathbf{x}}, \hat{\mathbf{x}}') = \delta(\hat{\mathbf{x}} - \hat{\mathbf{x}}'). \quad (5.33)$$

Substituting (5.32) into (5.19), we obtain a closed system of linear integral equations for $\delta\rho^{(k)}$,

$$\delta\rho^{(k)}(\mathbf{x}) + \sum_{l=0}^{m} w^{(l)}\int d\mathbf{x}' g_\Lambda^{kl}(\mathbf{x}, \mathbf{x}')\delta\rho^{(l)}(\mathbf{x}') = 0 \quad (5.34)$$

where we define the replica space density correlation functions

$$g_\Lambda^{kl}(\mathbf{x}, \mathbf{x}') \equiv \int d\hat{\mathbf{x}} \, \varphi_{mf}(\varsigma)\delta(\mathbf{x} - \mathbf{x}^{(k)})\int d\hat{\mathbf{x}}' \, \varphi_{mf}(\varsigma')\delta(\mathbf{x}' - \mathbf{x}'^{(l)})D(\Lambda; \hat{\mathbf{x}}, \hat{\mathbf{x}}'). \quad (5.35)$$

The problem can be further simplified by Fourier transforming Eq. (5.34). For this we have to calculate the Fourier transform of the functions $g_\Lambda^{kl}(\mathbf{x}, \mathbf{x}')$,

$$g_\Lambda^{kl}(\mathbf{q}^{(k)}, \mathbf{q}^{(l)}) \equiv \int d\hat{\mathbf{x}} \, \varphi_{mf}(\varsigma)\int d\hat{\mathbf{x}}' \, \varphi_{mf}(\varsigma')D(\Lambda; \hat{\mathbf{x}}, \hat{\mathbf{x}}') \exp(i\hat{\mathbf{q}}^{(k)} \cdot \hat{\mathbf{x}} - i\hat{\mathbf{q}}^{(l)} \cdot \hat{\mathbf{x}}'). \quad (5.36)$$

Changing the integration $d\hat{\mathbf{x}} \to d\mathbf{x}_L d\mathbf{x}_T$ (and $d\hat{\mathbf{x}}' \to d\mathbf{x}'_L d\mathbf{x}'_T$) and performing the integrations over the longitudinal coordinates using Eq. (5.27), yields

$$g^{kl}_\Lambda(\mathbf{q}^{(k)}, \mathbf{q}^{(l)}) = \delta(\mathbf{q}^{(k)}_L - \mathbf{q}^{(l)}_L)g^{kl}_\Lambda(\mathbf{q}^{(k)}_L), \tag{5.37}$$

$$g^{kl}_\Lambda(\mathbf{q}^{(k)}_L) = \int d\mathbf{x}_T \, \varphi_{mf}(s) \int d\mathbf{x}'_T \, \varphi_{mf}(s')$$
$$\times D(\Lambda - a^2(\mathbf{q}^{(k)}_L)^2; \mathbf{x}_T, \mathbf{x}'_T) \exp(i\mathbf{q}^{(k)}_T \cdot \mathbf{x}_T - i\mathbf{q}^{(l)}_T \cdot \mathbf{x}'_T). \tag{5.38}$$

In deriving Eq. (5.38), we used the replacement $\hat{\nabla}^2 \to \nabla^2_T - (\mathbf{q}^{(k)}_L)^2$, which introduced the shift $\Lambda \to \Lambda - a^2(\mathbf{q}^{(k)}_L)^2$ (Eq. (5.12)) into the longitudinal Fourier transform of Eq. (5.33). In calculating the above integrals we only consider the usual *continuous* description of a solid in which one only considers wavelengths that are much larger than the characteristic microscale (in our case, this microscale corresponds to the average spatial distance between cross-links, $a\overline{N}^{1/2}$). Since this distance is the characteristic length scale for the decay of the classical solution ($\varphi_{mf}(s \gg a^2\overline{N}) \to 0$), in evaluating the integrals we can expand the exponentials and keep only terms to second order in $\mathbf{q}^{(k)}_T \cdot \mathbf{x}_T$. Furthermore, since $\mathbf{q}^{(k)}_T$ and $\mathbf{q}^{(k)}_L$ can differ only by a factor of order unity (i.e., by a multiplicative factor of λ), the functions $g^{kl}_\Lambda(\mathbf{q}^{(k)}_L)$ have to be calculated also only to order $(\mathbf{q}^{(k)}_L)^2$.

Fourier transforming Eq. (5.34) and eliminating the integrations using the δ-functions in (6.21) yields a set of algebraic relations between the Fourier coefficients of the solutions $\delta\rho^{(0)}(\mathbf{x}) = \rho^{(0)}_{\mathbf{q}(0)}\exp(i\mathbf{q}^{(0)} \cdot \mathbf{x})$ and $\delta\rho^{(k)}(\mathbf{x}) = \rho^{(k)}_\mathbf{q} \exp(i\mathbf{q} \cdot \mathbf{x})$, with $\mathbf{q}^{(0)} = \lambda \star \mathbf{q} = \mathbf{q}_L$:

$$[1 + w^{(0)}g^{00}_\Lambda(\mathbf{q}_L)]\rho^{(0)}_{\mathbf{q}(0)} + wg^{01}_\Lambda(\mathbf{q}_L)\sum_{k=1}^m \rho^{(k)}_\mathbf{q} = 0$$

$$[1 + w(g^{11}_\Lambda(\mathbf{q}_L) - g^{12}_\Lambda(\mathbf{q}_L))]\rho^{(k)}_\mathbf{q} + w^{(0)}g^{10}_\Lambda(\mathbf{q}_L)\rho^{(0)}_{\mathbf{q}(0)} + wg^{12}_\Lambda(\mathbf{q}_L)\sum_{k=1}^m \rho^{(k)}_\mathbf{q} = 0 \tag{5.39}$$

where we used the identity of the $k \neq 0$ replicas to replace the general replica indices (k, l) in the functions (5.38), by those of the first and the second replicas. This symmetry can be used in Eq. (5.34) to recast it into a system of equations for the fields $\rho^{(0)}_{\mathbf{q}(0)}$ and $\eta_\mathbf{q} \equiv \Sigma_{k=1}^m \rho^{(k)}_\mathbf{q}$:

$$[1 + w^{(0)}g^{00}_\Lambda(\mathbf{q}_L)]\rho^{(0)}_{\mathbf{q}(0)} + wg^{01}_\Lambda(\mathbf{q}_L)\eta_\mathbf{q} = 0,$$

$$[1 + w(g^{11}_\Lambda(\mathbf{q}_L) - g^{12}_\Lambda(\mathbf{q}_L) + mwg^{12}_\Lambda(\mathbf{q}_L))] \eta_\mathbf{q} + mw^{(0)}g^{10}_\Lambda(\mathbf{q}_L)\rho^{(0)}_{\mathbf{q}(0)} = 0. \tag{5.40}$$

The eigenvalues $\Lambda (\mathbf{q}_L)$ are obtained from the condition of solvability of this linear system of equations,

$$[1 + w^{(0)}g^{00}_\Lambda(\mathbf{q}_L)][1 + w(g^{11}_\Lambda(\mathbf{q}_L) - g^{12}_\Lambda(\mathbf{q}_L) + mwg^{12}_\Lambda(\mathbf{q}_L))] = mw^{(0)}w[g^{01}_\Lambda(\mathbf{q}_L)]^2. \tag{5.41}$$

Strictly speaking, we can set $m \to 0$ in the above equation and find the two eigenvalues from the condition that one of the two terms in the square brackets vanishes. Extra care must be taken in calculating the corresponding eigenfunctions since Eq. (5.41) admits three different types of eigenmodes, two of which become degenerate in the limit $m \to 0$.

Density Modes in Initial State

The first type of eigenmodes corresponds to eigenvalues $\Lambda_D^{(0)}(\mathbf{q}_L)$ which, in the limit $m \to 0$, are determined by setting to zero the term in the first square bracket on the left-hand side of Eq. (5.41),

$$1 + w^{(0)} g_\Lambda^{00}(\mathbf{q}_L) = 0. \tag{5.42}$$

The function g_Λ^{00} can be represented in the form (recall that $\mathbf{q}_T^{(0)} = 0$)

$$g_\Lambda^{00}(\mathbf{q}_L) = \int d\mathbf{x}_T \varphi_{mf}(\varsigma) \Phi(\Lambda - a^2 \mathbf{q}_L^2; \varsigma) \tag{5.43}$$

where Φ is defined by

$$\Phi(\Lambda - a^2 \mathbf{q}_L^2; \varsigma) \equiv \int d\mathbf{x}_T' \varphi_{mf}(\varsigma') D(\Lambda - a^2 \mathbf{q}_L^2; \mathbf{x}_T, \mathbf{x}_T'). \tag{5.44}$$

Due to the spherical symmetry of the differential operator in Eq. (5.33), Φ depends only on the scalar combination $\varsigma = \mathbf{x}_T^2/2$. The integral (5.43) is calculated in the same way as in Eqs. (4.26) and (4.27) (Section 3.4) in which we replace $\tilde{H}(\varsigma)$ by $\varphi_{mf}(\varsigma) \Phi(\Lambda - a^2 \mathbf{q}_L^2; \varsigma)$ and take the limit $m \to 0$ (note that in calculating $\int d\mathbf{x}_T \tilde{H}(\varsigma)$ we only used the property that $\varphi_{mf}(\varsigma)$ is an exponentially decreasing function of ς). This yields

$$g_\Lambda^{00}(\mathbf{q}_L) = \varphi_{mf}(0) \Phi(\Lambda - a^2 \mathbf{q}_L^2; 0). \tag{5.45}$$

To calculate the function Φ, we apply the differential operator in Eq. (5.33) to the left-hand side of Eq. (5.44) and remove the resulting δ-function by integrating over $\mathbf{x}_T'$ on the right-hand side of the resulting equation. This gives the following equation for Φ:

$$\left[\frac{1}{\overline{N}} - (\Lambda - a^2 \mathbf{q}_L^2) - 2a^2 \varsigma \frac{\partial^2}{\partial \varsigma^2} - \frac{3z_c}{2} \varphi_{mf}^2(\varsigma) \right] \Phi(\Lambda - a^2 \mathbf{q}_L^2; \varsigma) = \varphi_{mf}(\varsigma). \tag{5.46}$$

For $\varsigma = 0$, the equation reduces to an algebraic relation

$$\Phi(\Lambda - a^2 \mathbf{q}_L^2; 0) = \frac{\varphi_{mf}(0)}{-2/\overline{N} - \Lambda + a^2 \mathbf{q}_L^2} \tag{5.47}$$

and, upon inserting this expression into Eq. (5.45), we finally get

$$g_\Lambda^{00}(\mathbf{q}_L) = \frac{2\rho^{(0)}}{-2/\bar{N} - \Lambda + a^2\mathbf{q}_L^2}. \tag{5.48}$$

Substituting this expression for g_Λ^{00} into (5.42) we find

$$\Lambda_{\text{gap}}(\mathbf{q}_L) = (w^{(0)} - z_c)\varphi_{mf}^2 + a^2(\mathbf{q}_L)^2. \tag{5.49}$$

Since this eigenvalue does not vanish (in general) in the limit $q \to 0$, following the usual terminology we say that the corresponding solution is *massive* (i.e., has an energy gap). The gap vanishes at

$$z_c = \frac{1}{\rho^{(0)}\bar{N}^{\text{min}}} = w^{(0)} \tag{5.50}$$

which can be interpreted as the "cross-link saturation threshold" that defines the highest density of cross-links that can be achieved by instantaneous cross-linking of a polymer solution (see Eq. (2.8)).

Further inspection of Eqs. (5.39) and (5.40) leads to the conclusion that the above solution corresponds to the case of identical densities $\rho_\mathbf{q}^{(k)}$ in all the replicas of the final state ($k = 1, \ldots, m$). Furthermore, in the limit $m \to 0$ (taken in the above-defined sense), these modes obey the affine relation between the densities $\rho_\mathbf{q}^{(k)} = \rho_{\lambda \star \mathbf{q}}^{(0)}$ in the final and the initial states.

To construct the eigenfunctions of the density modes which correspond to the eigenvalues (5.49), we have to Fourier transform (see Eq. (5.11)) Eq. (5.32) for the functions $\psi(\hat{\mathbf{x}})$:

$$\psi_{\mathbf{q}_L}(\mathbf{x}_T) = \int d\mathbf{x}_T' \, \varphi_{mf}(s') D(\Lambda - a^2\mathbf{q}_L^2; \mathbf{x}_T, \mathbf{x}_T') \sum_{k=0}^{m} w^{(k)} \rho_\mathbf{q}^{(k)} \exp(i\mathbf{q}_T^{(k)} \cdot \mathbf{x}_T'). \tag{5.51}$$

Using the rapid decrease of $\varphi_{mf}(s')$ with $s' = (\mathbf{x}_T')^2/2$, we can expand the exponentials in the above expression to second order in $\mathbf{q}_T^{(k)} \cdot \mathbf{x}_T'$. To calculate the above integrals, it is convenient to introduce the function

$$\mathbf{x}_T \Psi(\Lambda - a^2\mathbf{q}_L^2; s) \equiv \int d\mathbf{x}_T' \, \varphi_{mf}(s') D(\Lambda - a^2\mathbf{q}_L^2; \mathbf{x}_T, \mathbf{x}_T')\mathbf{x}_T'. \tag{5.52}$$

Taking into account the definitions of the functions Φ, Eq. (5.44), and Ψ, we arrive at the general result

$$\psi_{\mathbf{q}_L}(\mathbf{x}_T) = \sum_{k=0}^{m} \rho_\mathbf{q}^{(k)}[\Phi(\Lambda - a^2\mathbf{q}_L^2; s) + i\mathbf{q}_T^{(k)} \cdot \mathbf{x}_T \Psi(\Lambda - a^2\mathbf{q}_L^2; s)]. \tag{5.53}$$

Explicit expressions for the eigenfunctions can be obtained from (5.53) by substituting the corresponding eigenvalues Λ and using the appropriate relations between the $\rho_{\mathbf{q}}^{(k)}$ coefficients in each of the modes, Eqs. (5.39) and (5.40). Since we are interested in the continuum limit, we can set $\mathbf{q}_L = 0$ in Φ in the above expression. The next step is to realize that for the massive mode ($\Lambda_{\text{gap}} \to$ const as $\mathbf{q} \to 0$), the function Ψ remains finite in the limit $\mathbf{q} \to 0$. In this case, the second term on the right-hand side of Eq. (5.53) is of order $\mathbf{q}_T^{(k)} \cdot \mathbf{x}_T$ and can be neglected with respect to the first term. This gives

$$\psi_{\text{gap}}(\hat{\mathbf{x}}) = \Phi[(w^{(0)} - z_c)\varphi_{mf}^2; \varsigma] \exp(i\mathbf{q}_L \cdot \mathbf{x}_L) \tag{5.54}$$

where the function Φ is defined as the solution of Eq. (5.46), and where we dropped the (constant) $\rho_{\mathbf{q}}^{(k)}$ terms that appear in (5.53) since the latter only affect the (arbitrary) normalization of the eigenfunctions. An analytic expression for Φ cannot be given, in general, since for arbitrary ς, Eq. (5.46) only can be solved numerically. There is, however, one important case of networks prepared at the cross-link saturation threshold, in which we can find a simple analytical solution:

$$\Phi(0;\varsigma) = -\frac{\partial \varphi_{mf}(\varsigma)}{\partial \mu} = \overline{N}^2 \frac{\partial \varphi_{mf}(\varsigma)}{\partial \overline{N}}. \tag{5.55}$$

This result can be checked by differentiating Eq. (4.1) for $\varphi_{mf}(\varsigma)$ with respect to μ and comparing the resulting expression with Eq. (5.46).

Density Modes in Final State

In the limit $m \to 0$, the eigenvalue equation for the density modes in the final state is obtained by setting to 0 the term in the second square bracket on the left-hand side of Eq. (5.41):

$$1 + w(g_\Lambda^{11}(\mathbf{q}_L) - g_\Lambda^{12}(\mathbf{q}_L)) = 0. \tag{5.56}$$

We now proceed to calculate the replica-space correlation function g_Λ^{kl} for general k, l. To obtain the $\mathbf{q}^2$ corrections to g^{kl}, we have to expand the exponential in Eq. (5.38) to fourth order in $\mathbf{q}_T$ (linear terms vanish due to angular integrations):

$$g_\Lambda^{kl}(\mathbf{q}_L) = g_\Lambda^{00}(\mathbf{q}_L) + \int d\mathbf{x}_T \varphi_{mf}(\varsigma) \int d\mathbf{x}_T' \varphi_{mf}(\varsigma') D(\Lambda - a^2(\mathbf{q}_L^{(k)})^2; \mathbf{x}_T, \mathbf{x}_T')$$

$$\times \left[-\frac{1}{2}(\mathbf{q}_T^{(k)} \cdot \mathbf{x}_T - \mathbf{q}_T^{(l)} \cdot \mathbf{x}_T')^2 + \frac{1}{24}(\mathbf{q}_T^{(k)} \cdot \mathbf{x}_T - \mathbf{q}_T^{(l)} \cdot \mathbf{x}_T')^4 \right]. \tag{5.57}$$

Returning to the definition of Φ, Eq. (5.44) and of Ψ, Eq. (5.52), and using the symmetry of the integrand in Eq. (5.57) under the replacement $\mathbf{x}_T \leftrightarrow \mathbf{x}_T'$, we can rewrite the latter in the form

$$g_\Lambda^{kl}(\mathbf{q}_L) = g_\Lambda^{00}(\mathbf{q}_L) + \int d\mathbf{x}_T \varphi_{mf}(s) \left\{ -\frac{1}{2} \Phi(\Lambda - a^2 \mathbf{q}_L^2; s). \right.$$

$$\times [(\mathbf{q}_T^{(k)} \cdot \mathbf{x}_T)^2 + (\mathbf{q}_T^{(l)} \cdot \mathbf{x}_T)^2] + \Psi(\Lambda - a^2 \mathbf{q}_L^2; s) \left[(\mathbf{q}_T^{(k)} \cdot \mathbf{x}_T)(\mathbf{q}_T^{(l)} \cdot \mathbf{x}_T) \right.$$

$$\left. \left. - \frac{1}{6} (\mathbf{q}_T^{(k)} \cdot \mathbf{x}_T)^3 (\mathbf{q}_T^{(l)} \cdot \mathbf{x}_T) - \frac{1}{6} (\mathbf{q}_T^{(k)} \cdot \mathbf{x}_T) (\mathbf{q}_T^{(l)} \cdot \mathbf{x}_T)^3 \right] \right\}. \tag{5.58}$$

The angular integrations over the direction of the vector $\mathbf{x}_T$ in Eq. (5.58) are performed with the aid of the formulas

$$\int d\Omega x_{Ti} x_{Tj} = S_{3m} \frac{\mathbf{x}_T^2}{3m} \delta_{ij},$$

$$\int d\Omega \, x_{Ti} x_{Tj} x_{Tl} x_{Tk} = S_{3m} \frac{(\mathbf{x}_T^2)^2}{3m(3m+2)} (\delta_{ij}\delta_{kl} + \delta_{ik}\delta_{jl} + \delta_{il}\delta_{jk}) \tag{5.59}$$

where $S_{3m} = 2\pi^{3m/2}/\Gamma(3m/2)$ is the surface area of a unit $3m$-dimensional sphere (Γ is the gamma function). In the limit $m \to 0$, this yields

$$g^{kl}(\mathbf{q}_L) = g^{00}(\mathbf{q}_L) - \frac{1}{2} \int_0^\infty ds \, \varphi_{mf}(s)[(\mathbf{q}_T^{(k)})^2 + (\mathbf{q}_T^{(l)})^2]$$

$$\times \Phi(\Lambda - a^2 \mathbf{q}_L^2; s) + \int_0^\infty ds \, \varphi_{mf}(s)(\mathbf{q}_T^{(k)} \cdot \mathbf{q}_T^{(l)}) \Psi(\Lambda - a^2 \mathbf{q}_L^2; s)$$

$$- \frac{1}{2} \int_0^\infty ds \, s\varphi_{mf}(s)[(\mathbf{q}_T^{(k)})^2 + (\mathbf{q}_T^{(l)})^2](\mathbf{q}_T^{(k)} \cdot \mathbf{q}_T^{(l)}) \Psi(\Lambda - a^2 \mathbf{q}_L^2; s). \tag{5.60}$$

We now calculate the functions g^{kl}, defined by Eq. (5.60), in the limit $q \to 0$. From the definition of $\mathbf{q}_T^{(k)}$, Eq. (5.26),

$$\mathbf{q}_T^{(k)} \cdot \mathbf{q}_T^{(l)} = \hat{\mathbf{q}}^{(k)} \cdot \hat{\mathbf{q}}^{(l)} - \mathbf{q}_L^{(k)} \cdot \mathbf{q}_L^{(l)}$$

$$= (\lambda^{-1} \star \mathbf{q}_L)^2 \delta_{kl} - \mathbf{q}_L^2, \quad k \neq 0, \quad l \neq 0 \tag{5.61}$$

and, consequently, $(\mathbf{q}_T^{(k)})^2 = (\mathbf{q}_T^{(l)})^2$ for $k, l \neq 0$. Using these relations, we express the combination $g_\Lambda^{11}(\mathbf{q}_L) - g_\Lambda^{12}(\mathbf{q}_L)$ in Eq. (5.56) in terms of the function Ψ.

$$g_\Lambda^{11}(\mathbf{q}_L) - g_\Lambda^{12}(\mathbf{q}_L) = (\lambda^{-1} \star \mathbf{q}_L)^2 \int_0^\infty ds \, \varphi_{mf}(s) \Psi(\Lambda - a^2 \mathbf{q}_L^2; s). \tag{5.62}$$

The function Ψ can be calculated from the following equation (which can be obtained by applying the differential operator in Eq. (5.33) to both sides of Eq. (5.52)):

$$\left[\frac{1}{\overline{N}} - (\Lambda - a^2\mathbf{q}_L^2) - 2a^2\varsigma\frac{\partial^2}{\partial\varsigma^2} - 2a^2\frac{\partial}{\partial\varsigma} - 3\frac{z_c}{2}\varphi_{mf}^2(\varsigma)\right]$$

$$\times \Psi(\Lambda - a^2\mathbf{q}_L^2; \varsigma) = \varphi_{mf}(\varsigma). \tag{5.63}$$

In general, the function $\Psi(\Lambda - a^2\mathbf{q}_L^2; \varsigma)$ has to be calculated numerically. Note, however, that the differential operator in Eq. (5.63) has an eigenvalue $a^2\mathbf{q}_L^2 - \Lambda$, with an eigenfunction of the form $\partial\varphi_{mf}(\varsigma)/\partial\varsigma$. (This can be checked by differentiating Eq. (4.16) with respect to ς and comparing to Eq. (5.63).) Thus, when looking for the solution of Eq. (5.63) in the form of an expansion over the eigenfunctions of this operator, in the limit $\Lambda - a^2\mathbf{q}_L^2 \to 0$ we can retain only the contribution of this eigenfunction (ground state dominance):

$$\Psi(\Lambda - a^2\mathbf{q}_L^2; \varsigma) \to \frac{C_0}{a^2\mathbf{q}_L^2 - \Lambda}\frac{\partial\varphi_{mf}(\varsigma)}{\partial\varsigma} \tag{5.64}$$

where C_0 is obtained from the normalization condition, which yields

$$C_0 = \int_0^\infty d\varsigma\,\varphi_{mf}(\varsigma)\frac{\partial\varphi_{mf}(\varsigma)}{\partial\varsigma}\bigg/\int_0^\infty d\varsigma\left(\frac{\partial\varphi_{mf}(\varsigma)}{\partial\varsigma}\right)^2. \tag{5.65}$$

Substituting this expression into Eq. (5.62), we obtain

$$g_\Lambda^{11}(\mathbf{q}_L) - g_\Lambda^{12}(\mathbf{q}_L) = C\frac{(\lambda^{-1}\star\mathbf{q}_L)^2}{a^2\mathbf{q}_L^2 - \Lambda} \tag{5.66}$$

where the constant C is given by

$$C = \left(\int_0^\infty d\varsigma\,\varphi_{mf}(\varsigma)\frac{\partial\varphi_{mf}(\varsigma)}{\partial\varsigma}\right)^2\bigg/\int_0^\infty d\varsigma\left(\frac{\partial\varphi_{mf}(\varsigma)}{\partial\varsigma}\right)^2. \tag{5.67}$$

The numerator and the denominator in this expression are readily calculated by integrating by parts

$$\int_0^\infty d\varsigma\,\varphi_{mf}(\varsigma)\frac{\partial\varphi_{mf}(\varsigma)}{\partial\varsigma} = \frac{\varphi_{mf}^2(\varsigma)}{2}\bigg|_0^\infty = -\frac{\varphi_{mf}^2}{2} \tag{5.68}$$

and

$$\begin{aligned}
\int_0^\infty d\varsigma\left(\frac{\partial\varphi_{mf}(\varsigma)}{\partial\varsigma}\right)^2 &= -2\int_0^\infty d\varsigma\,\frac{\partial\varphi_{mf}(\varsigma)}{\partial\varsigma}\varsigma\frac{\partial^2\varphi_{mf}(\varsigma)}{\partial\varsigma^2}\\
&= -\int_0^\infty d\varsigma\,\frac{\partial\varphi_{mf}(\varsigma)}{\partial\varsigma}\frac{[1/\overline{N} - (z_c/2)\varphi_{mf}^2(\varsigma)]\varphi_{mf}(\varsigma)}{a^2}\\
&= (\varphi_{mf}^2/(2\overline{N}) - z_c\varphi_{mf}^4/8)/a^2\\
&= \varphi_{mf}^2/(4a^2\overline{N}) \tag{5.69}
\end{aligned}$$

where, in obtaining the second equality, we have used Eq. (4.16) for $\varphi_{mf}(\varsigma)$. Substituting the above into Eq. (5.67) gives

$$C = 2\rho^{(0)}a^2\overline{N}. \tag{5.70}$$

The eigenvalues corresponding to the gapless density modes are found by substituting the expression for $g_\Lambda^{11}(\mathbf{q}_L) - g_\Lambda^{12}(\mathbf{q}_L)$ (Eqs. (5.66) and (5.69)) into the eigenvalue equation (5.56). This gives the eigenvalues

$$\Lambda_D(\mathbf{q}_L) = a^2\mathbf{q}_L^2 + 2w\rho^{(0)}a^2\overline{N}(\lambda^{-1} \star \mathbf{q}_L)^2. \tag{5.71}$$

Since $\Lambda_D(\mathbf{q}_L \to 0) \to 0$, the corresponding fluctuations are Goldstone modes. The condition $\Lambda_D(\mathbf{q}_L) > 0$ is always satisfied when the final state of the gel corresponds to good solvent conditions ($w > 0$). In fact, for large deformations the positivity condition can be satisfied even for moderately poor solvent ($w < 0$), since the network can be stabilized against collapse by the external forces applied to its surface.

The next step is to calculate the eigenfunctions corresponding to the above eigenvalue. We find that there are m degenerate eigenfunctions (in the limit $m \to 0$) and therefore, before taking this limit we have to consider the case of arbitrary integer m.

Consider modes for which $\rho_{\mathbf{q}(0)}^{(0)} = \eta_\mathbf{q} = 0$ but $\rho_\mathbf{q}^{(k)} \neq 0$ (for $k = 1, \ldots, m$) (see Eq. (5.40)). In this case we cannot use the secular Eq. (5.41) and have to return to Eq. (5.39). This gives the eigenvalue equation

$$J_m(\Lambda_D(\mathbf{q}_L)) = 0 \tag{5.72}$$

where we define

$$J_m(\Lambda) \equiv 1 + w(g_\Lambda^{11}(\mathbf{q}_L) - g_\Lambda^{12}(\mathbf{q}_L)).$$

The $(m - 1)$ degenerate eigenfunctions corresponding to this eigenvalue are obtained by calculating the integrals in Eq. (5.32),

$$\psi_j(\hat{\mathbf{x}}) = \sum_{k=1}^{m} S_j^{(k)}\left(\frac{i\mathbf{q}_T^{(k)}}{\rho^{(0)}q^2} \cdot \nabla_T\varphi_{mf}(\varsigma)\right)\exp(i\mathbf{q}_L \cdot \mathbf{x}_L), \qquad j = 1, \ldots, m - 1. \tag{5.73}$$

Here, $\rho^{(0)}$ is the mean density in the initial state. The arbitrary coefficients $S_j^{(k)}$ obey the relation

$$\sum_{k=1}^{m} S_j^{(k)} = 0 \tag{5.74}$$

which follows from the condition $\eta_\mathbf{q} = 0$. (This can be checked by substituting ψ_j into Eq. (5.25) and summing over k.) The condition of orthogonality of the eigenfunctions, $\int d\hat{\mathbf{x}} \, \psi_i(\hat{\mathbf{x}})\psi_j^*(\hat{\mathbf{x}}) = 0$ for $i \neq j$ (where * denotes a complex conjugate and where the volume of the system is kept finite during the integration over $\mathbf{x}_T$), imposes another relation between the $S_j^{(k)}$ coefficients:

$$\sum_{k,l=1}^{m} S_i^{(k)} S_j^{(l)} (\delta_{kl} - q_L^2/q^2) = 0, \quad \text{for } i \neq j. \tag{5.75}$$

To gain some insight into the physical meaning of the above eigenfunctions, we note that the gradient term describes the infinitesimal displacement $\mathbf{x}_T \to \mathbf{x}_T + \mathbf{u}_T(\mathbf{x}_L)$ of the coordinate $\mathbf{x}_T$ (see Eq. (5.31)), where

$$\mathbf{u}_T(\mathbf{x}_L) = \frac{i\mathbf{q}_T^{(k)}}{q^2 \rho^{(0)}} \exp(i\mathbf{q}_L \cdot \mathbf{x}_L). \tag{5.76}$$

The displacements $\mathbf{u}_T$ given by Eq. (5.76) are orthogonal to the displacements that lead to pure shear modes (in replica space) derived in a previous subsection. Unlike the former, they correspond to fluctuations in the densities of the replicas.

We now consider $\eta_\mathbf{q} \neq 0$ (and, consequently, $\rho_{\mathbf{q}(0)}^{(0)} \neq 0$), in which case all the $\rho_\mathbf{q}^{(k)}$ are equal and the eigenvalue spectrum is obtained from the secular Eq. (5.41), which can be rewritten in the form

$$J_m(\Lambda) = mw \left[\frac{w^{(0)} (g_\Lambda^{01}(\mathbf{q}_L))^2}{1 + w^{(0)} g_\Lambda^{00}(\mathbf{q}_L)} - g_\Lambda^{12}(\mathbf{q}_L) \right]. \tag{5.77}$$

Using this expression we obtain, to first order in m,

$$\Lambda(\mathbf{q}_L) = \Lambda_D(\mathbf{q}_L) + m\Lambda_D'(\mathbf{q}_L) \tag{5.78}$$

where

$$\Lambda_D'(\mathbf{q}_L) \equiv \frac{w}{\partial J_0/\partial \Lambda} \left[\frac{w^{(0)} (g_\Lambda^{01}(\mathbf{q}_L))^2}{1 + w^{(0)} g_\Lambda^{00}(\mathbf{q}_L)} - g_\Lambda^{12}(\mathbf{q}_L) \right] \tag{5.79}$$

should be evaluated in the limit $m \to 0$. The corresponding fluctuations do not affect the density in the initial state since in this limit $\rho_{\mathbf{q}(0)}^{(0)} \sim \eta_\mathbf{q} = m\rho_\mathbf{q}^{(l)} \to 0$.

The eigenfunction corresponding to the eigenvalue (5.78) is obtained by calculating the integrals in Eq. (5.32):

$$\psi_0(\hat{\mathbf{x}}) = \sum_{k=1}^{m} \left(\frac{i\mathbf{q}_T^{(k)}}{\rho^{(0)} q^2} \cdot \nabla_T \varphi_{mf}(s) - w \frac{\partial \varphi_{mf}(s)}{\partial \mu} \right) \exp(i\mathbf{q}_L \cdot \mathbf{x}_L). \tag{5.80}$$

Comparing this equation with the expression for the degenerate eigenmodes, Eq. (5.73), and using Eq. (5.74), we can write a general expression for all m density eigenmodes $(j = 0, \ldots, m - 1)$

$$\psi_j(\hat{\mathbf{x}}) = \sum_{k=1}^{m} S_j^{(k)} \left(\frac{i\mathbf{q}_T^{(k)}}{\rho^{(0)} q^2} \cdot \nabla_T \varphi_{mf}(s) - w \frac{\partial \varphi_{mf}(s)}{\partial \mu} \right) \exp(i\mathbf{q}_L \cdot \mathbf{x}_L) \tag{5.81}$$

where all $S_0^{(k)}$ are equal to each other. From Eq. (5.75) and the automatic orthogonality of eigenfunctions corresponding to different eigenvalues, we can write the general orthonormality condition

$$\sum_{k,l=1}^{m} S'^{(k)}_{i} S'^{(l)}_{j} (\delta_{kl} - \mathbf{q}_L^2/q^2) = \delta_{ij} \tag{5.82}$$

where a convenient choice of normalization was made.

3.5.5 SUMMARY

We end this section with a brief summary. We proved that the homogeneous solution is unstable against fluctuations that mix the replicas and does not correspond to a minimum of the Hamiltonian. We then proceeded to study the fluctuations about the inhomogeneous mean field solution $\varphi_{mf}(\varsigma)$. We showed that, for this solution, all the eigenvalues Λ of the second derivative operator $\overset{\leftrightarrow}{K}$ are positive for networks prepared in sufficiently good solvents.

Here we give a short classification of these eigenvalues.

1. Rotational modes of the n-vector model, which are related to excluded volume effects.

   The eigenvalues of the operator $K^{\perp}$ are

   $$\Lambda^{\perp}(\mathbf{q}_L) = a^2 \mathbf{q}^2. \tag{5.83}$$

   The eigenfunctions corresponding to these eigenvalues are associated with the rotations of the vector $\vec{n}$ in the abstract n-dimensional space and are gapless Goldstone modes; i.e., their eigenvalues are positive definite and vanish in the long-wavelength limit. The situation is equivalent to that of a ferromagnet (with $n \to 0$ spin components) where the Goldstone modes describe "soft" ($q \to 0$) rotations of the magnetization vector [28], although in our case these modes do not have a simple physical interpretation.

2. Shear modes that do not affect the densities in the replicas.

   The lowest-energy shear modes are Goldstone modes with eigenvalues

   $$\Lambda_S(\mathbf{q}_L) = a^2 \mathbf{q}_L^2 \tag{5.84}$$

   which describe the infinitesimal displacement $\mathbf{x}_T \to \mathbf{x}_T + \mathbf{u}_T(\mathbf{x}_L)$ of the coordinate $\mathbf{x}_T$ (in the abstract transverse $3m$-dimensional subspace), subject to the condition that it does not affect the densities in any of the replicas.

3. Density modes in the initial state.

Eigenmodes for which the density fluctuations $\delta\rho^{(0)}(\mathbf{x}^{(0)})$ are not identically zero have eigenvalues

$$\Lambda_{\text{gap}}(\mathbf{q}_L) = (w^{(0)} - z_c)\, \varphi_{mf}^2 + a^2(\mathbf{q}_L)^2. \tag{5.85}$$

Since these eigenvalues do not vanish (in general) in the limit $q \to 0$, following the usual terminology we say that the corresponding solution is *massive* (i.e., has an energy gap). The gap vanishes at

$$z_c^{\max} = \frac{1}{\rho^{(0)}\overline{N}^{\min}} = w^{(0)} \tag{5.86}$$

which is identical to the cross-link saturation threshold condition, Eq. (2.8), that determines the maximal attainable density of cross-links in our model. Our inhomogeneous mean field solution becomes unstable (and therefore unphysical) if this density exceeds the saturation threshold.

4. Density modes in the final state.

The eigenvalues of these modes (with $\delta\rho^{(k)}(\mathbf{x}^{(k)}) \neq 0$ in at least one of the replicas of the final state),

$$\Lambda_D(\mathbf{q}_L) = a^2\mathbf{q}_L^2 + 2w\rho^{(0)}a^2\overline{N}(\lambda^{-1} \star \mathbf{q}_L)^2 \tag{5.87}$$

vanish in the limit $\mathbf{q}_L \to 0$ and hence the corresponding fluctuations are Goldstone modes. The stability criterion $\Lambda_D(\mathbf{q}_L) > 0$ is always satisfied when the final state of the gel corresponds to good solvent conditions ($w > 0$). For large deformations, the positivity condition can be satisfied even for moderately poor solvents ($w < 0$) since then the network is stabilized against collapse by the external forces applied to its surface.

The appearance of Goldstone modes is the consequence of the fact that the ground state of the stretched network corresponds to a solution with spontaneously broken symmetry with respect to translations in replica space, which describes a disordered solid.

The above classification contains information about the quenched (frozen) versus annealed (thermal) character of the fluctuations described by the $3m$ physical modes. Shear modes mix both types of fluctuations. The degenerate ψ_j ($j \neq 0$) density modes correspond to independent density fluctuations in the replicas of the final state ($\rho^{(k)}$ are only constrained by the relation $\eta = 0$) and therefore describe purely annealed thermal fluctuations of the density in the final state. The nondegenerate modes ψ_0 and ψ_{gap} describe density fluctuations that are identical in the replicas of the final state ($\rho^{(1)} = \rho^{(2)} = \cdots = \rho^{(m)}$) and thus correspond to purely quenched fluctuations.

We conclude that all the fluctuations about the inhomogeneous mean field solution $\varphi_{mf}(s)$ (including those that break replica symmetry) increase the energy of the system and, therefore, the above solution corresponds to a true minimum of the Hamiltonian. This shows a fundamental difference between the mean field solution of our model and that of the Sherrington-Kirkpatrick model of spin glasses [29]. While our solution corresponds to a true minimum of the Hamiltonian (i.e., is stable against arbitrary small fluctuations), the solution of Sherrington-Kirkpatrick model gives only the saddle point of the corresponding Hamiltonian. Although the above argument does not prove that our solution is the true *global* minimum of the Hamiltonian and that no other minima with lower energy exist, in a previous paper [3] we have shown that, under given thermodynamic conditions, there is only one state of mechanical equilibrium in which the average force on all the cross-links vanishes. Further support for the conclusion that polymer networks do not belong to the class of spin glasses comes from the work of Goldbart and Zippelius [26], who use a variational approach to study the present model (however, neglecting excluded volume interactions) and find no solutions with RSB. Although the above authors argue that RSB should appear in a more complete treatment of the Edwards model of polymer networks that would account for excluded volume effects, we have demonstrated by now that replica symmetry is maintained even when these interactions are exactly taken into account by the collective coordinates method.

The existence of a unique ground state is of fundamental importance to the physics of polymer networks and means that *a gel with a given structure and subjected to given thermodynamic conditions has a unique microscopic state of equilibrium defined by the complete set of the average positions of all the cross-links and, therefore, of all the monomers.* (Note that a similar conclusion has been reached already by James [30] in his study of localized phantom networks.) In this sense, gels resemble crystalline solids and differ dramatically from spin glasses [23] and amorphous materials in which there are many distinct microscopic equilibrium states under given thermodynamic conditions. A corollary of this statement is the somewhat astonishing prediction that *if the thermodynamic parameters are changed in an arbitrary way (by changing the solvent, temperature, forces on the boundaries, etc., without rupturing the network) and then returned to their initial values, all the network monomers will first undergo some displacement from their initial average positions and then will return to their original locations.* This prediction can be tested by experiments that probe the static inhomogeneous density distribution of the gel, e.g., by checking the reproducibility of the observed (seemingly random) speckle patterns of the intensity of light scattered from the gel, following a cyclic variation ($A \rightarrow B \rightarrow A$) of thermodynamic parameters. In spin glasses such cyclic variations do, in general, change the microscopic state of the sample.

Note that our mean field solution, by its very nature, does not contain any information about the deviations of the monomer *density* from its average value. Such deviations occur due to static inhomogeneities introduced by the statistical character of the cross-linking process and due to thermal fluctuations about this inhomogeneous density distribution. In the next section we proceed to study these density fluctuations and, in the process, gain important insights about the frozen inhomogeneous structure of polymer networks.

3.6   Static Inhomogeneities and Thermal Fluctuations

We now proceed to obtain all the statistical information concerning density fluctuations in the network. Note that the mean field solution, in spite of its rich physical contents, tells us nothing about the static inhomogeneities and the thermal fluctuations of the monomer density in polymer networks. Although we did study fluctuations about the mean field solution (to test its stability), we have been working in an abstract replica space in which all fluctuations—those of the frozen structure of the network as well as thermal ones—have been treated on the same footing. (Recall that the original reason for posing the problem in the language of replica field theory was to avoid dealing with the complicated averaging over the ensemble of different network structures and to treat static inhomogeneities in real space as thermal fluctuations in replica space.) The price we had to pay is that we can no longer distinguish in a simple way between these two very physically distinct types of "fluctuations."

To recover the important information about the inhomogeneous structure of polymer networks and about thermal fluctuations of monomers in the gel, we first eliminate all the collective coordinates that do not affect the monomer density and then go back to real space in which the distinction between static inhomogeneities and thermal density fluctuations comes out naturally. This program will be carried out in the following subsections.

3.6.1   RPA FREE ENERGY DENSITY FUNCTIONAL

In principle, one can express the replica-space Hamiltonian (5.1) in terms of the shear and density modes and obtain the full information about all shear and density fluctuations in the gel. In a similar way, the above Hamiltonian governs the response of the stretched and swollen network to small (in the linear response regime) deformations, on top of the stretching and swelling described by the deformation ratios $\{\lambda_\alpha\}$. The continuum limit of such a theory will give rise to a generalized theory of elasticity of two-component systems in which one of the components is a solid and the other a liquid [31], [14]. Such an approach (using a variational method) was used in ref. [32], where elastic moduli describing small deformations of an unstretched network ($\lambda = 1$) were calculated. Although conceptually important, a mesoscopic (i.e., finite $\mathbf{q}$) description of stress and strain fluctuations is of somewhat limited interest from the experimental point of view, since most scattering experiments do not measure the strain–strain (or stress–stress) correlation functions that can be calculated in this framework. (Information about such quantities can be obtained, in principle, from studies on the propagation and scattering of shear waves in the gel.) To focus on issues directly relevant to neutron and light scattering experiments that probe static density inhomogeneities and thermal *density* fluctuations in the gel, we will eliminate (i.e., integrate over) all the other degrees of freedom and obtain a reduced description in terms of the density fluctuations only.

Throughout this section we will assume that deviations from the mean density due to both thermal fluctuations and static inhomogeneities are small and can be described on a Gaussian (quadratic in the fluctuations) level. For thermal fluctuations this simplifying assumption is equivalent to the *random phase approximation* (RPA), which works well for concentrated polymer systems (for example, for polymer blends) but breaks down in the semidilute regime of polymer solutions in good solvents. Note, however, that RPA also can be applied to the study of long-wavelength fluctuations in semidilute polymer solutions, provided that one uses renormalized (due to strong fluctuations on scales smaller than the correlation length) values instead of the bare values of the parameters (monomer size, second virial coefficient) in the RPA Hamiltonian. This program was carried out using the renormalization group methods in ref. [3].

While the physical reasons for the breakdown of RPA due to the existence of strong thermal fluctuations on length scales smaller than the thermal correlation length (blob size) are well understood, little is known about the limits of applicability of RPA to static inhomogeneities. It was conjectured in ref. [33] that the statistics of static inhomogeneities are strongly non-Gaussian, so that all moments of their distribution are equally important. Actually, such non-Gaussian statistics are encountered only in gels prepared very close to the gelation threshold, which have a fractal structure. The thermodynamics and elastic properties of such gels swollen in good solvent were studied in ref. [11]. The amplitude and the size of static inhomogeneities also grow in approaching the cross-link saturation threshold. We will show in the next section that, when gels are prepared far from the gel point and away from the cross-link saturation threshold (this regime of intermediate cross-link densities is realized in most experiments), static inhomogeneities of monomer density are limited to length scales comparable to the average mesh size of the network and are therefore "weak." Thus, as long as one does not approach the "critical" regime of cross-link densities near the saturation threshold, one can safely assume the applicability of RPA to the study of such inhomogeneities.

3.6.2   DERIVATION OF THE ENTROPY DENSITY FUNCTIONAL

Recall that the fluctuation Hamiltonian can be written as the sum of a rotational part ($\Delta H^\perp$) and a part that describes shear and density fluctuations ($\Delta H^\parallel$) (see Eq. (5.2)). Since rotational modes do not couple to the latter fluctuations, they do not contribute to the partition function of shear and density modes Ξ_m^{SD}, defined as

$$\Xi_m^{SD} = \int D[\delta\varphi] \exp(-\Delta H^\parallel[\delta\varphi]) \tag{6.1}$$

where the Hamiltonian of shear and density fluctuations in replica space has the form

$$\Delta H^\parallel[\delta\varphi] \equiv \frac{1}{2} \int d\hat{\mathbf{x}} \, d\hat{\mathbf{x}}' \, \delta\varphi(\hat{\mathbf{x}}) K^\parallel(\hat{\mathbf{x}}, \hat{\mathbf{x}}') \delta\varphi(\hat{\mathbf{x}}')$$

$$= \sum_{k=0}^{m} \frac{\Delta U^{(k)}[\delta\rho^{(k)}]}{T^{(k)}} - \Delta S[\delta\varphi]. \tag{6.2}$$

In the second equality we used the definition of the $K^\parallel$ operator, Eq. (5.4), to represent the fluctuation Hamiltonian (for the shear and the density modes) as a sum of excluded volume

$$\Delta U^{(k)}[\delta\rho^{(k)}] \equiv \frac{w^{(k)} T^{(k)}}{2} \int d\mathbf{x}^{(k)} [\delta\rho^{(k)}(\mathbf{x}^{(k)})]^2 \tag{6.3}$$

and "entropic"

$$\Delta S[\delta\varphi] = -\frac{1}{2} \int d\hat{\mathbf{x}} \, \delta\varphi(\hat{\mathbf{x}}) \left[\frac{1}{N} - a^2 \hat{\nabla}^2 - \frac{3z_c}{2} \varphi_{mf}^2(s) \right] \delta\varphi(\hat{\mathbf{x}}) \tag{6.4}$$

contributions. Here $T^{(k)}$ is the temperature ($T^{(0)}$ and T in replicas of the initial and the final state, respectively).

We proceed to eliminate the shear fluctuations and the density fluctuations in

the initial state, i.e., integrate over those fluctuations that do not affect the densities in each of the replicas of the *final* state. (We are only interested in density
fluctuations in the final state; those in the state of preparation can be obtained
from the latter by setting $T = T^{(0)}$, $w = w^{(0)}$ and $\{\lambda_\alpha = 1\}$.) This is done by inserting the following representation of the unity

$$1 = \prod_{k=1}^{m} \int D[\delta\rho^{(k)}] \delta[\delta\rho^{(k)}(\mathbf{x}) - \int d\hat{\mathbf{x}} \ \varphi_{mf}(\mathsf{s})\delta\varphi(\hat{\mathbf{x}})\delta(\mathbf{x} - \mathbf{x}^{(k)})] \qquad (6.5)$$

into the integrand in Eq. (6.1) and moving the integration over $\delta\rho^{(k)}$ to the leftmost side of the integral. We obtain

$$\Xi_m^{SD} = \prod_{k=1}^{m} \int D[\delta\rho^{(k)}] \exp\left(-\sum_{k=1}^{m} \frac{\Delta U^{(k)}[\delta\rho^{(k)}]}{T} + \Delta S_m[\{\delta\rho^{(k)}\}]\right) \qquad (6.6)$$

where T is the temperature in the final state, and

$$\exp(\Delta S_m[\{\delta\rho^{(k)}\}]) \equiv \int D[\delta\varphi] \exp\left(\Delta S[\delta\varphi] - \frac{\Delta U^{(0)}[\delta\rho^{(0)}]}{T^{(0)}}\right)$$

$$\times \prod_{k=1}^{m} \delta[\delta\rho^{(k)}(\mathbf{x}) - \int d\hat{\mathbf{x}} \ \varphi_{mf}(\mathsf{s})\delta\varphi(\hat{\mathbf{x}})\delta(\mathbf{x} - \mathbf{x}^{(k)})]. \qquad (6.7)$$

Note that although ΔS_m depends only on the replica densities $\delta\rho^{(k)}$ in the final
state, it includes exactly all the contributions of both the density fluctuations in
the initial state and the shear fluctuations.

Elimination of Shear Fluctuations and of Density Fluctuations in the Initial State

An explicit expression for the entropy in the final state ($k = 1, \ldots, m$),
$\Delta S_m[\{\delta\rho^{(k)}\}]$, Eq. (6.7), is found by introducing the Fourier representation of the
δ-function through the auxiliary fields $\{h^{(k)}(\mathbf{x})\}$ (the field $h^{(k)}$ is conjugate to
$\delta\rho^{(k)}$):

$$\delta[\delta\rho^{(k)}(\mathbf{x}) - \int d\hat{\mathbf{x}} \ \varphi_{mf}(\mathsf{s})\delta\varphi(\hat{\mathbf{x}})\delta(\mathbf{x} - \mathbf{x}^{(k)})]$$

$$= \int D[h^{(k)}] \exp\left(i \int d\mathbf{x} \ \delta\rho^{(k)}(\mathbf{x})h^{(k)}(\mathbf{x}) - i \int d\hat{\mathbf{x}} \ \varphi_{mf}(\mathsf{s})\delta\varphi(\hat{\mathbf{x}})h^{(k)}(\mathbf{x}^{(k)})\right) \qquad (6.8)$$

where we used $\delta(\mathbf{x} - \mathbf{x}^{(k)})$ to integrate over $\mathbf{x}$ in the second term on the right-hand
side of this expression. Substituting into Eq. (6.7) and moving the integration
over $\{h^{(k)}(\mathbf{x})\}$ to the leftmost side of the resulting expression gives

$$\exp(\Delta S_m[\{\delta\rho^{(k)}\}]) = \int D\{h^{(k)}\} \exp\left(i\sum_{k=1}^{m}\int d\mathbf{x}\ \delta\rho^{(k)}(\mathbf{x})h^{(k)}(\mathbf{x})\right)I[\{h^{(k)}\}] \quad (6.9)$$

where

$$I[\{h^{(k)}\}] \equiv \int D[\delta\varphi] \exp\left\{-i\int d\hat{\mathbf{x}}\ \varphi_{mf}(\mathsf{s})\delta\varphi\ (\hat{\mathbf{x}})\sum_{k=1}^{m}h^{(k)}(\mathbf{x}^{(k)})\right.$$

$$\left. -\frac{1}{2}\int d\hat{\mathbf{x}}\int d\hat{\mathbf{x}}'\ \delta\varphi\ (\hat{\mathbf{x}})K_0^{\|}(\hat{\mathbf{x}},\ \hat{\mathbf{x}}')\delta\varphi(\hat{\mathbf{x}}')\right\}. \quad (6.10)$$

The operator $K_0^{\|}$ is defined by expression (5.4) for $K^{\|}$, in which we take $w = 0$.

To perform the integration over $\delta\varphi$ in the functional I, it is convenient to remove the term linear in $\delta\varphi$ by shifting the integration variables, $\delta\varphi(\hat{\mathbf{x}}) \to \delta\varphi(\hat{\mathbf{x}}) + \phi_h(\hat{\mathbf{x}})$, and demanding that the field ϕ_h obeys the relation

$$\int d\hat{\mathbf{x}}'K_0^{\|}(\hat{\mathbf{x}},\ \hat{\mathbf{x}}')\phi_h(\hat{\mathbf{x}}') \equiv i\varphi_{mf}(\mathsf{s})\sum_{k=1}^{m}h^{(k)}(\mathbf{x}^{(k)}). \quad (6.11)$$

The remaining Gaussian integration can be easily carried out with the result

$$I[\{h^{(k)}\}] = [\det K_0^{\|}]^{-1/2}\exp\left[-\frac{i}{2}\int d\hat{\mathbf{x}}'\ \varphi_{mf}(\mathsf{s}')\phi_h(\hat{\mathbf{x}}')\sum_{k=1}^{m}h^{(k)}(\mathbf{x}^{(k)})\right]. \quad (6.12)$$

The determinant $\det K_0^{\|}$ is given by the product of the eigenvalues $\Lambda_0(\mathbf{q}_L)$ of the operator $K_0^{\|}$, which are obtained by substituting $w = 0$ in the eigenvalues of the operator $K^{\|}$. We have shown previously that all the eigenvalues are positive and therefore $\det K_0^{\|}$ is a well-defined quantity that introduces fluctuation corrections to the mean field free energy. These corrections were calculated in ref. [3] and we will omit them in the following.

Equation (6.11) for the response field $\phi_h(\hat{\mathbf{x}})$ can be rewritten as

$$[1/\bar{N} - a^2\hat{\nabla}^2 - 3(z_c/2)\varphi_{mf}^2(\mathsf{s})]\phi_h(\hat{\mathbf{x}}) + \varphi_{mf}(\mathsf{s})w^{(0)}\rho_h^{(0)}(\mathbf{x}^{(0)}) = i\varphi_{mf}(\mathsf{s})\sum_{k=1}^{m}h^{(k)}(\mathbf{x}^{(k)}) \quad (6.13)$$

where

$$\rho_h^{(0)}(\mathbf{x}^{(0)}) = \int d\hat{\mathbf{x}}'\ \delta(\mathbf{x}^{(0)} - \mathbf{x}'^{(0)})\varphi_{mf}(\mathsf{s}')\phi_h(\hat{\mathbf{x}}') \quad (6.14)$$

can be interpreted as the density variation in the zeroth replica (the initial state), induced by the fields $\{h^{(k)}\}$ in replicas of the final state. Since the $\{h^{(k)}\}$ fields were introduced as conjugate fields to the densities $\{\delta\rho^{(k)}\}$ in the replicas of the final state, the fact that the density in the initial state is also affected by $\{h^{(k)}\}$ appears surprising. Further reflection reveals that the above relation expresses the simple physical fact that the densities in the initial and the final state are strongly correlated, since the structure of the network is identical in all the replicas.

Returning to Eq. (6.13), we first express ϕ_h through $\rho_h^{(0)}$,

$$\phi_h(\hat{\mathbf{x}}) = -\int d\hat{\mathbf{x}}' D(\hat{\mathbf{x}}, \hat{\mathbf{x}}') \varphi_{mf}(s') \left[w^{(0)} \rho_h^{(0)}(\mathbf{x}'^{(0)}) - i \sum_{k=1}^{m} h^{(k)}(\mathbf{x}') \right] \qquad (6.15)$$

where D is defined by the equation

$$[1/\overline{N} - a^2 \hat{\nabla}^2 - 3(z_c/2)\varphi_{mf}^2(\hat{\mathbf{x}})] D(\hat{\mathbf{x}}, \hat{\mathbf{x}}') = \delta(\hat{\mathbf{x}} - \hat{\mathbf{x}}'). \qquad (6.16)$$

Substituting expression (6.15) into Eq. (6.12) yields

$$I[\{h^{(k)}\}] = \exp\left\{ -\frac{1}{2} \int d\mathbf{x} \int d\mathbf{x}' \left[\sum_{k,l=1}^{m} g^{kl}(\mathbf{x}, \mathbf{x}') h^{(k)}(\mathbf{x}) h^{(l)}(\mathbf{x}') \right. \right.$$
$$\left. \left. + i w^{(0)} \rho_h^{(0)}(\mathbf{x}) g^{01}(\mathbf{x}, \mathbf{x}') \sum_{k=1}^{m} h^{(k)}(\mathbf{x}') \right] \right\} \qquad (6.17)$$

where we define the Green's function g^{kl} by

$$g^{kl}(\mathbf{x}, \mathbf{x}') \equiv \int d\hat{\mathbf{x}}\, \varphi_{mf}(s)\delta(\mathbf{x} - \mathbf{x}^{(k)}) \int d\hat{\mathbf{x}}'\, \varphi_{mf}(s')\delta(\mathbf{x}' - \mathbf{x}'^{(l)}) D(\hat{\mathbf{x}}, \hat{\mathbf{x}}'). \qquad (6.18)$$

The long-wavelength and short-wavelength limits of the Fourier transforms of these functions are calculated in Section 3.7.

We now proceed to eliminate $\rho_h^{(0)}$ in the expression for $I[\{h^{(k)}\}]$. Inserting Eq. (6.15) into Eq. (6.13), we obtain a linear integral equation for the function $\delta\rho_h^{(0)}(\mathbf{x})$:

$$\delta\rho_h^{(0)}(\mathbf{x}) + w^{(0)} \int d\mathbf{x}' g^{00}(\mathbf{x}, \mathbf{x}')\delta\rho_h^{(0)}(\mathbf{x}') = i \int d\mathbf{x}' g^{01}(\mathbf{x}, \mathbf{x}') \sum_{k=1}^{m} h^{(k)}(\mathbf{x}'). \qquad (6.19)$$

The problem can be further simplified by Fourier transforming Eq. (6.19). For this we have to calculate the Fourier transform of the functions $g^{kl}(\mathbf{x}, \mathbf{x}')$,

$$g^{kl}(\mathbf{q}^{(k)}, \mathbf{q}^{(l)}) \equiv \int d\hat{\mathbf{x}}\, \varphi_{mf}(s) \int d\hat{\mathbf{x}}'\, \varphi_{mf}(s') D(\hat{\mathbf{x}}, \hat{\mathbf{x}}') \exp(i\hat{\mathbf{q}}^{(k)} \cdot \hat{\mathbf{x}} - i\hat{\mathbf{q}}^{(l)} \cdot \hat{\mathbf{x}}'). \qquad (6.20)$$

Changing the integration $d\hat{\mathbf{x}} \to d\mathbf{x}_L d\mathbf{x}_T$ (and $d\hat{\mathbf{x}}' \to d\mathbf{x}_L' d\mathbf{x}_T'$) and performing the integrations over the longitudinal coordinates using Eq. (5.27) yields

$$g^{kl}(\mathbf{q}^{(k)}, \mathbf{q}^{(l)}) = \delta(\mathbf{q}_L^{(k)} - \mathbf{q}_L^{(l)}) g^{kl}(\mathbf{q}_L^{(k)}), \qquad (6.21)$$

$$g^{kl}(\mathbf{q}_L^{(k)}) = \int d\mathbf{x}_T \varphi_{mf}(s) \int d\mathbf{x}_T' \varphi_{mf}(s') D_T(\mathbf{x}_T, \mathbf{x}_T') \exp(i\mathbf{q}_T^{(k)} \cdot \mathbf{x}_T - i\mathbf{q}_T^{(l)} \cdot \mathbf{x}_T') \qquad (6.22)$$

where the kernel D_T is defined by Eq. (5.12) with the replacement $1/\overline{N} \to 1/\overline{N} + a^2(\mathbf{q}_L^{(k)})^2$, which is introduced by the shift $\hat{\nabla}^2 \to \nabla_T^2 - (\mathbf{q}_L^{(k)})^2$ (Eq. (5.12)) into the longitudinal Fourier transform of Eq. (6.16). In calculating the above integrals we note that to construct the usual *continuous* description of a solid we have to consider only wavelengths that are much larger than the characteristic microscale. (In our case, this microscale corresponds to $a\overline{N}^{1/2}$, which is the average monomer fluctuation radius.) Since this distance is the characteristic length scale for the decay of the classical solution $(\varphi_{mf}(\varsigma \gg a^2\overline{N}) \to 0)$, in evaluating the integrals we can expand the exponentials and keep only terms to second order in $\mathbf{q}_T^{(k)} \cdot \mathbf{x}_T$. Furthermore, since $\mathbf{q}_T^{(k)}$ and $\mathbf{q}_L^{(k)}$ can differ only by a factor of order unity (i.e., by a multiplicative factor of λ), the functions $g_\Lambda^{kl}(\mathbf{q}_L^{(k)})$ have to be calculated also only to order $(\mathbf{q}_L^{(k)})^2$. This calculation is carried out in Section 3.7.

Fourier transforming Eq. (6.19), we find (analogously to Eq. (5.39))

$$[1 + w^{(0)}g^{00}(\mathbf{q}^{(0)})]\rho_{\mathbf{q}^{(0)}}^{(0)} = ig^{01}(\mathbf{q}^{(0)})\sum_{k=1}^{m} h_{\mathbf{q}}^{(k)} \tag{6.23}$$

where the wave vectors $\mathbf{q}^{(0)}$ and $\mathbf{q}$ in the initial and the final states, respectively, are related by the affine transformation

$$\mathbf{q}^{(0)} = \lambda \star \mathbf{q}. \tag{6.24}$$

Substituting this solution into Eq. (6.12) yields an explicit expression for $I[\{h^{(k)}\}]$ in terms of the Fourier components $\{h_{\mathbf{q}}^{(k)}\}$:

$$I[\{h^{(k)}\}] = \exp\left\{-\int \frac{d\mathbf{q}}{(2\pi)^3}\left[\frac{g_{\mathbf{q}}}{2}\sum_{k=1}^{m} h_{\mathbf{q}}^{(k)}h_{-\mathbf{q}}^{(k)} + \frac{\nu_{\mathbf{q}}}{2}\sum_{k=1}^{m} h_{\mathbf{q}}^{(k)}\sum_{l=1}^{m} h_{-\mathbf{q}}^{(l)}\right]\right\} \tag{6.25}$$

with the coefficients

$$g_{\mathbf{q}} \equiv \frac{g^{11}(\lambda \star \mathbf{q}) - g^{12}(\lambda \star \mathbf{q})}{\lambda_x \lambda_y \lambda_z} \tag{6.26}$$

and

$$\nu_{\mathbf{q}} \equiv \frac{1}{\lambda_x \lambda_y \lambda_z}\left[g^{12}(\lambda \star \mathbf{q}) - \frac{w^{(0)}[g^{01}(\lambda \star \mathbf{q})]^2}{1 + w^{(0)}g^{00}(\lambda \star \mathbf{q})}\right]. \tag{6.27}$$

The fact that the wave vectors transform as $\mathbf{q}^{(0)} = \lambda \star \mathbf{q}$ automatically implies that the coordinates transform affinely as $\mathbf{x} = \lambda \star \mathbf{x}^{(0)}$, where $\mathbf{x}$ and $\mathbf{x}^{(0)}$ are the coordinates of a point in the deformed and the undeformed network, respectively. The factor $\lambda_x \lambda_y \lambda_z = V/V^{(0)}$ in Eqs. (6.26) and (6.27) reflects the change of volume in going from the initial to the final (deformed) state of the network.

Diagonalization in the Replicas

Expression (6.25) contains nondiagonal terms in the replicas. The above coupling between the replicas reflects the fact that all replicas have identical network structure. This observation is made more transparent if we diagonalize these terms using the identity (Eq. (A.6) in Appendix 3.A, with the replacement $\psi \to n$)

$$\exp\left\{-\int \frac{d\mathbf{q}}{(2\pi)^3} \frac{\nu_\mathbf{q}}{2} \sum_{k=1}^m h_\mathbf{q}^{(k)} \sum_{l=1}^m h_{-\mathbf{q}}^{(l)}\right\} \equiv \left\langle \exp\left\{-i\int \frac{d\mathbf{q}}{(2\pi)^3} n_\mathbf{q} \sum_{l=1}^m h_{-\mathbf{q}}^{(l)}\right\}\right\rangle_n \quad (6.28)$$

where $n_\mathbf{q}$ is a Fourier component of a random Gaussian field whose correlator (see Eq. (A.8), in Appendix 3.A)

$$\langle n_\mathbf{q} n_{-\mathbf{q}}\rangle_n = \nu_\mathbf{q}. \quad (6.29)$$

The probability distribution of this random field can be obtained from Eq. (A.7)

$$P[n] = \exp\left\{\frac{1}{2}\int \frac{d\mathbf{q}^{(0)}}{(2\pi)^3}\left[\ln(2\pi\nu_\mathbf{q}) - \frac{n_\mathbf{q} n_{-\mathbf{q}}}{\nu_\mathbf{q}}\right]\right\}. \quad (6.30)$$

We can now use identity (6.28) to represent Eq. (6.25) in the diagonal in the replicas form

$$I[\{h^{(k)}\}] = \left\langle \prod_{k=1}^m \exp\left\{-\int \frac{d\mathbf{q}}{(2\pi)^3}\left[\frac{g_\mathbf{q}}{2} h_\mathbf{q}^{(k)} h_{-\mathbf{q}}^{(k)} - i h_\mathbf{q}^{(k)} n_{-\mathbf{q}}\right]\right\}\right\rangle_n. \quad (6.31)$$

Upon inserting this relation into Eq. (6.9) and performing the Gaussian integrals over $\{h^{(k)}\}$, we get

$$\exp\Delta S_m[\{\rho^{(k)}\}] = \left\langle \prod_{k=1}^m \exp\Delta S[n, \delta\rho^{(k)}]\right\rangle_n \quad (6.32)$$

where the entropy functional is given by the simple expression

$$\Delta S[n, \delta\rho^{(k)}] = -\int \frac{d\mathbf{q}}{(2\pi)^3} \frac{(\rho_\mathbf{q}^{(k)} - n_\mathbf{q})(\rho_{-\mathbf{q}}^{(k)} - n_{-\mathbf{q}})}{2g_\mathbf{q}}. \quad (6.33)$$

The entropy $\Delta S[n, \delta\rho^{(k)}]$ is a functional of the density in the kth replica and of a random field n (the physical meaning of which will be discussed later). The averaging, $\langle \ \rangle_n$, in Eq. (6.32) is taken with respect to the distribution function, Eq. (6.30), with the correlator $\nu_\mathbf{q} = \langle n_\mathbf{q} n_{-\mathbf{q}}\rangle_n$.

To derive the grand canonical partition function of the replica system, we substitute Eq. (6.32) into Eq. (6.6) and use the independence of the n and the $\delta\rho^{(k)}$ fields to change the order of averaging (with respect to n) and integration (over $\rho^{(k)}$):

$$\Xi_m^{SD} = \left\langle \prod_{k=1}^m \int D[\delta\rho^{(k)}] \exp\left(-\frac{\Delta U^{(k)}[\delta\rho^{(k)}] - T\Delta S_n[\delta\rho^{(k)}]}{T}\right)\right\rangle_n. \quad (6.34)$$

Using the identity of the contributions from the different replicas of the final state and introducing the free energy functional

$$\mathcal{F}_n[\delta\rho] \equiv \Delta U[\delta\rho] - T\Delta S_n[\delta\rho] \tag{6.35}$$

where

$$\Delta U[\delta\rho] \equiv \frac{wT}{2} \int \frac{d\mathbf{q}}{(2\pi)^3} \rho_{\mathbf{q}}\rho_{-\mathbf{q}} \tag{6.36}$$

and

$$\Delta S_n[\delta\rho] = -\int \frac{d\mathbf{q}}{(2\pi)^3} \frac{(\rho_{\mathbf{q}} - n_{\mathbf{q}})(\rho_{-\mathbf{q}} - n_{-\mathbf{q}})}{2g_{\mathbf{q}}} \tag{6.37}$$

we can rewrite Eq. (6.34) as

$$\Xi_m^{SD} = \left\langle \left[\int D[\delta\rho] \exp\left(-\frac{\mathcal{F}_n[\delta\rho]}{T}\right) \right]^m \right\rangle_n. \tag{6.38}$$

Substituting this expression into Eq. (2.31), we obtain the final expression for the fluctuation contribution to the free energy of the deformed network, $\Delta\mathcal{F}\{\lambda_\alpha\}$ $= \mathcal{F}\{\lambda_\alpha\} - \mathcal{F}_{mf}\{\lambda_\alpha\}$ ($\mathcal{F}_{mf}$ is defined in Eq. (4.34)),

$$\Delta\mathcal{F}\{\lambda_\alpha\} = -T\int Dn\, P[n] \ln\left[\int D[\delta\rho] \exp\left(-\frac{\mathcal{F}_n[\delta\rho]}{T}\right) \right] \tag{6.39}$$

where $P[n]$ is given by Eq. (6.30), and

$$\frac{\mathcal{F}_n[\delta\rho]}{T} = \frac{1}{2} \int \frac{d\mathbf{q}}{(2\pi)^3} \left[\frac{(\rho_{\mathbf{q}} - n_{\mathbf{q}})(\rho_{-\mathbf{q}} - n_{-\mathbf{q}})}{g_{\mathbf{q}}} + w\rho_{\mathbf{q}}\rho_{-\mathbf{q}} \right]. \tag{6.40}$$

Note that since all the functional integrals in Eq. (6.39) are Gaussian (see Eqs. (6.30) and (6.40)), they can be calculated exactly and one can obtain a complete expression for the thermodynamic free energy in which all fluctuation corrections are included on the RPA level. To carry out this program, one has to calculate Fourier integrals over combinations of the functions $\nu_{\mathbf{q}}$ and $g_{\mathbf{q}}$ for which only asymptotic (i.e., in the long- and short-wavelength limits) results are available at present. Since the resulting corrections are small (otherwise, our RPA calculation is inconsistent), we will not consider them here. (See ref. [3], where the effects of strong thermal fluctuations in semidilute gels in good solvents were studied by renormalization group and scaling methods.)

What is the physical meaning of the random field n? Comparison between the definition of the free energy functional $\mathcal{F}_n[\delta\rho]$, defined in Eq. (6.40), and the expression for the thermodynamic free energy, Eq. (2.13), shows that the probability functional $P[n]$, defined in Eq. (6.30), gives the probability of realization of a given network structure $\mathbf{S}$, which is characterized by the value of n (in general, n is a complicated functional of $\mathbf{S}$ and depends on the deformation of the network)

$$P[n] = \int d\mathbf{S}\, \mathcal{P}(\mathbf{S})\delta[n - n(\mathbf{S})] \qquad (6.41)$$

where $\mathcal{P}(\mathbf{S})$ is defined in Eq. (2.12). As long as we are interested in observable quantities such as density correlation functions, which can be represented as statistical averages, we do not need the explicit functional form of n($\mathbf{S}$) and it is sufficient to specify the probability functional $P[n]$ or, alternatively, the function ν_q. The function ν_q depends on the frozen structure of the network and on the deformation but does not depend on the quality of solvent in the final state.

3.6.3   THERMAL AND STRUCTURE AVERAGES

Although, in principle, a given realization of network structure is described by the value of the field $n(\mathbf{x})$ at each point $\mathbf{x}$ in the gel, if we are interested in statistical information we only have to know the probability $P[n(\mathbf{x})]$ (defined in Eq. (6.30)) of observing this particular value at this point in space. Thus, the only meaningful information about the density profile one can obtain from our statistical mechanical description involves *averages* of moments of the density field (e.g., density correlation functions).

We now introduce two important relations. The first is the consequence of our demonstration that a gel with a given network structure and thermodynamic parameters has a unique microscopic equilibrium state characterized by a density distribution ρ_q^{eq}. When the network is deformed, a new equilibrium state results that is characterized by a different equilibrium density distribution. Gels are, of course, nonergodic in the sense that the configurational space available to the cross-linked network is smaller than that of the pre-cross-linked polymer solution. Nevertheless, the existence of a single state of equilibrium under given thermodynamic conditions implies that they possess *restricted ergodicity;* i.e., if we were to prepare an ensemble of gels with *identical* structures, averaging over this ensemble with respect to the Gibbs distribution is the same as measuring time averages in a single gel. We therefore conclude that thermal (annealed) averaging is equivalent to time averaging.

The second relation is based on the fact that, as long as the cross-linking is done away from the cross-link saturation threshold, the resulting static density inhomogeneities will be uncorrelated over macroscopic distances. We can, therefore, mentally decompose the entire network into small but still macroscopic domains, each of which is characterized by a different structure. In the thermodynamic limit, the probability of finding a region with a given structure S is the same as that of finding a network with this structure from the ensemble of networks with all possible structures (but prepared under the same conditions, i.e., temperature, density of cross-links, etc.). Thus, the *structure average* of a quantity over the latter ensemble is equivalent to the *spatial average* of this quantity over the volume of a single network. The averages we calculate are structure averages over the ensemble of different structures and the above equivalence implies that our results can be directly applied to scattering experiments that measure averages over the volume of a single gel with a unique structure.

We begin the calculation of the density correlation functions with the definition of thermal (time) and structure (spatial) averages. Thermal averages for a given network structure (which enters through the value of n) are taken with respect to the Gibbs probability distribution

$$P_n[\delta\rho] \equiv \frac{\exp(-\mathcal{F}_n[\delta\rho]/T)}{\int D[\delta\rho]\,\exp(-\mathcal{F}_n[\delta\rho]/T)} \tag{6.42}$$

where $\mathcal{F}_n[\delta\rho]$ is defined in Eq. (6.40)). The thermal average of a functional $A[\delta\rho]$ is denoted by

$$\langle A[\delta\rho]\rangle \equiv \int D[\delta\rho]A[\delta\rho]P_n[\delta\rho]. \tag{6.43}$$

The structure average of a functional $B[n]$ will be denoted by

$$\overline{B[n]} \equiv \int dS\,P(S)B[n(S)] = \int DnB[n]P[n] \equiv \langle B[n]\rangle_n \tag{6.44}$$

where we used Eq. (6.41) to replace the averaging over the structure with averaging with respect to n. Finally, the complete structure and thermal (spatial and time) average of the functional $C_n[\delta\rho]$ is defined as

$$\overline{\langle C_n[\delta\rho]\rangle} \equiv \int DnP[n]\int D[\delta\rho]\,P_n[\delta\rho]C_n[\delta\rho]. \tag{6.45}$$

3.6.4 DENSITY CORRELATION FUNCTIONS

One of the salient characteristics of gels is the presence of static spatial inhomogeneities of the density. Whereas in liquids the time average of the density fluctuations vanishes ($\langle\delta\rho\rangle = 0$), in polymer gels static spatial density inhomogeneities are always present due to the statistical nature of the process of cross-linking, which results in a unique equilibrium density distribution ρ_q^{eq}. Straightforward calculation (by Gaussian integration, using the probability distribution defined in Eq. (6.42)) of the thermally averaged Fourier component of the density fluctuations gives the amplitude of this density distribution:

$$\langle \rho_{\mathbf{q}} \rangle = \frac{n_{\mathbf{q}}}{1 + wg_{\mathbf{q}}} = \rho_{\mathbf{q}}^{eq}. \qquad (6.46)$$

Note that if we were to switch off the excluded volume ($w = 0$) while keeping the other thermodynamic parameters fixed (e.g., fixed volume V of the gel), we would have $\langle \rho_{\mathbf{q}} \rangle|_{w=0} = n_{\mathbf{q}}$ and, therefore, *the random field* n(x) *can be interpreted as the time-averaged density in the "elastic" reference state*, in which only elastic forces act in the network.

Since every given realization of the network is characterized by a unique equilibrium density profile $\rho^{eq}(\mathbf{x}) = \rho + \delta\rho^{eq}(\mathbf{x}) = \rho + \int [d\mathbf{q}/(2\pi)^3]\rho_{\mathbf{q}}^{eq} \exp(i\mathbf{q} \cdot \mathbf{x})$, the time-averaged density $\langle \rho(\mathbf{x}) \rangle$ fluctuates in space across the gel. This leads to the appearance of stationary speckle patterns and to the presence of a time-independent component in measurements of the temporal decay of intensity correlations in light scattering from gels [34], [35]. Finally, $\overline{\rho_{\mathbf{q}}^{eq}} \equiv \overline{\langle \rho_{\mathbf{q}} \rangle} = 0$, as expected for an average deviation from the mean density $\rho \equiv N_{tot}/V$.

Using the above equation we can introduce the amplitude of thermal density fluctuations $\delta\rho^{th}(\mathbf{x}, t)$ as the deviation of the instantaneous density $\rho(\mathbf{x}, t)$ from its equilibrium value,

$$\delta\rho^{th}(\mathbf{x}, t) \equiv \rho(\mathbf{x}, t) - \rho^{eq}(\mathbf{x}) \qquad (6.47)$$

which, according to Eq. (6.46), satisfies $\langle \delta\rho^{th}(\mathbf{x}) \rangle = 0$. (We replace time averages by ensemble averages.) We now proceed to calculate the different density correlators, evaluating Gaussian integrals by the method described in Appendix 3.A.

The correlator of the thermal density fluctuations is given by

$$G_{\mathbf{q}} \equiv \langle \rho_{\mathbf{q}}^{th}\rho_{-\mathbf{q}}^{th} \rangle = \frac{g_{\mathbf{q}}}{1 + wg_{\mathbf{q}}}. \qquad (6.48)$$

The above expression can be rewritten in a form that emphasizes the similarity with the RPA relation of the theory of polymer liquids [15], that gives the effective Hamiltonian of thermal fluctuations ($G_{\mathbf{q}}^{-1}$) as the sum of entropic and excluded volume contributions:

$$G_{\mathbf{q}}^{-1} = g_{\mathbf{q}}^{-1} + w. \qquad (6.49)$$

Thus, $g_{\mathbf{q}}$ *can be interpreted as the thermal structure factor of the gel, in the absence of excluded volume interactions* (i.e., in the elastic reference state).

The correlator of the static density inhomogeneities (the Fourier transform of the spatially averaged two-point correlation function $\overline{\delta\rho^{eq}(\mathbf{x})\delta\rho^{eq}(\mathbf{x}'))}$ can be found using the definition (6.46) and Eq. (6.29):

$$C_{\mathbf{q}} \equiv \overline{\rho_{\mathbf{q}}^{eq}\rho_{-\mathbf{q}}^{eq}} = \frac{\nu_{\mathbf{q}}}{(1 + wg_{\mathbf{q}})^2}. \tag{6.50}$$

Setting $w = 0$ in the above expression we conclude that $\nu_{\mathbf{q}}$ can be interpreted as the spatially averaged (structure-averaged) equilibrium density correlator, in the absence of excluded volume interactions (in the elastic reference state).

Using Eq. (6.47) we arrive at the following expression for the total structure factor, which includes the contributions of both static inhomogeneities and thermal fluctuations and which is a measure of the total deviation of the density from its mean value, N_{tot}/V:

$$S_{\mathbf{q}} \equiv \langle\rho_{\mathbf{q}}\rho_{-\mathbf{q}}\rangle = G_{\mathbf{q}} + C_{\mathbf{q}}. \tag{6.51}$$

This structure factor is proportional to the scattered intensity at a wave vector $\mathbf{q}$, measured in static scattering experiments.

3.7 Analytical Expressions for the Correlators

3.7.1 FINAL STATE, LONG-WAVELENGTH LIMIT

Simple analytical formulas for the functions $g_{\mathbf{q}}$ and $\nu_{\mathbf{q}}$ which enter the expressions (6.48)–(6.51) for the structure factor of the gel in the final state, can be obtained in the continuum limit $\mathbf{q} \to 0$. (These formulas were derived in Section 3.5.)

The long-wavelength limit of the correlator of thermal fluctuations in the elastic reference state, $g_{\mathbf{q}}$, is given by

$$g_{\mathbf{q}\to 0} = \frac{2\rho\overline{N}\mathbf{q}^2}{(\lambda \star \mathbf{q})^2}. \tag{7.1}$$

Note that this function retains its angular dependence (associated with the direction of the wave vector $\mathbf{q}$) for anisotropic deformations, even in the limit $\mathbf{q} \to 0$. The presence of the $\mathbf{q}^2/(\lambda \star \mathbf{q})^2$ term is related the fact that $g_{\mathbf{q}\to 0}$ is a response function that governs thermal fluctuations about the anisotropic deformed equilibrium state of the network (in the elastic reference state). Analysis of the connection between the long-wavelength limit of the present theory and the continuum theory of elasticity of gels shows that the elastic moduli of the gel are proportional to $g_{\mathbf{q}\to 0}^{-1}$. These moduli depend on the applied deformation and are anisotropic, and this anisotropy results in the above dependence [3].

We now consider the correlator of the static inhomogeneities in the elastic reference state, ν_q. In the long-wavelength limit ν_q approaches a constant value:

$$\nu_{q\to 0} = \lim_{q\to 0} \overline{n_q n_{-q}} = \rho\overline{N}\left(6 + \frac{9}{w^{(0)}\rho^{(0)}\overline{N} - 1}\right) \tag{7.2}$$

where ρ is the density of monomers in the *final* deformed state of the network. The quantity $\nu_{q\to 0}$ diverges at the cross-link saturation threshold, $\overline{N}^{min} = 1/(w^{(0)}\rho^{(0)})$, at which the characteristic size of static spatial inhomogeneities of cross-link (and hence of monomer) density diverges. We would like to emphasize that the finite value of $\nu_{q\to 0}$ (away from the saturation threshold) does not mean that there are frozen clusters of macroscopic dimensions; instead, this is a trivial consequence of the Fourier representation (*all* length scales contribute to the Fourier transform of the density in the limit $q \to 0$).

We can now estimate the amplitude of static density inhomogeneities and check the applicability of RPA. Away from the saturation threshold one can neglect the second term in the square brackets in Eq. (7.2) and write

$$\overline{(\delta n(x))^2} = \int \frac{dq}{(2\pi)^3}\,\nu_q \approx \frac{\rho\overline{N}}{(aN^{1/2})^3} \tag{7.3}$$

where we have used the fact that there are no static density inhomogeneities on length scales smaller than the mesh size $a\overline{N}^{1/2}$. The RPA assumption that the amplitude of inhomogeneities is much smaller than the mean density ρ corresponds to the condition

$$\rho a^3 \overline{N}^{1/2} \gg 1 \tag{7.4}$$

which is equivalent to the assumption that *the volume of an average chain between cross-links is permeated by many other chains*. In the following we will assume that this condition is always satisfied and that static inhomogeneities are correctly described in the RPA approximation.

It is interesting to consider the limiting case of cross-linking in the melt. (Strictly speaking, this case is outside the domain of applicability of our model since excluded volume in a melt cannot be described by a second virial coefficient.) In this limit, $w^{(0)} \sim 1/\rho^{(0)} \sim a^3$ and $\nu_{q\to 0} \simeq 6\rho\overline{N}$. The finite value of $\nu_{q\to 0}$ tells us that even though there are no density fluctuations in the melt, there are still finite inhomogeneities of the network structure that can be revealed upon swelling.

An explicit form for the correlators in the *final deformed state* of the network can be given in the continuum limit ($q \to 0$), using expressions (7.1) and (7.2) for the functions g_q and ν_q in Eqs. (6.48) and (6.50):

$$G_{\mathbf{q}\to 0} = \frac{2\rho\overline{N}}{\left(\lambda \star \dfrac{\mathbf{q}}{|\mathbf{q}|}\right)^2 + 2w\rho\overline{N}} \tag{7.5}$$

$$C_{\mathbf{q}\to 0} = \frac{\left(\lambda \star \dfrac{\mathbf{q}}{|\mathbf{q}|}\right)^4 \rho\overline{N}\left(6 + \dfrac{9}{w^{(0)}\rho^{(0)}\overline{N} - 1}\right)}{\left[\left(\lambda \star \dfrac{\mathbf{q}}{|\mathbf{q}|}\right)^2 + 2w\rho\overline{N}\right]^2}. \tag{7.6}$$

Several general comments can be made regarding the properties of the RPA correlators in the continuum limit.

1. For anisotropic deformations both correlators have an angular singularity associated with the direction of the scattering wave vector $\mathbf{q}$, even in the limit $\mathbf{q} \to 0$. In the important case of uniaxial deformations, *thermal fluctuations are suppressed along the extension axis and enhanced normal to it and static inhomogeneities exhibit the reverse behavior.* Suppression of thermal fluctuations along the stretching direction was predicted in ref. [36].

2. The only explicit dependence on the conditions of preparation of the gel (apart from the trivial dependence on the density of cross-links) appears in the correlator of the static inhomogeneities $C_{\mathbf{q}\to 0}$, which diverges at the cross-link saturation threshold. The control parameter that measures the "strength" of these inhomogeneities is the *heterogeneity parameter,* $X_{RPA} = (w^{(0)}\rho^{(0)}\overline{N} - 1)^{-1}$.

3. Static inhomogeneities are more sensitive than the thermal ones to deformation and swelling. Away from the Θ-point ($w\rho\overline{N} \gg 1$), $G_{\mathbf{q}\to 0} \sim$ $1/w$ is nearly independent of the deformation and of the density of cross-links. The intensity of scattering from static inhomogeneities, $C_{\mathbf{q}\to 0} \sim (\lambda \star \mathbf{q}/|\mathbf{q}|)^4/(w^2\rho\overline{N})$, increases with the degree of cross-linking. When the gel is uniformly swollen ($\lambda \sim 1/\rho^{1/3}$), scattering from static inhomogeneities increases with swelling (RPA predicts $1/\rho^{7/4}$ dependence). Under uniaxial extension it grows rapidly with the deformation ratio in the stretching direction (as λ^4) and decreases normal to it (as $1/\lambda^2$). In the vicinity of the Θ-point, the intensity of thermal fluctuations $G_{\mathbf{q}\to 0}$ decreases as $1/\lambda^2$ along the stretching direction (and increases as λ normal to it) and $C_{\mathbf{q}\to 0}$ approaches a constant (independent of λ).

3.7.2   FINITE STATE, MESOSCOPIC RANGE (FINITE q)

The explicit expressions for the density correlators, Eqs. (7.5) and (7.6), are valid only in the continuum limit ($\mathbf{q} \to 0$). To relate our predictions to scattering experiments that probe the mesoscopic range of length scales (50–5000 Å), we have to include terms of higher order in $\mathbf{q}$. This can be done by calculating the correlators in the elastic reference state $g_{\mathbf{q}}$ and $\nu_{\mathbf{q}}$ to order $a^2 \overline{N} \mathbf{q}^2$.

Long-Wavelength Limit, q^2 Corrections

We proceed to calculate the $\mathbf{q}^2$ corrections to the functions g^{kl}, defined by Eq. (5.60). Since all we need are the functions D,Φ, and Ψ evaluated at $\Lambda = 0$, we can simplify the notation by *omitting* the argument Λ. The function Ψ is defined by Eq. (5.52) (with $\Lambda = 0$),

$$\left[\frac{1}{\overline{N}} + a^2 \mathbf{q}_L^2 - 2a^2 \varsigma \frac{\partial^2}{\partial \varsigma^2} - 2a^2 \frac{\partial}{\partial \varsigma} - \frac{3z_c}{2} \varphi_{mf}^2(\varsigma) \right] \Psi(\varsigma) = \varphi_{mf}(\varsigma). \tag{7.7}$$

To solve this equation, we rewrite its right-hand side as the sum of two terms,

$$\varphi_{mf}(\varsigma) = C_0 \frac{\partial \varphi_{mf}(\varsigma)}{\partial \varsigma} + \tilde{\varphi}_{mf}(\varsigma), \qquad \int_0^\infty d\varsigma \, \tilde{\varphi}_{mf}(\varsigma) \frac{\partial \varphi_{mf}(\varsigma)}{\partial \varsigma} = 0 \tag{7.8}$$

the first of which is "parallel" to the eigenfunction $\partial \varphi_{mf}(\varsigma)/\partial \varsigma$ of the differential operator in Eq. (7.7) and the second of which is orthogonal to it. The constant C_0 was calculated earlier (Eq. (5.65)). The solution of the equation can be represented in the same form as in Eq. (7.8):

$$\Psi(\varsigma) = \frac{C_0}{a^2 \mathbf{q}_L^2} \frac{\partial \varphi_{mf}(\varsigma)}{\partial \varsigma} + \tilde{\Psi}(\varsigma), \qquad \int_0^\infty d\varsigma \, \tilde{\Psi}(\varsigma) \frac{\partial \varphi_{mf}(\varsigma)}{\partial \varsigma} = 0. \tag{7.9}$$

The function $\tilde{\Psi}(\varsigma)$ is the solution of Eq. (7.7), where on the right-hand side we replace the function $\varphi_{mf}(\varsigma)$ with $\tilde{\varphi}_{mf}(\varsigma)$. Since we are interested only in terms of order $\mathbf{q}^2$, we can omit the $a^2 \mathbf{q}_L^2$ term in this equation.

The next step is to introduce the dimensionless function

$$\tilde{\Psi}(\varsigma) = -\sqrt{\frac{2\overline{N}}{z_c}} \vartheta(t), \qquad t = \frac{\varsigma}{2a^2 \overline{N}} \tag{7.10}$$

which obeys the equation

$$(1 - 3\chi^2(t)) \vartheta(t) - t\vartheta''(t) - \vartheta'(t) = \chi(t) + \chi'(t) \tag{7.11}$$

with $\chi(t)$ defined in Eq. (4.17). Although we do not know the explicit analytical expression for the function $\chi(t)$, the general solution of the differential equation can be found, since we know one of the solutions ($\chi'(t)$) of the corresponding homogeneous equation. This general solution can be written as

$$\vartheta(t) = \frac{1}{2}[\chi(t) + t\chi'(t)] + \chi'(t)\left[C_1 + C_2 \int^t \frac{dt'}{t'(\chi'(t))^2}\right] \quad (7.12)$$

where C_1 and C_2 are integration constants. Since the integral in Eq. (7.12) diverges, we have to take $C_2 = 0$. Also, we must have $C_1 = 0$ because of the orthogonality condition, Eq. (7.9).

We now turn to the calculation of the functions g^{kl}, Eq. (5.60). Substituting the expression $\Phi(s) = \overline{N}^2 \partial\varphi_{mf}(s)/\partial\overline{N}$, Eq. (5.44), in the first integral, we find

$$\int_0^\infty ds\, \varphi_{mf}(s)\Phi(s) = \overline{N}^2 \frac{\partial}{\partial\overline{N}} \int_0^\infty ds\, \frac{\varphi_{mf}^2(s)}{2} = \frac{2a^2\overline{N}^2}{z_c} \frac{\partial}{\partial\overline{N}} \int_0^\infty dt\, \chi^2(t) = 0. \quad (7.13)$$

Substituting the expressions (7.9) and (7.10) into Eq. (5.60), we find

$$g^{kl}(\mathbf{q}_L) = g^{00}(\mathbf{q}_L) + \int_0^\infty ds\, \tilde{\Psi}(s)\tilde{\varphi}_{mf}(s)(\mathbf{q}_T^{(k)} \cdot \mathbf{q}_T^{(l)})$$

$$+ \frac{(\mathbf{q}_T^{(k)} \cdot \mathbf{q}_T^{(l)})}{a^2\mathbf{q}_L^2}\left\{C + \frac{C_0 I_1}{4}[(\mathbf{q}_T^{(k)})^2 + (\mathbf{q}_T^{(l)})^2]\right\} \quad (7.14)$$

where

$$I_1 \equiv \int_0^\infty dt\, \chi^2(t) = 0.524. \quad (7.15)$$

Using the definition (7.10), we can rewrite the corresponding integral in Eq. (7.14) in the form

$$\int_0^\infty ds\, \tilde{\Psi}(s)\tilde{\varphi}_{mf}(s) = 2\rho^{(0)}\overline{N}I_2 \quad (7.16)$$

$$I_2 \equiv \int_0^\infty dt(\chi(t) + t\chi'(t))(\chi(t) + \chi'(t)). \quad (7.17)$$

Let us estimate the integrals that appear in Eq. (7.17). Multiplying Eq. (4.18) for the function $\chi(t)$ by $t\chi'(t)$ and $\chi(t)$ and integrating by parts, we obtain

$$\int_0^\infty dt\, t(\chi'(t))^2 = \frac{1}{2}\int_0^\infty dt\, \chi^2(t) - \frac{1}{4}\int_0^\infty dt\, \chi^4(t), \quad (7.18)$$

$$\frac{1}{2} - \int_0^\infty dt\, t(\chi'(t))^2 = \int_0^\infty dt\, \chi^2(t) - \int_0^\infty dt\, \chi^4(t) \quad (7.19)$$

respectively. Combining these two equations, we find

$$\int_0^\infty dt \, t(\chi'(t))^2 = \frac{1}{5}\int_0^\infty dt \, \chi^2(t) - \frac{1}{10}. \tag{7.20}$$

Multiplying Eq. (4.18) by $\chi'(t)$ and integrating by parts yields

$$\int_0^\infty dt(\chi'(t))^2 = \frac{1}{2}. \tag{7.21}$$

Using these relations we finally obtain

$$I_2 = \frac{2}{5} - \frac{7}{10}\int_0^\infty dt \, \chi^2(t) = 0.033. \tag{7.22}$$

Short-Wavelength Limit

We now calculate the functions $g^{kl}(\mathbf{q}_L)$, defined in Eq. (5.35) with $\Lambda = 0$, in the short-wavelength limit $a^2\overline{N}(\lambda \star \mathbf{q})^2 \gg 1$. In this limit the Laplacian gives the leading contribution to the differential operator in Eq. (5.33). Its eigenfunctions are plane waves, $\exp(i\mathbf{q}_T \cdot \mathbf{x}_T)$, and the corresponding eigenvalues are

$$\varepsilon(\mathbf{q}_T^2) = \frac{1}{\overline{N}} + a^2(\mathbf{q}_L^2 + \mathbf{q}_T^2) - \frac{3z_c}{2V^m}\int d\mathbf{x}_T \varphi_{mf}^2(\mathbf{s}). \tag{7.23}$$

In the limit $m \to 0$, this takes the form

$$\varepsilon(\mathbf{q}_T^2) = a^2(\mathbf{q}_L^2 + \mathbf{q}_T^2) - 2/\overline{N}. \tag{7.24}$$

The short-wavelength limit of the function D is

$$D(\mathbf{x}_T, \mathbf{x}_T') = \int \frac{d\mathbf{q}_T}{(2\pi)^{3m}} \frac{e^{i\mathbf{q}_T \cdot (\mathbf{x}_T - \mathbf{x}_T')}}{\varepsilon(\mathbf{q}_T^2)} \tag{7.25}$$

where, for the sake of simplicity, we omit the argument Λ.

Substituting this function into Eq. (5.35), we find

$$g^{kl}(\mathbf{q}_L) = \int \frac{d\mathbf{q}_T}{(2\pi)^{3m}} \frac{\varphi(\mathbf{q}_T - \mathbf{q}_T^{(k)})\varphi(-\mathbf{q}_T + \mathbf{q}_T^{(l)})}{\varepsilon(\mathbf{q}_T^2)} \tag{7.26}$$

where the function $\varphi(\mathbf{q}_T)$ is the Fourier transform (in the transverse subspace) of the function $\varphi_{mf}(\mathbf{s})$:

$$\varphi(\mathbf{q}_T) \equiv \int d\mathbf{x}_T \varphi_{mf}(\mathbf{s})e^{-i\mathbf{q}_T \cdot \mathbf{x}_T}. \tag{7.27}$$

Since this function has a very sharp peak at the origin and falls down exponentially at infinity, we can estimate the integral (7.27) in the limit of large $\mathbf{q}_*^2 \equiv |\mathbf{q}_T^{(k)}| = |\mathbf{q}_T^{(l)}|$ as

$$g^{kl}(\mathbf{q}_L) \approx \frac{1}{\varepsilon(\mathbf{q}_*^2)} \int \frac{d\mathbf{q}_T}{(2\pi)^{3m}} \varphi(\mathbf{q}_T - \mathbf{q}_T^{(k)}) \varphi(-\mathbf{q}_T + \mathbf{q}_T^{(l)})$$

$$= \frac{1}{\varepsilon(\mathbf{q}_*^2)} \int d\mathbf{x}_T \varphi_{mf}^2(\varsigma) \exp[i(\mathbf{q}_T^{(k)} - \mathbf{q}_T^{(l)})\mathbf{x}_T]. \tag{7.28}$$

We now split the $\mathbf{x}_T$ integration into an angular integration and an integration over $|\mathbf{x}_T|$. In calculating the former, we use the equality

$$\int d\Omega e^{i\mathbf{q}_T\mathbf{x}_T} = S_{3m}\Gamma\left(1 + \frac{3m}{2}\right)\left(\frac{2^{1/2}}{|\mathbf{q}_T|\varsigma^{1/2}}\right)^{3m/2} J_{3m/2}(|\mathbf{q}_T|\sqrt{2\varsigma}). \tag{7.29}$$

Substituting it into the last expression (7.26), we find (in the limit $m \to 0$)

$$g^{kl}(\mathbf{q}_L) \approx -\frac{1}{\varepsilon(\mathbf{q}_*^2)} \int_0^\infty d\varsigma \frac{d\varphi_{mf}^2(\varsigma)}{d\varsigma} J_0\left(|\mathbf{q}_T^{(k)} - \mathbf{q}_T^{(l)}|\sqrt{2\varsigma}\right). \tag{7.30}$$

In the short-wavelength limit this equation reduces to

$$g^{kl}(\mathbf{q}_L) = \frac{2\rho^{(0)}}{a^2(\lambda^{-1} \star \mathbf{q}_L)^2 - 2/\overline{N}}, \quad \text{for } k = 1. \tag{7.31}$$

In the case $k \neq l$, we use the equality

$$(\mathbf{q}_T^{(k)} - \mathbf{q}_T^{(l)})^2 = (\hat{\mathbf{q}}^{(k)} - \hat{\mathbf{q}}^{(l)})^2 = 2(\lambda^{-1} \star \mathbf{q}_L)^2 \tag{7.32}$$

and find from equation (7.30)

$$g^{kl}(\mathbf{q}_L) \approx -\frac{1}{\varepsilon(\mathbf{q}_*^2)} \frac{d\varphi_{mf}^2(\varsigma)}{d\varsigma}\bigg|_{\varsigma=0} \int_0^\infty d\varsigma \, \theta(\varsigma)J_0(2\sqrt{(\lambda^{-1} \star \mathbf{q}_L)^2\varsigma}) \tag{7.33}$$

where $\theta(0) = 1$ and, according to Eqs. (7.12) and (4.19), the function $\theta(\varsigma)$ decreases when $\varsigma \to \infty$ as $\exp\{-4[\varsigma/(2a^2\overline{N})]^{1/2}\}$. The asymptotics under consideration do not depend on the form of this function and, therefore, we can use this exponent for $\theta(\varsigma)$ in the entire region of variation of ς. This gives

$$g^{kl}(\mathbf{q}_L) = \frac{\rho^{(0)}|\chi'(0)|}{a^2(\lambda^{-1} \star \mathbf{q}_L)^2[a^2\overline{N}(\lambda^{-1} \star \mathbf{q}_L)^2 - 2]}, \quad \text{for } k \neq 1 \tag{7.34}$$

where numerical calculations give $\chi'(0) = -1.21$.

Notice that in the derivation of expressions (7.32) and (7.34) we neglected terms of order of $\mathbf{q}_L^{-6}$.

3.7.3 ASYMPTOTIC EXPRESSIONS FOR THE CORRELATORS $g_{\mathbf{q}}$ AND $\nu_{\mathbf{q}}$

We begin with the summary of the main results for the correlation functions $g_{\mathbf{q}}^{kl}$ of the replica system (with $\Lambda = 0$). Then we derive exact asymptotic expressions for the long- and short-wavelength limits of the correlators $g_{\mathbf{q}}$ and $\nu_{\mathbf{q}}$.

An exact (to all orders in $\mathbf{q}^2$) expression for g^{00} is given in Eq. (5.48). Substituting $\Lambda = 0$ into the above expression, we find

$$g^{00}(\lambda \star \mathbf{q}) = \frac{2\rho^{(0)}\overline{N}}{-2 + Q^2} \tag{7.35}$$

where we introduced the dimensionless wave vector Q, defined as $Q^2 \equiv a^2\overline{N}\mathbf{q}^2$. Due to the symmetry of the replica Hamiltonian with respect to permutations of all the replicas of an undeformed network ($\{\lambda_\alpha = 1\}$), we find that all g^{kl} are equal for $k \neq l$ and that $g^{kk} = g^{00}$ for $k = 1, \ldots, m$. Substituting Eqs. (6.26) and (6.27) into Eqs. (6.48)–(6.51) and using these symmetry relations, we find, after some algebra, that all dependence of the total structure factor on the nondiagonal elements of g^{kl} drops out, and we obtain

$$S_{\mathbf{q}}^{(0)} = \frac{g^{00}(\mathbf{q})}{1 + w^{(0)}g^{00}(\mathbf{q})}. \tag{7.36}$$

In the general case, $\{\lambda_\alpha \neq 1\}$, the replica space correlation functions have been calculated only in the long-wavelength ($Q \ll 1$) and the short-wavelength ($Q \gg 1$) limits (see Subsection 3.7.2). Combining the expressions for the functions g^{kl} obtained above, we write

$$g^{11}(\lambda \star \mathbf{q}) - g^{12}(\lambda \star \mathbf{q}) = \begin{cases} \dfrac{2\rho^{(0)}\overline{N}}{(\lambda \star \check{\mathbf{q}})^2}\,[1 + \alpha\,(\check{\mathbf{q}})Q^2], & \text{for } Q \ll 1 \\[3mm] \dfrac{2\rho^{(0)}\overline{N}}{Q^2}\left(1 + \dfrac{I_3}{Q^2}\right), & \text{for } Q \gg 1 \end{cases} \tag{7.37}$$

where we introduced the unit vector $\check{\mathbf{q}} = \mathbf{q}/|\mathbf{q}|$ in the direction of the wave vector $\mathbf{q}$, and

$$\alpha\,(\check{\mathbf{q}}) = 2I_1[(\lambda \star \check{\mathbf{q}})^2 - 1] + I_2(\lambda \star \check{\mathbf{q}})^2 \tag{7.38}$$

with the constants $I_1 \simeq 0.524$, $I_2 \simeq 0.033$, and $I_3 \simeq 1.395$.

3.8 Nonaffine Deformation of Network Chains

In this section we describe the average deformation of a chain of a given length
N (which may deviate from the average chain length in the network, $\overline{N}$) for a giv-
en deformation of the macroscopic gel. In our derivation we will not assume, the
way it is usually done in classical theories of rubber elasticity, that the deformation is *affine*, i.e., that when the entire network is deformed by the factors $\{\lambda_\alpha\}$,
the individual chains in the network deform by the same ratios. Since the chain is
connected to the network through its cross-links, the deformation of the chain is
not necessarily affine because the response of a network chain to externally imposed stress depends on its local environment (i.e., on whether the considered
chain is connected to long or short chains). The local density of cross-links
varies through the network because of the randomness of the cross-linking
process. Thus, the deformation of the chain has to be considered as a random value that can be characterized by the end-to-end distribution function.

3.8.1 CHAIN DISTRIBUTION FUNCTIONS

There are two types of information that characterize a given chain in the network: the statistical distribution of chain lengths in the network and the conformations of chains of given length. Thus, we have to consider two types of distribution functions.

1. The distribution function of chain lengths, $f(N)$, which depends on the
   conditions of preparation of the network but not on the particular realization of network structure **S** (due to the self-averaging property). This
   function is normalized by the condition

$$\int_0^{N_{\text{tot}}} dN f(N) = 1 \qquad (8.1)$$

and defines the average chain length between neighboring cross-links

$$\overline{N} \equiv \int_0^{N_{\text{tot}}} dN N f(N). \qquad (8.2)$$

2. The probability that a chain of contour length aN has an end-to-end distance vector **R**, given that the network is stretched by factors λ_α ($\alpha = x$,
   y, z) with respect to the state in which it was synthesized. This distribution function is defined by

$$W_N(\mathbf{R}|\lambda) \equiv \langle \delta(\mathbf{R} - \mathbf{R}_l) \rangle_N \qquad (8.3)$$

where $\mathbf{R}_l$ is the end-to-end distance of the chains l of length $N_l = N$. $\langle \cdots \rangle_N$ means averaging over all such chains and the usual thermodynamic averaging with Gibbsian weight (corresponding to a particular realization of network structure $\mathbf{S}$):

$$\mathcal{P}[\mathbf{x}(s), \mathbf{S}] \equiv [Z(\mathbf{S})]^{-1} \exp(-\mathcal{H}[\mathbf{x}(s)]/T) \prod_{\{i,j\}} \delta[\mathbf{x}(s_i) - \mathbf{x}(s_j)]. \tag{8.4}$$

This equation is the generalization of Eq. (2.10) for the state of preparation. The Edwards Hamiltonian $\mathcal{H}$ of the deformed gel is defined in Eq. (2.19) and its partition function $Z(\mathbf{S})$ is given by Eq. (2.20). According to the definition of the distribution function, Eq. (8.3), it is normalized by the condition

$$\int d\mathbf{R}\, W_N(\mathbf{R}|\lambda) = 1. \tag{8.5}$$

To calculate the above distribution functions, we introduce the following generating function

$$\Phi_N(\mathbf{R}|\lambda) = \left\langle \sum_l \delta_{N,N_l} \delta(\mathbf{R} - \mathbf{R}_l) \right\rangle$$

$$\equiv \int D\mathbf{x}(s)\mathcal{P}[\mathbf{x}(s), \mathbf{S}] \sum_l \delta_{N,N_l} \delta(\mathbf{R} - \mathbf{R}_l) \tag{8.6}$$

where the sum is over all the chains of the network and the angle brackets denote the usual thermodynamic averaging with the Gibbsian weight $\mathcal{P}$, Eq. (8.4). This function can be split into the product of the above distribution functions, $f(N)$ and $W_N(\mathbf{R}|\lambda)$:

$$\Phi_N(\mathbf{R}|\lambda) = N_{ch} f(N) W_N(\mathbf{R}|\lambda) \tag{8.7}$$

where N_{ch} is the total number of subchains in the network. The functions $f(N)$ and $W_N(\mathbf{R}|\lambda)$ can be obtained from this product using the normalization condition, Eq. (8.5):

$$f(N) = \frac{1}{N_{ch}} \int d\mathbf{R}\, \Phi_N(\mathbf{R}|\lambda). \tag{8.8}$$

We will show in the following subsection that the mean field solution obtained in Section 3.4 contains statistical information about the local properties of network structure and the response of network chains to imposed deformation.

3.8.2   FIELD THEORETICAL FORMULATION

We proceed to calculate the function $\Phi_N(\mathbf{R}|\lambda)$ [11]. Up to this point we were only interested in thermodynamic quantities and therefore could treat the network as a single polymer chain in replica space, with effective attractions induced by the presence of cross-links. Since now we want to study the behavior of individual chains, we have to label the subchains by their length N (chains with different lengths cannot be considered as identical) and our homopolymer model has to be replaced by a model that describes a *heteropolymer* made of labeled chain segments l of variable length $N_l \equiv s_{l+1} - s_l$.

Since we are interested in the properties of individual network chains, it is convenient to rewrite the Gibbsian weight (8.4) in the form of the averaged product of contributions of all the network chains l:

$$P[\mathbf{x}(s), \mathbf{S}] = (Z(\mathbf{S}))^{-1} \left\langle \prod_l P_l[\mathbf{x}(s), ih(\mathbf{x})] \prod_{\{i,j\}} \delta[\mathbf{x}(s_i) - \mathbf{x}(s_j)] \right\rangle_h \qquad (8.9)$$

where the averaging is taken over the random Gaussian field $h(\mathbf{x})$ with the correlator

$$\langle h(\mathbf{x})h(\mathbf{x}') \rangle_h = w\delta(\mathbf{x} - \mathbf{x}'). \qquad (8.10)$$

The contribution P_l of the lth Gaussian subchain to the product is

$$P_l[\mathbf{x}(s), ih(\mathbf{x})] \equiv \exp\left(-\int_{s_l}^{s_{l+1}} ds \left\{ \frac{1}{2a^2}\left(\frac{d\mathbf{x}(s)}{ds}\right)^2 + ih[\mathbf{x}(s)] \right\}\right). \qquad (8.11)$$

The idea of representing the excluded volume interactions by a random field goes back to Edwards and coworkers [17, 15, 18]. This approach appears to be very convenient when treating effects associated with density inhomogeneities in heteropolymer systems because it reduces the problem of monomer interactions to the consideration of noninteracting polymer systems in an external field. What is the physical meaning of the field $ih(\mathbf{x})$? In the mean field approximation this is the well-known self-consistent field which acts on each of the monomers. The fluctuations of this field represent the random character of monomer collisions in the process of thermodynamic fluctuations. The random field $h(\mathbf{x})$ is δ-correlated (see Eq. (8.10)) because of the local character of such interactions.

Our method of calculation of the generating function defined in Eq. (8.6)) is based on the consideration of a network with labeled chains. We label the chains of N monomers by the index ε and introduce the distribution function

$$P_l^\varepsilon[\mathbf{x}(s), ih(\mathbf{x})] \equiv P_l[\mathbf{x}(s), ih(\mathbf{x})][1 + \varepsilon\delta(\mathbf{x}(s_{l+1}) - \mathbf{x}(s_l) - \mathbf{R})]. \qquad (8.12)$$

Note that in the limit $\varepsilon \to 0$ we have $P_l^\varepsilon \to P_l$. Unlabeled chains of $N_l \neq N$ monomers are characterized by the distribution function $P_l^\varepsilon \equiv P_l$ (see Eq. (8.11)). The partition function of such a labeled network is defined as

$$Z^\varepsilon(\mathbf{S}) \equiv \left\langle \int D\mathbf{x}(s) \prod_l P_l^\varepsilon[\mathbf{x}(s), ih(\mathbf{x})] \prod_{\{i,j\}} \delta[\mathbf{x}(s_i) - \mathbf{x}(s_j)] \right\rangle_h. \qquad (8.13)$$

In the limit $\varepsilon \to 0$ it reduces to the partition function $Z(\mathbf{S})$ of the unlabeled network.

Substituting Eqs. (8.12) into (8.13) and differentiating the latter with respect to ε, we can rewrite Eq. (8.6) for the distribution function Φ_N, as

$$\Phi_N(\mathbf{R}|\lambda) = \frac{1}{Z(\mathbf{S})} \frac{\partial Z^\varepsilon(\mathbf{S})}{\partial \varepsilon}\bigg|_{\varepsilon=0}. \tag{8.14}$$

We can further recast Eq. (8.14) in the form

$$\Phi_N(\mathbf{R}|\lambda) = \frac{\partial \ln Z^\varepsilon(\mathbf{S})}{\partial \varepsilon}\bigg|_{\varepsilon=0} = -\frac{1}{T} \frac{dF^\varepsilon(\lambda)}{d\varepsilon}\bigg|_{\varepsilon=0}. \tag{8.15}$$

In writing the second equality we took into account the fact that in the thermodynamic limit the free energy is independent of the choice of the network structure $\mathbf{S}$ (see Section 3.2). Using the self-averaging property of the free energy F^ε, we can write

$$F^\varepsilon(\lambda) \equiv -T\int d\mathbf{S} P(\mathbf{S}) \ln Z^\varepsilon(\mathbf{S}). \tag{8.16}$$

Equation (8.15) reduces the calculation of the function Φ_N to the problem of evaluating the averaged free energy F^ε of the network with the labeled chains. Since the analogous problem for a nonlabeled network was analyzed earlier in this work, we shall not repeat the details of this calculation. Using the replica trick and transforming into the grand canonical ensemble (Section 3.2.3), we can recast Eq. (8.16) in the form

$$F^\varepsilon(\lambda) = -T\frac{\partial \ln \Xi_m^\varepsilon(\mu, z_c)}{\partial m}\bigg|_{m=0}. \tag{8.17}$$

Using the methods of Section 3.3, we arrive at the following field theoretical representation of the grand canonical partition function

$$\Xi_m^\varepsilon(\mu, z_c) = \left\langle \int D\vec{\varphi}(\hat{\mathbf{x}}) \left[\int d\hat{\mathbf{x}} \, \varphi_1(\hat{\mathbf{x}}) \right]^2 \exp\{-H^\varepsilon[\vec{\varphi}, i\hat{h}]\} \right\rangle_{\hat{h}} \tag{8.18}$$

where the averaging is performed over the random field $\hat{h}(\hat{\mathbf{x}})$ (in replica space), which can be represented as a sum of random Gaussian fields $h^{(k)}(\mathbf{x}^{(k)})$ in each of the replicas:

$$\hat{h}(\hat{\mathbf{x}}) \equiv \sum_{k=0}^{m} h^{(k)}(\mathbf{x}^{(k)}). \tag{8.19}$$

These fields are uncorrelated in different replicas, $\langle h^{(k)}(\mathbf{x}) h^{(l)}(\mathbf{x}')\rangle_h = 0$ for $k \neq l$, since there are no excluded volume interactions between monomers in different replicas. The correlations of the field $h^{(k)}(\mathbf{x})$ in each of the replicas $k = 0, \ldots, m$

are described by the correlation function (8.10) where we have to substitute $w \rightarrow w^{(0)}$ for zeroth replica.

The effective dimensionless Hamiltonian in Eq. (8.18) has the form

$$H^\varepsilon[\vec{\varphi}, i\hat{h}] = \int d\hat{x} \left[\frac{1}{2} \int d\hat{x}' [\hat{G}^\varepsilon]^{-1}(\hat{x}, \hat{x}', [i\hat{h}]) (\vec{\varphi}(\hat{x}) \cdot \vec{\varphi}(\hat{x}')) - \frac{z_c}{4} (\vec{\varphi}^{\,2}(\hat{x}))^2 \right]. \quad (8.20)$$

The operator $[\hat{G}^\varepsilon]^{-1}$ is the inverse of the operator $\hat{G}^\varepsilon$ (see Eq. A.3) of Appendix 3.A),

$$\int d\hat{x}' [\hat{G}^\varepsilon]^{-1}(\hat{x}, \hat{x}', [i\hat{h}]) \hat{G}^\varepsilon(\hat{x}', \hat{x}'', [i\hat{h}]) = \delta(\hat{x} - \hat{x}'') \quad (8.21)$$

with the kernel

$$\hat{G}^\varepsilon(\hat{x}, \hat{x}'; [i\hat{h}]) = \int_0^\infty dM \, e^{-\mu M} \hat{G}_M\{\hat{x}, \hat{x}'; [i\hat{h}]\}$$

$$\times \prod_{k=1}^m [1 + \varepsilon \delta(M - N)\delta(\mathbf{x}^{(k)} - \mathbf{x}^{(k)'} - \mathbf{R})]. \quad (8.22)$$

Here $\hat{G}_M$ is the partition function of a Gaussian chain with M monomer units, defined in Eq. (B.1) of Appendix 3.B. For $\varepsilon = 0$, the grand canonical partition function $\hat{G}^\varepsilon = \hat{G}$ of such a chain with ends fixed at points $\hat{x}$ and $\hat{x}'$ (in a $3(1 + m)$-dimensional replica space) turns into that of Eq. (3.19). Equations (8.17)–(8.22) give the field theoretical representation of the free energy of a network with labeled chains. It can be further simplified in the limit $\varepsilon \rightarrow 0$, which should be taken to calculate the distribution function $\Phi_N(\mathbf{R}|\lambda)$ (see Eq. (8.15)).

Note that according to Eq. (8.15) we can keep only terms up to first order in the parameter ε in the product in Eq. (8.22):

$$\hat{G}^\varepsilon = \hat{G} + \varepsilon \sum_{k=1}^m \hat{G}^{(k)},$$

$$\hat{G}^{(k)}\{\hat{x}, \hat{x}'; [i\hat{h}]\} = e^{-\mu N} \hat{G}_N(\hat{x}, \hat{x}'; [i\hat{h}]) \delta(\mathbf{x}^{(k)} - \mathbf{x}^{(k)'} - \mathbf{R}). \quad (8.23)$$

The inversion of the operator $\hat{G}^\varepsilon$ (solution of Eq. (8.21), to first order in the small parameter ε) yields

$$[\hat{G}^\varepsilon]^{-1} = \hat{G}^{-1} - \varepsilon \sum_{k=1}^m \hat{G}^{-1} \hat{G}^{(k)} \hat{G}^{-1}. \quad (8.24)$$

Substituting this expression in the effective Hamiltonian (8.20), we find

$$H^\varepsilon = H - \varepsilon \sum_{k=1}^m H^{(k)} \quad (8.25)$$

where the effective Hamiltonians are given by

$$H[\vec{\varphi}, i\hat{h}] = \int d\hat{\mathbf{x}} \left[\frac{\mu + i\hat{h}(\hat{\mathbf{x}})}{2} \vec{\varphi}^2(\hat{\mathbf{x}}) + \frac{a^2}{2}(\hat{\nabla}\vec{\varphi}(\hat{\mathbf{x}}))^2 - \frac{z_c}{8}(\vec{\varphi}^2(\hat{\mathbf{x}}))^2 \right],$$

$$H^{(k)}[\vec{\varphi}, i\hat{h}] = \frac{1}{2}\int d\hat{\mathbf{x}} \int d\hat{\mathbf{x}}' \hat{G}^{(k)}(\hat{\mathbf{x}}, \hat{\mathbf{x}}'; [i\hat{h}])(\vec{\psi}(\hat{\mathbf{x}}) \cdot \vec{\psi}(\hat{\mathbf{x}}')). \tag{8.26}$$

Here the function $\vec{\psi}(\hat{\mathbf{x}})$ is defined as

$$\vec{\psi}(\hat{\mathbf{x}}) \equiv \hat{G}^{-1}\vec{\varphi}(\hat{\mathbf{x}}) = (\mu + i\hat{h}(\hat{\mathbf{x}}) - a^2\hat{\nabla}^2)\vec{\varphi}(\hat{\mathbf{x}}). \tag{8.27}$$

Substituting Eq. (8.25) back into Eq. (8.18) and performing the differentiation over ε in Eq. (8.17), we arrive at the following expression for the "free energy" (valid to first order in the parameter ε)

$$-T \ln \Xi_m^\varepsilon(\mu, z_c) = F_m - \varepsilon T \sum_{l=1}^m \langle H^{(k)}[\vec{\varphi}, i\hat{h}] \rangle_{\vec{\varphi}, \hat{h}}. \tag{8.28}$$

Here the averaging is taken both over the field $\vec{\varphi}$ (with the weight $\exp\{-H[\vec{\varphi}, i\hat{h}]\}$, Eq. (8.26)) and over the field $\hat{h}$ (Eq. (8.19)). Substituting Eq. (8.28) into Eq. (8.15) and using the symmetry of the Hamiltonian with respect to the permutations of the replicas of the final system, we finally get

$$\Phi_N(\mathbf{R}|\lambda) = \langle H^{(1)}[\vec{\varphi}, i\hat{h}] \rangle_{\vec{\varphi}, \hat{h}}|_{m=0}. \tag{8.29}$$

3.8.3   MEAN FIELD THEORY

In the framework of the mean field approach, we neglect the fluctuations of the fields $\vec{\varphi}$ and $\hat{h}$ in Eq. (8.29), substituting the mean field values for these fields into this equation. In the limit $m \to 0$ the mean field value of the field $\hat{h}$ (Eq. (8.19)) is $i\hat{h} = ih^{(0)} = w^{(0)}\rho^{(0)}$, and the function $\vec{\varphi}$ is given by expressions (4.14) and (4.17). Substituting the classical solution (4.17) into the integrand of expression (8.27), we find from Eqs. (8.26)–(8.29)

$$\Phi_N(\mathbf{R}|\lambda) = \frac{1}{2}\int d\hat{\mathbf{x}}\psi(s)\int d\hat{\mathbf{x}}' \hat{G}^{(1)}\{\hat{\mathbf{x}}, \hat{\mathbf{x}}', [w^{(0)}\rho^{(0)}]\}\psi(s'). \tag{8.30}$$

The function $\psi(s)$ is related to the mean field solution, $\varphi_{mf}(s)$, through Eq. (8.27). Using the mean field equation (4.1), we can recast this function in the form

$$\psi(s) \equiv (z_c/2)\varphi_{mf}^3(s). \tag{8.31}$$

Expression (8.30) can be further simplified by taking the Fourier transform:

$$\Phi_N(\mathbf{R}|\lambda) = \frac{V^{(0)}}{2} \int \frac{d\mathbf{q}_T}{(2\pi)^{3m}} |\psi_{\mathbf{q}_T}|^2 \hat{G}_{\mathbf{q}_T}^{(1)} \tag{8.32}$$

where $\hat{G}_{\mathbf{q}_T}^{(1)}$ and $\psi_{\mathbf{q}_T}$ are Fourier transforms of the functions $\hat{G}^{(1)}$ (Eq. (8.23)) and $\psi(s)$, respectively:

$$\hat{G}_{\mathbf{q}_T}^{(1)} \equiv \int d\hat{\mathbf{x}} \hat{G}^{(1)}\{\hat{\mathbf{x}}, 0; [w^{(0)}\rho^{(0)}]\} e^{i\mathbf{q}_T \cdot \mathbf{x}_T}, \tag{8.33}$$

$$\psi_{\mathbf{q}_T} \equiv \int d\mathbf{x}_T \psi(s) e^{i\mathbf{q}_T \cdot \mathbf{x}_T}. \tag{8.34}$$

The remainder of this section is devoted to the calculation of the integral in Eq. (8.34). We split the $\mathbf{x}_T$-integration into an angular integration and that over $|\mathbf{x}_T|$. It is convenient to perform first the angular integration with the aid of the equality (7.29). Substituting that into Eq. (8.34), we find in the limit $m \to 0$

$$\psi_{\mathbf{q}_T} = -\int_0^\infty ds \frac{d\psi(s)}{ds} J_0(|\mathbf{q}_T|\sqrt{2s}). \tag{8.35}$$

According to Eqs. (4.17) and (8.31), the function $\psi(s)$ can be written in the dimensionless form

$$\psi(s) = \frac{2^{1/2}}{z_c^{1/2} \overline{N}^{3/2}} \chi^3(t), \qquad t = \frac{s}{2a^2\overline{N}} \tag{8.36}$$

and the function $\psi_{\mathbf{q}_T}$ (Eq. (8.35)) has the dimensionless representation

$$\psi_{\mathbf{q}_T} = \frac{2^{1/2}}{z_c^{1/2} \overline{N}^{3/2}} \kappa(a^2\overline{N}\mathbf{q}_T^2) \tag{8.37}$$

where the function $\kappa(p)$ is related to $\chi^3(t)$ by the equation

$$\kappa(p) = -\int_0^\infty dt \frac{d\chi^3(t)}{dt} J_0(\sqrt{2pt}). \tag{8.38}$$

To calculate this integral, we use the theorem of Laplace analysis:

$$\kappa(p) = \int_0^\infty ds\, q(s) \exp\left(-\frac{p}{2s}\right), \qquad \chi^3(t) = \int_0^\infty ds\, q(s) e^{-st} \tag{8.39}$$

where $q(s)$ is the Laplace transform of the function $\chi^3(t)$. The asymptotic behavior $s \to 0$ of this function,

$$q(s) \sim s^{-3} \exp(-9/s) \tag{8.40}$$

can be found from the asymptotic behavior at $t \to \infty$ of the function $\chi(t)$ (Eq. (4.1)). We find from the second equation (8.39) that this function is normalized to unity and, therefore, can be considered as the distribution function of the parameter s. This parameter characterizes the stiffness of the network at the point where the labeled chain is connected to it: The larger the s, the stiffer the local environment of the network. Eq. (8.40) shows that both very stiff ($s \to \infty$) and very soft ($s \to 0$) regions are rarely encountered in the network.

Substituting Eqs. (8.37) and (8.39) into Eq. (8.32), we get

$$\Phi_N(\mathbf{R}|\lambda) = \frac{V^{(0)}}{z_c \overline{N}^3} \int_0^\infty ds\, q(s) \int_0^\infty ds'\, q(s')$$

$$\times \int \frac{d\mathbf{q}_T}{(2\pi)^{3m}} \hat{G}^{(1)}_{\mathbf{q}_T} \exp\left[-\frac{a^2 \overline{N} \mathbf{q}_T^2}{2}\left(\frac{1}{s} + \frac{1}{s'}\right)\right] \tag{8.41}$$

Note that two arguments s and s' (and two distribution functions $q(s)$ and $q(s')$) appear in this expression because the chain is connected to the network by its two ends, and the local stiffness at each end is characterized by the distribution function $q(s)$. It is clear from Eq. (8.41) that it is convenient to introduce the new variable

$$\xi \equiv a\overline{N}^{1/2}\left(\frac{1}{s} + \frac{1}{s'}\right)^{1/2} \tag{8.42}$$

which is an average measure of the local softness of the network in the neighborhood of a given chain. Eq. (8.41) can be rewritten in the form

$$\Phi_N(\mathbf{R}|\lambda) = \frac{V^{(0)}}{z_c \overline{N}^3} \int_0^\infty d\xi\, P(\xi) \int \frac{d\mathbf{q}_T}{(2\pi)^{3m}} \hat{G}^{(1)}_{\mathbf{q}_T} \exp\left(-\frac{\xi^2 \mathbf{q}_T^2}{2}\right) \tag{8.43}$$

where we have introduced the function

$$P(\xi) \equiv \int_0^\infty ds\, q(s) \int_0^\infty ds'\, q(s')\delta\left[\xi - a\overline{N}^{1/2}\left(\frac{1}{s} + \frac{1}{s'}\right)^{1/2}\right] \tag{8.44}$$

which depends on the dimensionless variable $\xi/(a\overline{N}^{1/2})$ only. Using the second Eq. (8.39), we can check that this function is normalized,

$$\int_0^\infty d\xi\, P(\xi) = 1 \tag{8.45}$$

and thus it can be considered as the distribution function of the variable ξ defined by Eq. (8.42). The universal plot of $P(\xi)$ as a function of $\xi/(a\overline{N}^{1/2})$ is shown in Figure 3.7. Substituting the definition (8.33) of the function $\hat{G}^{(1)}_{\mathbf{q}_T}$

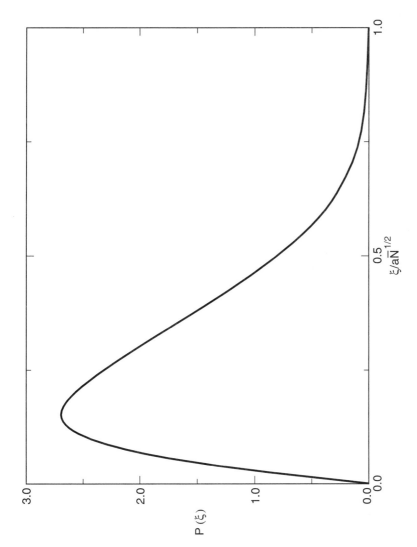

Figure 3.7 Plot of the probability distribution $P(\xi)$ of the random length ξ versus the dimensionless variable $\xi/(a\overline{N}^{1/2})$.

173

into Eq. (8.43) and performing the integration over $\mathbf{q}_T$, we arrive at the expression

$$\Phi_N(\mathbf{R}|\lambda) = \frac{V^{(0)}}{z_c \overline{N}^3} \int_0^\infty d\xi\, P(\xi) \int d\hat{\mathbf{x}}\, \hat{G}^{(1)}\{\hat{\mathbf{x}}, 0; [w^{(0)}\rho^{(0)}]\}\, \exp[-\mathbf{x}_T^2/(2\xi^2)]. \quad (8.46)$$

Using Eq. (8.23) and equality $\mu + w^{(0)}\rho^{(0)} = 1/\overline{N}$, we find the following expression for the function $G^{(1)}$:

$$\hat{G}^{(1)}\{\hat{\mathbf{x}}, 0; [w^{(0)}\rho^{(0)}]\} = \delta(\mathbf{x}^{(1)} - \mathbf{R}) \frac{e^{-N/\overline{N}}}{(2\pi a^2 N)^{3(1+m)/2}} \exp\left[-\sum_{k=0}^m \frac{(\mathbf{x}^{(k)})^2}{2a^2 N}\right]. \quad (8.47)$$

Substituting this into Eq. (8.46) and calculating Gaussian integrals over $\mathbf{x}^{(k)}$, we finally find

$$\Phi_N(\mathbf{R}|\lambda) = \frac{N_{ch}}{\overline{N}} e^{-N/\overline{N}} W_N\{R_\alpha|\lambda_\alpha\} \quad (8.48)$$

where the function W_N is given by the expression

$$W_N(\mathbf{R}|\lambda) = \int_0^\infty d\xi\, P(\xi) \prod_\alpha (2\pi\langle R_\alpha^2(N, \xi|\lambda_\alpha)\rangle)^{-1/2} \exp\left[-\frac{R_\alpha^2}{2\langle R_\alpha^2(N, \xi|\lambda_\alpha)\rangle}\right]. \quad (8.49)$$

The average projection of the α-th component squared of the end-to-end distance vector for a given value of ξ is defined by

$$\langle R_\alpha^2(N, \xi|\lambda_\alpha)\rangle = a^2 N \frac{1 + \lambda_\alpha^2 a^2 N/\xi^2}{1 + a^2 N/\xi^2}. \quad (8.50)$$

Note that this is the sum of a two contributions: The first one, proportional to λ_α^2, reflects the affine deformation of the average distance between the chain ends; the second one (independent of λ_α) is the contribution of thermal fluctuations that is not affected by the macroscopic deformation.

Substituting expression (8.48) into Eq. (8.8), we find the distribution function of chain lengths in the network

$$f(N) = \overline{N}^{-1} e^{-N/\overline{N}}. \quad (8.51)$$

Comparing Eqs. (8.48) and (8.51) with Eq. (8.7), we conclude that the function W_N (Eq. (8.49)) is the probability that a chain of contour length aN has components of the end-to-end distance vector R_α ($\alpha = x, y, z$).

3.8.4   SUMMARY

What is the physical interpretation of the length ξ? We know that ξ is a random variable characterized by the distribution function $P(\xi)$ and that its average value is of the order of the unperturbed size of the average chain (mesh size), $\overline{\xi} \simeq 0.27 a \overline{N}^{1/2}$ (see Figure 3.7). Further insight can be obtained from Eq. (8.50), which describes the response of a chain of contour length aN to a macroscopically induced deformation. The fact that the chain under consideration is coupled to a network, the local properties of which (e.g., density of cross-links) may vary from point to point due to the random character of the cross-linking process, is reflected in the appearance of the random length ξ. This length characterizes the local stiffness of the network at the points where the chain is coupled to it. We can distinguish between the following cases depending on the relation between the local stiffness of the network and the "spring constant" of the chain [12].

A. In the direction of elongation, $\lambda_\alpha > 1$, there are three regimes depending on N, ξ, and λ_α (these regimes become well-separated only in the limit $\lambda_\alpha \gg 1$).

　1. For $N \ll [\xi/(a\lambda_\alpha)]^2$, the size of the chain is essentially unaffected by the macroscopic elongation. For such short chains, macroscopic deformation can only lead to disinterpenetration of undeformed chains.

　2. Chains with number of monomers in the range $[\xi/(a\lambda_\alpha)]^2 \ll N \ll (\xi/a)^2$ are stretched nonaffinely, with $R_\alpha^2 = (a^2 N/\xi^2)\, a^2 N \lambda_\alpha^2 \ll a^2 N \lambda_\alpha^2$.

　3. Finally, when $N \gg (\xi/a)^2$, the chain stretches affinely with the applied strain, i.e., the condition $R_\alpha^2 = a^2 N \lambda_\alpha^2$ is satisfied. In this case, the rms distance between the ends of the chain is determined by the applied deformation and the fluctuational contribution is negligible.

B. In the direction of compression, $\lambda_\alpha < 1$, there are also three regimes of chain length.

　1. Chains with N smaller than $(\xi/a)^2$ interpenetrate without changing their undeformed size.

　2. When $[\xi/a\lambda_\alpha)]^2 \gg N \gg (\xi/a)^2$, the distance between chain ends is unaffected by compression and remains pinned down at ξ.

　3. For $[\xi/a\lambda_\alpha)]^2 \ll N$, the chains are compressed affinely with the network.

The most important geometric characteristic of the network is the average mesh size, which can be defined as the rms end-to-end distance of an average chain (of $\overline{N}$ monomers). It follows from our analysis that the average mesh size deforms affinely under macroscopic *stretching or swelling,* i.e.,

$$\overline{R}_{\text{mesh}}^{\text{str}} \approx a\overline{N}^{1/2}\lambda \tag{8.52}$$

but does not change under *compression* of the network (compression leads to enhanced interpenetration of the meshes [37], [38]):

$$\overline{R}_{\text{mesh}}^{\text{comp}} \approx a\overline{N}^{1/2}. \tag{8.53}$$

The above consideration leads to a deeper understanding of the deformation of individual network chains. We have seen that our mean field solution tells us that *the average positions of all the monomers change affinely with the macroscopic deformation of the network.* It also gives us the statistical information about the local stiffness of the network, which determines both the fluctuations of monomers about these mean positions and the fluctuations of the end to end distance of chains of a given contour length. *Deviations from affine displacement of average monomer positions occur only due to thermal fluctuations and take place only on length scales smaller than the mesh size.*

3.9 Discussion

In this work we presented a comprehensive statistical mechanical analysis of the Edwards model of arbitrarily deformed polymer networks, formed by instantaneous cross-linking of semidilute polymer solutions. The model takes into account both excluded volume interactions between the monomers and the random character of the process of cross-linking, but neglects permanent (topological) entanglements. Starting from a "microscopic" Hamiltonian, we have used replica field theory to account for the heterogeneous structure of polymer networks and obtained extensive statistical information about the macroscopic, mesoscopic, and microscopic behavior of polymer gels. Our solution of the statistical mechanics of this problem is on the same level of mathematical rigor as that of the well-established theory of polymer solutions.

We showed that the thermodynamic conditions—such as quality of solvent, degree of cross-linking, monomer concentration, and temperature—under which the network is prepared determine the statistical properties of its disordered structure. The inhomogeneous distribution of cross-links has a characteristic length scale that depends on the conditions of preparation and can vary from microscopic to macroscopic dimensions, depending on whether the gel is prepared away from or close to the cross-link saturation threshold. We found that *for each choice of thermodynamic conditions, a given network has a unique state of microscopic equilibrium* in which the average position of each cross-link and of each monomer is uniquely determined by the thermodynamic parameters. When these parameters (temperature, quality of solvent, degree of swelling, forces applied to the boundary of the gel) change, the new balance of elastic and excluded volume forces produces a new state of equilibrium. We have given a complete statistical characterization of this state, in terms of the moments of the equilibrium density profile. These monomer density correlation functions (correlators) can be directly measured in scattering experiments. The total structure factor can be represented as the sum of two terms: the correlator of static inhomogeneities, which characterizes the statistical properties of the inhomogeneous equilibrium density profile of the gel, and the correlator of thermal fluctuations about this equilibrium. The presence of static inhomogeneities gives rise to the observed stationary speckle patterns in light scattering from gels [34]. When the gel is stretched, the anisotropy of the inhomogeneous equilibrium density profile leads to enhanced scattering in the direction of the stretching and to the appearance of butterfly patterns in iso-intensity plots.

Throughout this chapter we dealt with the case of well-cross-linked networks and did not consider the transition between the liquid and amorphous solid states, which occurs at cross-link densities about one per chain (gel point, gelation threshold). Some aspects of this solidification transition are reviewed in [33] on the basis of the order parameter approach, and it is claimed that this "solidification" is a continuous thermodynamic phase transition. Actually, none of the thermodynamic parameters of the system can have any singularities at the gel point [39]. Gels prepared close enough to the gelation point have a percolating structure and the gel-formation transition is not a thermodynamic but rather a percolation-type transition. A general mean field consideration of gels prepared both close and far from the gel point was developed in [14]. The scaling description of such percolation networks was given in ref. [40] (for equilibrium physical gelation, when cross-links form and break up reversibly) and in ref. [11] (for chemical gels with permanent cross-links).

In this chapter we presented an RPA-level description of polymer networks that fails to account for strong short-wavelength thermal fluctuations of a network embedded in a good, low molecular weight solvent [41, 42]. (We have shown that such a problem does not arise for static inhomogeneities, which can be always described on the RPA level for networks prepared away from the cross-link saturation threshold.) A renormalization group approach [43] for dealing with such problems in the case of polymers with fixed structure was developed in ref. [44]. We adapted the methods in refs. [45], [46], and [3] (see also refs.

[47] and [48]) to the present case and obtained a scaling description of the thermodynamics of gels swollen in good solvents. We found that the equilibrium swelling state of gels cross-linked away from the cross-link saturation threshold is much more concentrated than the c^* state [4] and that there are many other network chains in the volume spanned by a chain that connects two adjacent cross-links. This means that the intuitive picture suggested by the c^* theorem, in which there is only one chain in the volume occupied by the average mesh, has to be replaced by one in which there are many chains per mesh volume and which corresponds to a semidilute solution of interpenetrating network chains.

Density correlation functions of gels in good solvent were calculated in ref. [3], both in the short- and long-wavelength limits. The application of semidilute solution ideas to cross-linked gels is based on the fact that whereas static density inhomogeneities take place only on scales comparable to or larger than the characteristic mesh size of the network, thermal fluctuations are dominated by small scale phenomena that are quite similar to those in semidilute polymer solutions. The absence of static inhomogeneities on these small length scales means that, in deriving the long-wavelength description (effective free energy) of the gel, the only contributions from the small scales will come from thermal fluctuations. These fluctuations can be taken into account within the framework of mean field theory if the system is described in terms of de Gennes' blobs instead of monomers.

Although replica field theory is a powerful tool for the study of the statistical mechanics of polymer gels, it has several drawbacks—the most important of which is its mathematical complexity. A more intuitive approach that captures the main physical ingredients of the more rigorous method is developed in ref. [49]. This approach has several important advantages, the foremost of which is that it does not require the use of advanced methods of mathematical physics (some of which are unfamiliar to the mainstream polymer community). Furthermore, although we have presented here and in ref. [3] the complete formal solution for the density correlators, explicit analytical results were obtained only in the long- and the short-wavelength limits. Thus, we were unable to describe the interesting phenomena associated with the transition from liquidlike to solidlike behavior that takes place on length scales of the order of the monomer fluctuation radius R (the typical length scale over which a monomer fluctuates about its mean position in the network). At length scales greater than R, the gel behaves as a usual elastic solid, but at shorter length scales its properties resemble those of a polymer liquid. The approach of ref. [49] is based on the separation of liquid- and solidlike degrees of freedom and gives a qualitatively correct picture of the behavior of gels on arbitrary spatial scales. Unlike a usual liquid, which is described by a single collective variable (the density ρ), a network must be characterized *by two independent thermodynamic variables: the density ρ and the displacement vector* **u**. This vector defines the displacement of a point of a macroscopic body with respect to its position in the undeformed reference state. The usual geometrical relation between these variables, $\delta\rho/\rho = -\nabla\mathbf{u}$, is not valid for the description of small-scale deformations and is restored only in the limit of deformations that vary infinitely slowly in space. The displacement vector **u** describes only the solidlike degrees of freedom of the network due to the displacements of the cross-links relative to their positions in the undeformed reference state. The liquidlike degrees of freedom are described by the monomer density $\rho(\mathbf{x})$ of the network subchains. The interplay between liquid- and solidlike degrees of the freedom is responsible for the richness of the finite-scale physics of polymer gels.

This approach has been extended to the description of thermal fluctuations and frozen heterogeneities in weakly charged gels [50]. We showed that the monomer density correlation functions of charged gels can be obtained from their neutral counterparts simply by replacing the second virial coefficients in the state of preparation and in the final observed state with effective interaction coefficients that contain a wavelength-dependent electrostatic contribution due to screened Coulomb interactions. We presented explicit analytic expressions for the total structure factor, for the correlator of static inhomogeneities and for the correlator of thermal fluctuations, in terms of the thermodynamic parameters in the state of preparation and in the final state of the gel. Some of our predictions, including the nonmonotonous dependence of the scattered intensity on the degree of cross-linking (for gels prepared in good solvent and studied in poor solvent), have been recently confirmed by light scattering experiments [51], [52].

Variation of thermodynamic conditions to which charged gels are subjected can result in dramatic macroscopic effects on their volume, shape, and permeability and in the appearance of complex patterns on the gel surface [53]. Recent neutron scattering experiments [54] support our prediction [49] that charged gels undergo microphase separation, which leads to the reorganization of their density profile on microscopic length scales.

We analyzed the thermodynamics of phase transitions in charged gels in poor solvents in [55] and demonstrated that *two types of phase transitions are possible in gels*. A gel can undergo a *volume phase transition* into a new homogeneous and isotropic phase by expelling the solvent and changing its volume. This type of transition is intimately related to the fact that the gel is a solid that has a unique volume under given thermodynamic conditions and has no analog in binary liquids. The second type of transition, which can take place in gels as well as in liquids, is *phase separation* into two coexisting phases of different densities. However, the presence of long-range elastic forces results in important differences between the thermodynamics and the kinetics of phase separation in gels and in binary liquids. Phase separation in liquids proceeds through nucleation or spinodal decomposition and results in the formation of coexisting bulk phases, each of which is isotropic and homogeneous. In gels, phase separation proceeds through the formation of an anisotropic phase on the surface of the gel [56]. Upon further change in the quality of solvent, this layer will initially grow at the expense of the bulk phase and this process will continue until both coexisting phases become strongly inhomogeneous. The above analysis was generalized in ref. [55] to the case of gels subjected to osmotic pressure and forces on their surface, and the corresponding phase diagrams were constructed.

Although the physical picture that emerges from our work is very different from that of the classical theories of polymer gels, many of our thermodynamic results (on the RPA level) agree qualitatively with the classical theories of elasticity of polymer networks [8, 10] and give rise to similar stress–strain relations. Such theories give a good description of the elasticity of swollen gels [57] but fail to predict the elastic response of dense networks for which Mooney-Rivlin corrections [58] have to be introduced. These corrections are usually attributed to entanglement effects, which are important for concentrated, sparsely cross-linked ($\overline{N} \gtrsim N_e$, where N_e is the entanglement length) networks not considered in this work.

Although our theory does not include some of the features of real polymer gels (such as entanglement contributions to elasticity [59] and the non-Gaussian character of real chains [60]), it does capture what we consider to be the most important characteristics of polymer gels: the frozen randomness of their structure introduced by the statistical character of their preparation and the interplay between short-range ("liquid") osmotic and long-range ("solid") elastic forces. Moreover, entanglement effects can be included by a proper generalization of the present model, which accounts for the effective "tube" introduced by the topological constraints [13, 61]. This approach predicts a transition from Mooney-Rivlin to Flory-type elasticity, with progressive swelling from the dense state of preparation to the semidilute, equilibrium swelling regime. Although our model strictly applies only to semidilute gels in which the second virial approximation holds (with the exception of the case of free chains dissolved in the network, where strong screening gives rise to a broader range of applicability of mean field arguments [3]), the generalization to the concentrated regime by replacing the second virial approximation to the osmotic free energy by an empirical concentration-dependent expression is straightforward.

Acknowledgments

S.P. would like to acknowledge the hospitality of the Department of Physics of Bar-Ilan University, where this work was done, and financial support from the Soros Foundation. This research was supported by grants from the Israeli Academy of Sciences and Humanities, the Israeli Ministry of Science and Technology, and the Research Authority of Bar-Ilan University.

Appendix 3.A   Functional Integrals

We start with the following fundamental identity for a *Gaussian* field $\psi(\mathbf{x})$:

$$\left\langle \exp\left\{\int d\mathbf{x} h(\mathbf{x})\psi(\mathbf{x})\right\}\right\rangle_\psi \equiv \int D\psi(\mathbf{x}) P[\psi(\mathbf{x})] \exp\left\{\int d\mathbf{x} h(\mathbf{x})\psi(\mathbf{x})\right\}$$

$$= \exp\left\{\frac{1}{2}\int d\mathbf{x}\int d\mathbf{x}' h(\mathbf{x}) g(\mathbf{x}, \mathbf{x}') h(\mathbf{x}')\right\} \tag{A.1}$$

where the averaging is performed with the weight

$$P[\psi] = \frac{\exp\left\{-\frac{1}{2}\int d\mathbf{x}\int d\mathbf{x}' \psi(\mathbf{x}) g^{-1}(\mathbf{x}, \mathbf{x}') \psi(\mathbf{x}')\right\}}{\int D\psi(\mathbf{x})\exp\left\{-\frac{1}{2}\int d\mathbf{x}\int d\mathbf{x}' \psi(\mathbf{x}) g^{-1}(\mathbf{x}, \mathbf{x}')\ \psi(\mathbf{x}')\right\}}. \tag{A.2}$$

In Eq. (A.1), $h(\mathbf{x})$ is an auxiliary vector field and g is an arbitrary positive definite operator. Its inverse, g^{-1}, is defined by

$$\int d\mathbf{x}'' g^{-1}(\mathbf{x}, \mathbf{x}'') g(\mathbf{x}'', \mathbf{x}') = \delta(\mathbf{x} - \mathbf{x}'). \tag{A.3}$$

Identity (A.1) can be proved by introducing a shift

$$\psi(\mathbf{x}) \rightarrow \psi(\mathbf{x}) + \int d\mathbf{x}' g(\mathbf{x}, \mathbf{x}') h(\mathbf{x}')$$

into the second term in Eq. (A.1). Differentiating the third term in this equation with respect to the field h and taking the limit $h = 0$ yields

$$g(\mathbf{x}, \mathbf{x}') = \frac{\delta^2}{\delta h(\mathbf{x})\delta h(\mathbf{x}')}\left\langle \exp\left\{\int d\mathbf{x} h(\mathbf{x})\psi(\mathbf{x})\right\}\right\rangle_\psi\Big|_{h=0}$$

$$= \int D\psi \psi(\mathbf{x})\psi(\mathbf{x}') P[\psi(\mathbf{x})] \equiv \langle \psi(\mathbf{x})\psi(\mathbf{x}')\rangle_\psi. \tag{A.4}$$

Here we have used the usual definition of a functional derivative

$$\delta I[h] \equiv I[h + \delta h] - I[h] = \int d\mathbf{x}\, \delta h(\mathbf{x}) \frac{\delta I[h]}{\delta h(\mathbf{x})} \tag{A.5}$$

which holds to first order in the arbitrarily small variation δh.

In the case when the function g depends only on the difference of its arguments, $g(\mathbf{x}, \mathbf{x}') = g(\mathbf{x} - \mathbf{x}')$, it is convenient to write Eqs. (A.1)–(A.4) in terms of the Fourier coefficients. We obtain

$$\left\langle \exp\left\{ \int \frac{d\mathbf{q}}{(2\pi)^3} h_{\mathbf{q}} \psi_{\mathbf{q}} \right\} \right\rangle_\psi = \exp\left\{ \frac{1}{2} \int \frac{d\mathbf{q}}{(2\pi)^3} \frac{h_{\mathbf{q}} h_{-\mathbf{q}}}{g_{\mathbf{q}}} \right\}. \tag{A.6}$$

The averaging is performed with the weight

$$P[\psi] = \exp\left\{ \frac{1}{2} \int \frac{d\mathbf{q}}{(2\pi)^3} \left[\ln\left(2\pi g_{\mathbf{q}}\right) - \frac{\psi_{\mathbf{q}} \psi_{-\mathbf{q}}}{g_{\mathbf{q}}} \right] \right\} \tag{A.7}$$

where the $\ln\left(2\pi g_{\mathbf{q}}\right)$ contribution comes from integrating the denominator in Eq. (A.2). In deriving the above equation we used the fact that Eq. (A.3) becomes a trivial identity in the Fourier representation, $g_{\mathbf{q}}^{-1} = 1/g_{\mathbf{q}}$. Eq. (A.4) transforms into

$$g_{\mathbf{q}} = \langle \psi_{\mathbf{q}} \psi_{\mathbf{q}} \rangle_\psi. \tag{A.8}$$

Appendix 3.B   Field Representation for Gaussian Chains

Equation (A.4) can be used to construct the field theoretical representation of the partition function of a Gaussian chain of N monomers with ends fixed at points $\mathbf{x}$ and $\mathbf{x}'$, in an external field $h(\mathbf{x})$. This function is given by the functional integral

$$G_N\{\mathbf{x}, \mathbf{x}'; [h]\} = \int_{\mathbf{x}}^{\mathbf{x}'} D\mathbf{x}(s)\, \exp\left\{ -\int_0^N ds\left[\frac{1}{2a^2}\left(\frac{d\mathbf{x}}{ds} \right)^2 + h(\mathbf{x}(s)) \right] \right\} \tag{B.1}$$

which is the solution of the diffusion—like equation

$$\left[\frac{\partial}{\partial N} - a^2 \nabla^2 + h(\mathbf{x}) \right] G_N\{\mathbf{x}, \mathbf{x}'; [h]\} = 0 \tag{B.2}$$

with the "initial" condition

$$G_N\{\mathbf{x}, \mathbf{x}'; [h(\mathbf{x})]\}|_{N=0} = \delta(\mathbf{x} - \mathbf{x}'). \tag{B.3}$$

It is convenient to introduce the grand canonical analog of this partition function by the Laplace transform:

$$G\{\mathbf{x}, \mathbf{x}'; [h]\} \equiv \int_0^\infty dN e^{-\mu N} G_N\{\mathbf{x}, \mathbf{x}'; [h]\} \qquad \text{(B.4)}$$

where μ is the chemical potential of monomers. Using the Laplace transform of Eq. (B.2) with the initial condition (B.3), it can be shown that this function obeys the equation

$$[\mu - a^2 \nabla^2 + h(\mathbf{x})] G\{\mathbf{x}, \mathbf{x}'; [h]\} = \delta(\mathbf{x} - \mathbf{x}'). \qquad \text{(B.5)}$$

Comparison of Eqs. (A.3) and (B.5) shows that the inverse operator $G^{-1}\{\mathbf{x}, \mathbf{x}'; [h]\}$ is defined by

$$G^{-1}\{\mathbf{x}, \mathbf{x}'; [h]\} = \delta(\mathbf{x} - \mathbf{x}')[\mu - a^2 \nabla^2 + h(\mathbf{x})]. \qquad \text{(B.6)}$$

We proceed to derive an explicit field theoretical expression for the grand canonical partition function of a Gaussian chain. Substituting the operator G^{-1} into the second term in Eq. (A.4), performing the integration over $\mathbf{x}'$ with the aid of the δ-function, and transforming the Laplacian into a square gradient by integration in parts, we finally obtain

$$G\{\mathbf{x}, \mathbf{x}'; [h]\} = \frac{\int D\varphi \varphi_1(\mathbf{x}) \varphi(\mathbf{x}') \exp\{-H_0[h, \varphi]\}}{\int D\varphi \exp\{-H_0[h, \varphi]\}} \qquad \text{(B.7)}$$

where the effective (dimensionless) Hamiltonian H_0 is defined as

$$H_0[h, \varphi] = \int d\mathbf{x} \left[\frac{1}{2} (\mu + h(\mathbf{x})) \varphi^2(\mathbf{x}) + \frac{a^2}{2} (\nabla \varphi(\mathbf{x}))^2 \right]. \qquad \text{(B.8)}$$

To avoid dealing with the denominator in Eq. (B.7), we introduce de Gennes' $n = 0$ model. The trick consists of introducing an n-component vector field $\vec{\varphi}(\mathbf{x})$ with components $\varphi_i(\mathbf{x})$; $i = 1, \ldots, n$ and noticing that Eq. (B.7) can be formally written as

$$G\{\mathbf{x}, \mathbf{x}'; [h]\} = \lim_{n \to 0} \left[\int D\varphi \exp\{-H_0[h, \varphi]\} \right]^{n-1}$$
$$\times \int D\varphi_1 \varphi_1(\mathbf{x}) \varphi_1(\mathbf{x}') \exp\{-H_0[h, \varphi_1]\}. \qquad \text{(B.9)}$$

Since, for integer n we can write

$$\left[\int D\varphi \exp\{-H_0[h, \varphi]\} \right]^{n-1} = \prod_{i=2}^n \int D\varphi_i \exp\{-H_0[h, \varphi_i]\}. \qquad \text{(B.10)}$$

Equation (B.9) can be recast into the simple expression

$$G\{\mathbf{x}, \mathbf{x}'[h]\} = \int D\vec{\varphi}\varphi_1(\mathbf{x})\varphi_1(\mathbf{x}') \exp\{-H_0[h, \vec{\varphi}]\} \tag{B.11}$$

where an analytic continuation over the number of components of the field $\vec{\varphi}$, from integer values of n to the limit $n = 0$, is implied. Here, $H_0[h, \vec{\varphi}]$ is defined by replacing φ by $\vec{\varphi}$ in $H_0[h, \varphi]$ and using

$$\vec{\varphi}^2(\mathbf{x}) \equiv \sum_{i=1}^{n} \varphi_i^2(\mathbf{x}), \qquad (\nabla\vec{\varphi}(\mathbf{x}))^2 = \sum_{i=1}^{n} (\nabla\varphi_i(\mathbf{x}))^2. \tag{B.12}$$

References

1. R. T. Deam and S. F. Edwards, *Philos. Trans. R. Soc. London* Ser. A **280**: 317 (1976).
2. M. Warner and S. F. Edwards, *J. Phys. A* **11**: 1649 (1978).
3. S. Panyukov and Y. Rabin, *Phys. Rep.* **269**: 1 (1996).
4. P.-G. de Gennes *Scaling Concepts in Polymer Physics,* Cornell University Press, Ithaca, NY, 1979.
5. P. W. Anderson, *Basic Notions of Condensed Matter Physics,* Benjamin, Reading, MA, 1984.
6. S. V. Panyukov, *JETP Lett.* **55**: 608 (1992).
7. S. F. Edwards and P. W. Anderson *J. Phys. (Paris)* **F5**: 1965 (1975).
8. P. J. Flory and J. Rehner, *J. Chem. Phys.* **2**: 521 (1943).
9. P. J. Flory, *Principles of Polymer Chemistry,* Cornell University Press, Ithaca, NY, 1971.
10. H. M. James and E. Guth, *J. Chem. Phys.* **11**: 455 (1943).
11. S. V. Panyukov, *JETP Lett.* **58**: 119 (1993).
12. S. Panyukov, Y. Rabin, and A. Feigel, *Europhys. Lett.* **28**: 149 (1994).
13. S. V. Panyukov, *Sov. Phys. JETP* **67**: 2274 (1988); S. V. Panyukov, *Sov. Phys. JETP* **69**: 342 (1989).
14. S. V. Panyukov, *Sov. Phys. JETP* **76**: 808 (1993).
15. M. Doi and S. F. Edwards, *The Theory of Polymer Dynamics* Clarendon Press, Oxford, 1986.
16. G. S. Grest, K. Kremer, and E. R. Duering, *Physica A* **194**: 330 (1993).
17. R. C. Ball and S. F. Edwards, *Macromolecules* **13**: 748 (1980).
18. S. F. Edwards and T. A. Vilgis, *Rep. Prog. Phys.* **51**: 243 (1988).
19. P. Goldbart and N. Goldenfeld, *Phys. Rev. A* **39**: 1412 (1989).
20. H. E. Castillo, P. M. Goldbart. and A. Zippelius, *Europhys. Lett.* **28** 519 (1994).
21. J. des Cloizeaux and G. Jannink, *Polymers in Solution. Their Modelling and Structure,* Clarendon Press, Oxford, 1990.
22. A. Zippelius, P. M. Goldbart, and N. Goldenfeld, *Europhys. Lett.* **23**: 451 (1993).
23. M. Mezard, G. Parisi, and M. Virasoro, *Spin Glass Theory and Beyond,* World Scientific, Singapore, 1987.
24. P. Goldbart and N. Goldenfeld, *Phys. Rev. Lett.* **58**: 2676 (1987).

25. W. Paul, K. Binder, D. W. Heerman, and K. Kremer, *J. Chem. Phys.* **95:** 7726 (1991).

26. P. M. Goldbart and A. Zippelius, *J. Phys. A: Math. Gen.* **27:** 6375 (1994).

27. L. D. Landau and E. M. Lifshitz, *Quantum Mechanics,* Pergamon, Oxford, 1984.

28. D. Forster, *Hydrodynamic Fluctuations, Broken Symmetry and Correlation Functions,* Addison-Wesley, New York, 1983.

29. D. Sherrington and S. Kirkpatrick, *Phys. Rev. Lett.* **32:** 1792 (1975).

30. H. M. James, *J. Chem. Phys.* **15:** 651 (1947).

31. Y. Rabin and R. Bruinsma, *Europhys. Lett.* **20:** 79 (1992); R. Bruinsma and Y. Rabin, *Phys. Rev. E.* **49:** 554 (1994).

32. P. M. Goldbart and A. Zippelius, *Phys. Rev. Lett.* **71:** 2256 (1993).

33. P. M. Goldbart, H. E. Castillo, A. Zippelius, *Adv. in Phys.* **45:** 393 (1996).

34. J. G. Joosten, J. L. McCarthy, and P. N. Pusey, *Macromolecules* **24:** 6690 (1991).

35. M. J. Orkisz, Ph. D. thesis (MIT, 1994).

36. A. Onuki, *J. de Phys. II* **2:** 45 (1992).

37. J. Bastide, C. Picot, and S. Candau, *J. Macromol. Sci. Phys.* **19:** 13 (1981).

38. M. Daoud, E. Bouchaud, and G. Jannink, *Macromolecules* **19:** 1955 (1986).

39. S. V. Panyukov, *Sov. Phys. JETP* **61:** 1065 (1985).

40. S. V. Panyukov, *Sov. Phys. JETP* **66:** 829 (1987).

41. A. Y. Grosberg and A. R. Khokhlov, *Statistical Physics of Macromolecules,* AIP Press, New York, 1994.

42. K. F. Freed, *Renormalization Group Theory of Macromolecules,* Wiley, New York 1987.

43. S.-K. Ma *Modern Theory of Critical Phenomena,* Benjamin, Reading, MA, 1976.

44. S. V. Panyukov, *Sov. Phys. JETP* **67:** 930 (1988).

45. S. V. Panyukov, *JETP Lett.* **51:** 253 (1990); S. V. Panyukov, *Sov. Phys. JETP* **71:** 372 (1990).

46. S. V. Panyukov, *Sov. Phys. JETP* **71:** 372 (1990).

47. S. P. Obukhov, M. Rubinstein, and R. H. Colby, *Macromolecules* **27:** 3191 (1994).

48. P. Pekarski, A. Thachenko, and Y. Rabin, *Macromolecules* **27:**7192 (1994).

49. S. Panyukov and Y. Rabin, *Macromolecules* **29:** 7960 (1996).

50. Y. Rabin and S. Panyukov, *Macromolecules* **30:** 301 (1996).

51. F. Ikkai and M. Shibayama, *Phys. Rev. E* **56:** R51 (1997).

52. M. Shibayama, F. Ikkai, Y. Shiva, and Y. Rabin, *J. Chem. Phys.* , in press.

53. M. Tokita and T. Tanaka, *Science* **253:** 1121 (1991).

54. M. Shibayama, T. Tanaka, and C. C. Han, *J. Chem. Phys.* **97:** 6842 (1992).

55. S. Panyukov and Y. Rabin, *Macromolecules* **29:** 8530 (1996).

56. K. Sekimoto, *Phys. Rev. Lett.* **70:** 4154 (1993).

57. A.-M. Hecht et al., *Macromolecules* **24:** 4183 (1991).

58. M. Mooney, *J. Appl. Phys.* **19:** 434 (1948); R. S. Rivlin, *Philos. Trans. R. Soc. London Ser. A* **241:** 379 (1948).

59. P. J. Flory and B. Erman, *Macromolecules* **15:** 801 (1982).

60. J. E. Mark and J. G. Curro, *J. Chem. Phys.* **79:** 5705 (1983).

61. S. V. Panyukov and I. I. Potemkin, *J. Phys. I (France)* **7:** 273 (1997).

Chapter 4 | Winding Angle Distributions for Directed Polymers

Barbara Drossel and Mehran Kardar

Department of Physics
Massachusetts Institute of Technology
Cambridge, Massachusetts

Abstract

In this chapter we study analytically and numerically the winding of directed polymers of length t *around each other or around a rod. Unconfined polymers in pure media have exponentially decaying winding angle distributions; the decay constant depends on whether the interaction is repulsive or neutral, but not on microscopic details. In the presence of a chiral asymmetry, the exponential tails become nonuniversal. In all these cases, the mean winding angle is proportional to* ln t. *When the polymer is confined to a finite region around the winding center—e.g., due to an attractive interaction—the winding angle distribution is Gaussian, with a variance proportional to* t. *We also examine the windings of polymers in random systems. Our results suggest that randomness reduces entanglements, leading to a narrow (Gaussian) distribution with a mean winding angle of the order of* $\sqrt{\ln t}$.

4.1 Introduction

The topological constraints produced by the windings of polymers [1] strongly affect the dynamics of polymer solutions. As a consequence of polymer entanglement, the viscosity of a solution of polymers above the overlap concentration is many orders of magnitude higher than the viscosity of the solvent. An analytical treatment of these topological constraints is extremely difficult. Theoretical efforts therefore focus on the limit of high polymer concentrations, where effective medium theories and the tube model successfully describe several aspects of the dynamics of the polymer solution [2], or on the limit of only one or two polymers, where the different possible configurations can be studied explicitly [3].

In this chapter, we take the latter approach, focusing on the winding of a directed polymer (DP) around a rod or of two DPs around each other, as shown in Fig. 4.1. DPs have a preferred direction $\hat{\tau}$, and their configuration can be described by the function $[\vec{r}(\tau)]$, with $\tau \in [0, t]$, where $\vec{r} = (x_1, x_2)$ is the coordinate

THEORETICAL AND MATHEMATICAL
MODELS IN POLYMER RESEARCH

Copyright © 1998 by Academic Press.
All rights of reproduction in any form reserved.
ISBN 0-12-304140-6/$25.00.

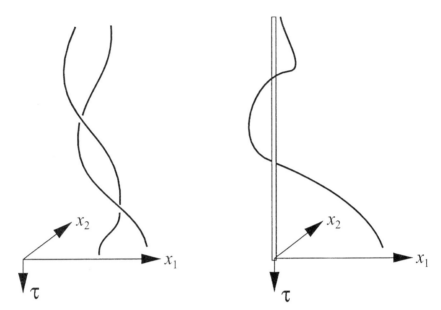

Figure 4.1 (a) Two directed polymers winding around each other. (b) A directed polymer
winding around a rod.

in the plane perpendicular to the preferred direction. Going to relative coordinates $\vec{r}_2(\tau) - \vec{r}_1(\tau)$, the winding of two DPs around each other can be mapped to the winding of a single DP around a rod (see Section 4.2). Since DPs cannot have knots, their main topological constraints are windings.

Although less common than flexible polymers, DPs are a good model of several semiflexible and rigid polymers. Examples are biological macromolecules such as DNA or liquid crystals composed of stacks of disk-shaped molecules. These polymers are aligned parallel to each other when their concentration is sufficiently large, forming crystalline and liquid crystalline phases (for a review on statistical mechanics of DPs, see ref. [4]). Isolated DPs can be realized by embedding a long polymer in a nematic solvent [5]. Another important class of directed "polymers" are magnetic flux lines in high-T_c superconductors that are oriented parallel to the direction of the external magnetic field. Due to the high temperature in the system and the weak coupling between different layers in the superconductor, thermal fluctuations of the flux lines are considerable, leading to entanglements [6, 7] and windings around columnar pins [8].

To calculate the winding angle distribution of DPs, we map them onto two-dimensional walks $[\vec{r}(\tau)]$, where the arc length τ plays the role of the time coordinate. The winding angle distribution depends on the interaction between the polymer and the winding center and on the properties of the embedding medium. In the following three sections, we discuss three different classes of winding angle distributions. The scaling variable is a different combination of the winding angle θ and the polymer length t for each class. In Section 4.2, we consider DPs in an infinitely large pure medium. These polymers can be mapped on ideal random walks, where the mean horizontal distance $\langle|\vec{r}(\tau) - \vec{r}(0)|\rangle$ from the starting point increases with the square root of τ. The number of returns to the winding center is proportional to $\ln t$ for such a random walk. We will see that because of the finite return probability to the winding center even for large times, the winding angle distribution depends on properties of the winding center. We will find different winding angle distributions depending on whether the interaction between the winding center and the polymer is a hard-core repulsion or is neutral, and on whether the winding center shows a chiral asymmetry. We will also see that the winding angle distribution does not depend on microscopic details such as the shape of the winding center or a possible underlying lattice structure.

In the limit $t \to \infty$, the probability distribution for the winding angle depends only on the combination $x = 2\theta/\ln(t)$ of the winding angle and the length of the walk, not on each variable separately. For large $|x|$, all three mentioned winding angle distributions decay exponentially in $|x|$. The scaling variable proportional to $\theta/\ln(t)$ can be explained as follows: After time t, the walker has a typical distance $r(t) \propto \sqrt{t}$ from the starting point, which is chosen to be close to the winding center. Assuming that $r(t)$ is the only relevant length scale, dimensional arguments—combined with the Markovian property—suggest that $dr/d\theta = rf(\theta)$. The rotational invariance of the system implies that $f(\theta)$ must be a constant; i.e., the increase in winding angle cannot depend on the number of windings or angular position. Hence,

$$d\theta \propto \frac{dr}{r} \propto \frac{dt}{t} = d(\ln t), \tag{1}$$

leading to a scaling variable proportional to $\theta/\ln t$.

If the interaction between the polymer and the winding center is attractive, the polymer can be bound to the winding center and its transverse wandering is limited. Polymers can also be confined by a finite container or by neighboring polymers. In all these cases, polymer segments of length Δt that are small compared to the total length t, but large compared to the length needed to make a winding, have identical winding angle distributions. Applying the central limit theorem, we conclude that the total winding angle distribution is Gaussian with a scaling variable θ^2/t. This situation will be discussed in Section 4.3.

Finally, we discuss in Section 4.4 certain DPs that cannot be described by ideal random walks since they do not satisfy a Markovian property. Polymers that are embedded in a random medium (e.g., a gel) have an energy that depends on the polymer configuration. Similarly, the energy of magnetic flux lines in high-T_c superconductors with point defects depends on their configuration. The mean distance from the starting point for these polymers in random media increases faster than for ideal walks since the line searches for low-energy configurations. The number of returns to the winding center remains finite in the limit of infinite length. Consequently, the interaction of the line with the winding center does not affect the winding angle distribution as long as it is not strong enough to bind the polymer.

We will see that the winding angle distribution of polymers in random media is Gaussian with a variance proportional to ln t, leading to a scaling variable $x \propto \theta/\sqrt{\ln(t)}$. This means that the pinning to randomness decreases the mean winding angle from the order of $\sqrt{\ln t}$ to the order of $\sqrt{\ln t}$. Interestingly, this winding angle distribution is similar to that of two-dimensional self-avoiding random walks [9]. The following scaling argument explains why the winding angle distribution in both situations is Gaussian with a variance proportional to ln t: Starting from the origin, divide the walk into segments of $1, 2, \ldots, 2^n \approx t/2$ steps. Since the αth segment is at a distance of roughly $2^{\alpha\nu}$ from the center (since $r \propto t^\nu$) and has a characteristic size of the same order, it is reasonable to assume that each segment spans a random angle θ_α of order 1. Under the mild assumption that the sum $\theta = \sum_{\alpha=1}^{n} \theta_\alpha$ satisfies the central limit theorem, we then conclude that θ is Gaussian distributed with a variance proportional to $n \propto \ln t$. Since this argument relies on the irrelevance of the winding center, it cannot be applied to the distributions in Section 4.2.

Many results of this chapter have been reported previously in ref. [10]. They point out the rich behavior already present in the simplest of problems involving topological defects. Properties of the winding center, interactions, and various types of randomness are all potentially relevant, leading to different universal distribution functions. In Section 4.5 we give an outlook on possible further universality classes and on the winding of nondirected polymers.

4.2   Winding Angle Distributions in an Infinite Homogeneous Medium

In this section we study winding angle distributions of DPs in infinite homogeneous media, all characterized by a scaling variable $x = 2\theta/\ln t$ and exponential tails. We keep the initial point of the polymers fixed, but otherwise allow them to move freely. The precise form of the winding angle distribution depends on the interaction with the winding center. In Subsection 4.2.1, we consider two DPs with hard-core repulsion, or, equivalently, one DP winding around a repulsive rod, leading to the distribution in Eq. (14). For neutral winding centers, the corresponding winding angle distribution given in Eq. (17) has a decay constant that is smaller by one-half (Subsection 4.2.2). These two distributions occur under fairly general conditions (see Subsection 4.2.3). However, when the symmetry with respect to the sign of the winding angle is broken, new (asymmetric) distributions occur (Subsection 4.2.4), with the decay constants of the exponential tails depending on the degree of chirality.

4.2.1   WINDING IN THE PRESENCE OF HARD-CORE REPULSION

Mapping to a Random Walk with Absorbing Boundary Conditions

The energy of a given configuration of two DPs $\vec{r}_1(\tau)$ and $\vec{r}_2(\tau)$ of length t is given by

$$E[\vec{r}_1(\tau), \vec{r}_2(\tau)] = \int_0^t d\tau \left[c\left(\frac{d\vec{r}_1}{d\tau}\right)^2 + c\left(\frac{d\vec{r}_2}{d\tau}\right)^2 + V(\vec{r}_1 - \vec{r}_2) \right]. \quad (2)$$

The potential $V(\vec{r})$ has a hard core, $V(r) = \infty$ for $r < a$ and $V(r) = 0$ for $r > a$. The first two terms are the elastic energies of the polymers, where the parameter c is related to their stiffness. Introducing the relative coordinate $\vec{r} = \vec{r}_1 - \vec{r}_2$ and the center-of-mass coordinate $\vec{R} = (\vec{r}_1 + \vec{r}_2)/2$, Eq. (2) becomes

$$E[\vec{R}(\tau), \vec{r}(\tau)] = \int_0^t d\tau \left[\frac{c}{2}\left(\frac{d\vec{r}}{d\tau}\right)^2 + 2c\left(\frac{d\vec{R}}{d\tau}\right)^2 + V(\vec{r}) \right].$$

The partition function for the two polymers is

$$Z = \int \mathcal{D}[\vec{R}(\tau)]\mathcal{D}[\vec{r}(\tau)] \, \exp\{-E[\vec{R}(\tau), \vec{r}(\tau)]/k_B T\}, \quad (3)$$

where the integral is taken over all possible configurations $[\vec{R}(\tau)]$ and $[\vec{r}(\tau)]$. The expression $\mathcal{D}[\vec{r}(\tau)]$ denotes a path integral and is the continuum limit of $\Pi_{i=1}^n (\int d\vec{r}(\tau_i))$, k_B is the Boltzmann constant, and T is the temperature.

   As long as we are only interested in quantities related to the relative coordinate (like the winding angle), we can integrate out the center-of-mass variations and focus on the partition function for the relative coordinate alone. That is,

$$Z = \int \mathcal{D}[\vec{r}(\tau)] \, \exp\left\{-\int_0^t d\tau \left[\frac{c}{2}\left(\frac{d\vec{r}}{d\tau}\right)^2 + V(\vec{r})\right]\middle/k_B T\right\}. \quad (4)$$

This is identical to the partition function for a single DP winding around a rod. Due to the hard-core repulsion, all configurations where the polymer and the rod penetrate each other do not contribute to the partition function ($V = \infty$), while $V = 0$ for all other configurations.

A two-dimensional random walk can be described by the Langevin equation

$$\frac{d\vec{r}}{dt} = \vec{\eta}(t),$$ (5)

where η is a stochastic force with zero mean ($\langle \eta(t) \rangle = 0$) and the correlation function $\langle \vec{\eta}(t)\vec{\eta}(t') \rangle = 2D\delta(t - t')$. The probability distribution of $\vec{\eta}$ is Gaussian; i.e.,

$$P[\eta(t)] \propto \exp[-D(\vec{\eta}(t))^2].$$

With Eq. (5), we find that the probability for a given trajectory $[\vec{r}(\tau)]$ of the random walk is proportional to

$$\exp\left\{-\int_0^t d\tau \left[D\left(\frac{d\vec{r}}{d\tau}\right)^2\right]\right\}.$$

When all walks that enter a region of radius a around the origin get absorbed, the probability that the random walk has a trajectory $[\vec{r}(\tau)]$ is identical to the probability that the above DP has the configuration $[\vec{r}(\tau)]$ (compare to Eq. (4), with $D = c/k_BT$). This correspondence between DPs with hard-core repulsion and random walks with absorbing boundary conditions was first pointed out by Rudnick and Hu [11].

Conformal Mapping of the Random Walk

Since we are interested in the winding angle of the random walk, it is convenient to perform a transformation such that the winding angle becomes one of the co-ordinates. To this purpose, we represent the walk $\vec{r}(t) = (x_1(t), x_2(t))$ by the complex number

$$z(t) = x_1(t) + ix_2(t).$$

The time evolution of each random walker satisfies

$$dz = \eta(t)dt,$$ (6)

where $\eta(t)$ is now complex, with

$$\langle \eta(t)\eta^*(t') \rangle = 2D\delta(t - t').$$ (7)

We now introduce the new variable ζ by the transformation

$$\zeta(t) = \ln z(t) = \ln r(t) + i\theta(t), \tag{8}$$

where $r = \sqrt{x_1^2 + x_2^2}$. Since $d\zeta = \eta(t)dt/z(t)$, the stochastic motion of the walker in the new complex plane is highly correlated to its location; i.e., the walk is no longer random. This feature can be removed by defining a new time variable

$$d\tilde{t} = \frac{dt}{|z(t)|^2} \tag{9}$$

for each walker, which leads to

$$d\zeta = \mu(\tilde{t})d\tilde{t} \quad \text{with} \quad \mu(\tilde{t}) = z^*(t)\eta(t). \tag{10}$$

Since

$$\langle \mu(\tilde{t})\mu^*(\tilde{t}')\rangle = 2D|z(t)|^2\delta(t - t') = 2D\delta(\tilde{t} - \tilde{t}'), \tag{11}$$

the evolution of $\zeta(\tilde{t})$ is that of a random walk. Under the transformation in Eq. (8), the absorbing disc in the z-plane maps onto an absorbing wall in the ζ-plane (see Fig. 4.2).

For simplicity we choose the initial condition $\zeta(t = \tilde{t} = 0) = 0$; i.e., the original walker starts out at $z = 1$. We also set the diffusion constant to $D = 1/2$, so that the mean square distance over which the walker moves during a time t is $\langle r^2(t)\rangle = t$. Consequently, the probability that $r(t)$ is within an interval $[\sqrt{\pi t}^{(1-\epsilon)/2}, \sqrt{\pi t}^{(1+\epsilon)/2}]$ around its mean value of $\sqrt{\pi t}$ is

$$p(t, \epsilon) = \int_{\sqrt{\pi t}^{(1-\epsilon)/2}}^{\sqrt{\pi t}^{(1+\epsilon)/2}} \frac{\exp(-r^2/2t)}{2\pi t} 2\pi r\, dr$$

$$= \int_{\pi t^{-\epsilon/2}}^{\pi t^{\epsilon/2}} \exp(-s)ds, \tag{12}$$

and approaches unity in the limit $t \to \infty$. The effect of the absorbing disc on this probability can be neglected in the limit $t \to \infty$ since the disc becomes smaller when viewed from larger distances. In this limit, the distance r from the starting point $z = 1$ is identical to the distance from the origin, and $p(t,\epsilon)$ is identical to the probability that $\zeta(\tilde{t})$ is in the interval $[0.5(1 - \epsilon)\ln t, 0.5(1 + \epsilon)\ln t]$. So the endpoints of all walks (except for an infinitesimal fraction) that take a time t in the original plane map within a strip of width $\epsilon \ln t$ in the ζ-plane, as indicated in Fig. 4.2. If we shrink the complex plane ζ by a factor of $(\ln t)/2$, the walker is within a distance ϵ of the line with real value of unity. Thus, all walks of length t in the z-plane are mapped onto walks that end at the line with real value of unity,

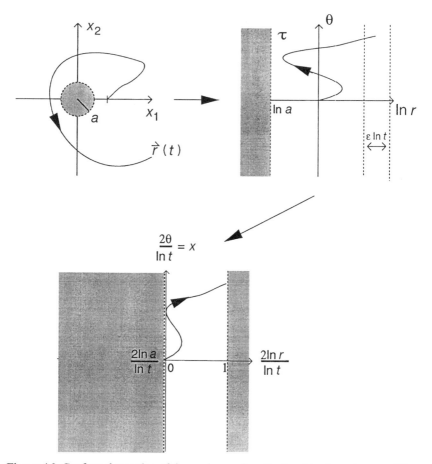

Figure 4.2 Conformal mapping of the random walk with absorbing boundary conditions and subsequent rescaling.

without having gone beyond (see Fig. 4.2). Since there is a separate transformation $\tilde{t}(t)$ for each walker, walks of the same length t map onto walks of different length $\tilde{t}$. (To be precise, we also have to shrink the time scale $\tilde{t}$ when shrinking the ζ-plane, but for simplicity we denote the new time again by $\tilde{t}$.)

Calculation of the Winding Angle Distribution

The imaginary (or vertical) coordinate x in the rescaled ζ-plane is related to the winding angle by $x = 2\theta(t)/\ln t$. To obtain the winding angle distribution, we have to determine the vertical position of a random walk starting at the origin, at the moment when it reaches for the first time the wall at distance 1 from the vertical axis, without going beyond the absorbing wall at distance $2|\ln a|/\ln t$ on the opposite side. Since we are interested in a walk only up to the moment when it reaches the right-hand wall, we can consider this wall also as absorbing. Since walks of length t in the original plane map onto walks of different length $\tilde{t}$ in the new plane, we need the probability that the walk is absorbed at this wall before time $\tilde{t}$.

We formulate this problem more generally and determine the probability $P_{\alpha,\beta}(y, \tilde{t})$ that a one-dimensional random walk starting at y between two absorbing points α and β at time 0 is absorbed at the point β before time $\tilde{t}$. Since for sufficiently small $\Delta\tilde{t}$ the walker is only a short distance Δy from its starting point, we have

$$P_{\alpha,\beta}(y, \tilde{t}) = \int_0^\infty d(\Delta y) \frac{1}{\sqrt{2\pi\Delta\tilde{t}}} \exp\left[-\frac{(\Delta y)^2}{2\Delta\tilde{t}}\right]$$

$$\times [P_{\alpha,\beta}(y + \Delta y, \tilde{t} - \Delta\tilde{t}) + P_{\alpha,\beta}(y - \Delta y, \tilde{t} - \Delta\tilde{t}].$$

Expanding the above equation to the order of $\Delta\tilde{t}$ indicates that $P_{\alpha,\beta}(y, \tilde{t})$ satisfies a diffusion equation. The appropriate boundary conditions are $P_{\alpha,\beta}(\alpha, \tilde{t}) = 0$ and $P_{\alpha,\beta}(\beta, \tilde{t}) = 1$ with the initial value $P_{\alpha,\beta}(y, 0) = 0$, resulting in [12]

$$P_{\alpha,\beta}(y, \tilde{t}) = \frac{y - \alpha}{\beta - \alpha} + \frac{2}{\pi} \sum_{v=1}^\infty \frac{(-1)^{v+1}}{v} \sin\left(\frac{\pi v(y - \alpha)}{\beta - \alpha}\right) \times \exp\left[-\frac{1}{2}\left(\frac{\pi v}{\beta - \alpha}\right)^2 \tilde{t}\right].$$

The probability that the walk is absorbed at the right-hand boundary during the time interval $[\tilde{t}, \tilde{t} + d\tilde{t}]$, is $d\tilde{t}\, \partial_{\tilde{t}} P_{\alpha,\beta}(y, \tilde{t})$.

Note, however, that

$$\int_0^\infty d\tilde{t}\; \partial_{\tilde{t}} P_{\alpha,\beta}(y, \tilde{t}) = (y - \alpha)/(\beta - \alpha);$$

i.e., it is equal to the total fraction of particles absorbed at the right-hand boundary (inversely proportional to the separations from the boundaries). To calculate the winding angle distribution $p_A(x)$, we need the fraction of these walks absorbed between $\tilde{t}$ and $\tilde{t} + d\tilde{t}$, equal to $((\beta - \alpha)/(y - \alpha))\partial_{\tilde{t}} P$. Hence (with $\alpha = 2 \ln a/\ln t$, $\beta = 1$ and $y = 0$),

$$p_A(x) = \int_0^\infty d\tilde{t}\; \frac{1 - \alpha}{-\alpha} \frac{\partial P_{\alpha,1}(0, \tilde{t})}{\partial \tilde{t}} \frac{\exp(-x^2/2\tilde{t})}{\sqrt{2\pi\tilde{t}}}$$

$$= \int_0^\infty d\tilde{t} \sum_{v=1}^\infty \frac{(-1)^{v+1}}{\sqrt{2\pi\tilde{t}}} \frac{\pi v}{\alpha(1 - \alpha)} \sin\left(\frac{\pi v\alpha}{1 - \alpha}\right) \times \exp\left[-\frac{1}{2}\left(\frac{\pi v}{1 - \alpha}\right)^2 \tilde{t} - \frac{x^2}{2\tilde{t}}\right]$$

$$= \sum_{v=1}^\infty \frac{(-1)^{v+1}}{\alpha} \sin\left(\frac{\pi v\alpha}{1 - \alpha}\right) \exp\left[-\frac{\pi v|x|}{(1 - \alpha)}\right].$$

The last step is achieved by first performing a Fourier transform with respect to x, followed by integrating over $\tilde{t}$, and finally inverting the Fourier transform. (Alternatively, the $\tilde{t}$ integration can be performed by the saddle point method.) In the limit of large t, the variable α is very small, and we can replace the sine function by its argument.

Taking the sum over ν, we find

$$p_A(x) = \frac{\pi}{(1 - \alpha)} \frac{\exp[\pi x/(1 - \alpha)]}{\{\exp[\pi x/(1 - \alpha)] + 1\}^2}. \tag{13}$$

Changing the variable from x to

$$\tilde{x} = \frac{x}{(1 - \alpha)} = \frac{2\theta}{\ln(t/a^2)},$$

and noting that $p_A(x)dx = p_A(\tilde{x})d\tilde{x}$, leads from Eq. (13) to

$$p_A\left(\tilde{x} = \frac{2\theta}{\ln(t/a^2)}\right) = \frac{\pi}{4\cosh^2(\pi\tilde{x}/2)}. \tag{14}$$

The above distribution, which is exact in the limit $t \to \infty$, has an exponential decay for large $\tilde{x}$, as first derived in ref. [11]. The complete form of Eq. (14) was first given in ref. [13], however, without derivation. The analogy to random walkers in the plane ζ, confined by the two walls, provides simple physical justifications for the behavior of the winding angle. In the presence of both walls, the diffusing particle is confined to a strip, and loses any memory of its starting position at long times. The probability that a particle that has already traveled a distance θ in the vertical direction proceeds a further distance $d\theta$ without hitting either wall is thus independent of θ, leading to the exponential decay.

Comparison to the Winding Angle Distribution Around a Point Center

The method described in this section was used earlier to derive the winding angle distribution for Brownian motion around a point center [14]. The resulting probability distribution for the winding angle in this case is [15]

$$\lim_{t \to \infty} p\left(x = \frac{2\theta}{\ln t}\right) = \frac{1}{\pi} \frac{1}{1 + x^2}, \tag{15}$$

leading to an infinite mean winding angle. Since there is no confining wall on the left-hand side, the particles may diffuse arbitrarily far in that direction, making it less probable to hit the wall on the right-hand side. In the original z-plane, the walker takes no time at all to make an infinitely small winding around the point center. This is clearly an unphysical feature, since real winding centers are finite and since real random walks (or polymers) need a finite time (a finite length segment) to make a winding. We therefore do not consider this situation any further.

Exercise: Derive Equation (15), repeating the calculation of this section, but with no absorbing wall (see ref. [10]).

4.2.2 WINDING OF DIRECTED POLYMERS AROUND NEUTRAL WINDING CENTERS

Instead of having first a rod and then inserting the DP into the system, we can also first have a free configuration of a DP and then insert a rod into it. If the polymer cannot relax to its thermal equilibrium distribution after insertion of the rod (e.g., because its ends are fixed or because its configuration is frozen), the resulting winding angle distribution will be different from that in the previous subsection. No configuration of the polymer is forbidden, but those configurations that interfere with the rod become *deformed*. The degree of deformation may depend on the diameter of the rod, but the winding angle does not. Alternatively, we could consider a winding center that has no interaction at all with the polymer (e.g., a light beam) or some structural defect in the solvent that is not felt by the polymer. In this case we would find the same winding angle distribution as in the case of a rod that deforms the polymer. In the language of a random walk, this situation corresponds to having a disc that reflects all walks that hit it. The walks that would go through the disc thus become deformed, but are not removed from the statistical ensemble.

We can obtain the winding angle distribution by repeating the calculations of the previous subsection, but replacing the absorbing boundary condition $P_{\alpha,\beta}(\alpha, \tilde{t}) = 0$ with the reflecting condition $\partial P_{\alpha,\beta}(y, \tilde{t})/\partial y|_{y=\alpha} = 0$, leading to

$$P_{\alpha,\beta}(y,\tilde{t}) = 1 - \frac{2}{\pi} \sum_{\nu=0}^{\infty} \frac{1}{\nu + 1/2} \sin\left(\frac{\pi(\nu + 1/2)(\beta - y)}{\beta - \alpha}\right) \times \exp\left[-\frac{1}{2}\left(\frac{\pi(\nu + 1/2)}{\beta - \alpha}\right)^2 \tilde{t}\right].$$

$$(16)$$

There is thus no current leaving the system at point α, and walkers that hit the winding center are reflected. We then find the winding angle distribution

$$p_R(\tilde{x}) = \frac{1}{2\cosh(\pi\tilde{x}/2)},$$

$$(17)$$

where again $\tilde{x} = 2\theta/\ln(t/a^2)$, and the limit $t \to \infty$ has been taken. For large $\tilde{x}$, where the walk has lost the memory of its initial distance from both walls, this probability decays exponentially as $\exp[-\pi\tilde{x}/2]$, i.e., exactly half as fast as for absorbing boundary conditions. A random walk confined between an absorbing and a reflecting wall that have a distance 1 can be mapped to a random walk confined between two absorbing walls at distance 2. After rescaling the wall distance and the $\tilde{x}$-coordinate by 2, this explains the factor 1/2 between the decay constants in the tails of the distributions in Eqs. (14) and (17).

Exercise: Derive Equation (17) (see ref. [10]).

4.2.3 UNIVERSALITY OF THE WINDING ANGLE DISTRIBUTION

The winding angle distribution in Eq. (17), which we derived in the previous subsection for Brownian motion around a reflecting disc, was obtained previously by several authors in different contexts. For example, Bélisle [16] calculated the winding angle distribution both for a random walk on a two-dimensional lattice around a point that is different from any lattice site and for a random walk with steps of finite size taken in arbitrary directions around a point in the two-dimensional plane, obtaining in both cases Eq. (17). The same result was obtained by Pitman and Yor [17] for the distribution of "big windings" of Brownian motion around two pointlike winding centers. Comtet *et al.* [18] divided the two-dimensional plane into three concentric sections and determined the contribution of each section to the winding angle for Brownian motion around a point, finding Eq. (17) for the contribution of the outer section.

This universality seems surprising since one might expect that the main increase in winding angle occurs when the walk is close to the winding center, where details like the lattice symmetry and the shape and size of the winding center determine how much time it takes to make one winding. However, a careful look at Fig. 4.2 reveals that this is not the case: The main increase in winding angle does not occur when the walker is within a small distance from the left-hand wall. Since all distances have been scaled by $1/\ln(t)$, a small distance from the left-hand wall corresponds to a large distance (of the order of $\ln t$) from the winding center. Therefore, almost all windings are made far from the winding center, where microscopic details do not matter. The properties that do affect the winding angle distribution are conservation laws (absorbing or reflecting boundary conditions), symmetries (with respect to the sign of the angle—see the following subsection), singularities (as for the winding of Brownian motion around a point center), and interactions (self-avoidance) or randomness (see Section 4.4).

To further test this universality hypothesis, we determined numerically the winding angle distribution for a random walk on a lattice with reflecting and absorbing boundary conditions. Reflecting boundary conditions are realized by choosing a winding center different from the vertices of the lattice, which is thus never crossed by the walker (this is exactly the situation treated analytically in ref. [16]). On the other hand, to model absorbing boundary conditions, the winding center is chosen as one of the lattice sites (say, the origin), but no walk is allowed to go through this point.

The winding angle distributions are most readily obtained using a transfer matrix method that calculates the number of all walks with a given winding angle and a given endpoint after t steps from the same information after $t-1$ steps. The winding center is at $(0.5, 0.5)$ for reflecting boundary conditions and at the origin for absorbing boundary conditions. The walker starts at $(1, 0)$, and the winding angle is increased or decreased by 2π every time it crosses the positive branch of the x_1-axis. Due to limitations in computer memory, we applied a cutoff in system size and winding angle for times $t > 120$, making sure that the results were not affected by this approximation. The largest times used, $t = 9728$, required approximately three days to run on a Silicon Graphics Indy Workstation.

Figures 4.3 and 4.4 show the results for the two cases. The asymptotic exponential tails predicted by theory can clearly be seen; deviations from the theoretical curve for smaller values of the scaling variable $x = 2\theta/\ln(2t)$ are due to the slow convergence to the asymptotic limit. Since the scaling variable depends logarithmically on time, the asymptotic limit is reached only for large $\ln t$. Note that

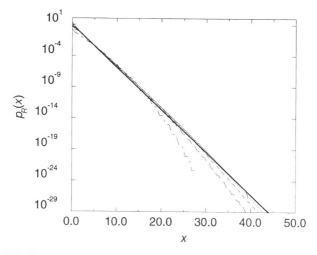

Figure 4.3 Winding angle distribution for random walks on a square lattice with *reflecting* boundary conditions for $t = 38$ (dot-dashed), 152 (long dashed), 608 (dashed), 2432 (dotted), and 9728 (solid). The horizontal axis is $x = 2\theta/\ln(2t)$. The thick solid line is the analytical result of Eq. (17).

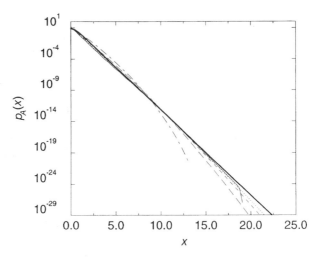

Figure 4.4 Winding angle distribution for random walks on a square lattice with *absorbing* boundary conditions. The symbols and the variable x are the same as in Figure 4.3. The thick solid line is the analytical result of Eq. (14).

the only free parameter in fitting to the analytical form is the characteristic timescale appearing inside the logarithm. With t measured in units of single steps on the lattice, we found that a factor 2 in the scaling variable provides the best fit. In the limit $t \to \infty$, different scales of t of course give the same asymptotic winding angle distribution.

Exercise: Perform the numerical calculations mentioned in this section. Study also the case of absorbing and reflecting winding centers that comprise several lattice points. How does the size of the winding center affect the convergence toward the asymptotic winding angle distribution?

We also studied the winding of a DP proceeding along the diagonal of a cubic lattice in three dimensions (see Fig. 4.5). The polymer starts at $(1, 0, 0)$ and at each step increases one of its three coordinates by 1. We determined the winding angle distribution around the diagonal $(1,1,1)$–direction, excluding from the walk all points that are on this diagonal (a repulsive columnar defect, corresponding to the case of an absorbing winding center). The excluded points lie on the origin when the polymer is projected in a plane perpendicular to the diagonal. In this plane, the polymer proceeds along the bonds of a triangular lattice, alternating between the three different sublattices. A cutoff of 243 in system size was imposed for the transfer matrix calculations. The winding angle distribution $p(x = 2\theta/\ln t)$ is shown in Fig. 4.6 for different times. As for the square lattice, an

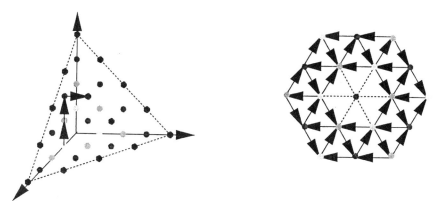

Figure 4.5 (a) Construction of a polymer directed along the (1, 1, 1)-diagonal of a cubic lattice. (b) Projection into the plane perpendicular to the (1, 1, 1)-diagonal. The three sublattices are indicated by different shades of gray.

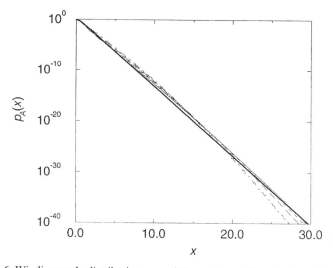

Figure 4.6 Winding angle distribution around the preferred direction for a flux line (directed path) in three dimensions for $t = 243$ (dot-dashed), 729 (long dashed), 2187 (dashed), 6561 (dotted), and 19,684 (solid). The horizontal axis is $x = 2\theta/\ln(2t)$. The thick solid line is the analytical result of Eq. (14).

exponential tail with decay constant of π can be seen. Our numerical results, as well as the analytical considerations, thus indicate clearly that the winding angle distributions for reflecting and absorbing boundary conditions are universal and do not depend on microscopic details.

Due to the special properties of directed paths along the diagonal of the cube, the case of reflecting boundary conditions leads to an asymmetry between windings in positive and negative directions. This is because it takes only three steps to make the smallest possible winding in one direction, but six steps in the opposite direction. This situation is discussed in detail in the following subsection.

4.2.4 WINDING CENTERS WITH CHIRAL ASYMMETRY

So far we have considered only situations that are symmetric with respect to the angles $\pm\theta$. For directed paths on certain lattices, however, this symmetry is broken. A directed walk that proceeds at each step along the $+x_1$, $+x_2$, or $+x_3$ direction on a cubic lattice can be mapped onto a random walker on a two-dimensional triangular lattice, as indicated in Figure 4.7(a). Each bond can be crossed in only one direction, and the winding center for reflecting boundary conditions must be different from the vertices of the lattice. It is apparent from this figure that the random walker can go around the center in three steps in one angular direction, but in no less than six steps in the other direction.

An alternative description is obtained by examining the position of the walker after every three time steps. The resulting coarse-grained random walk takes

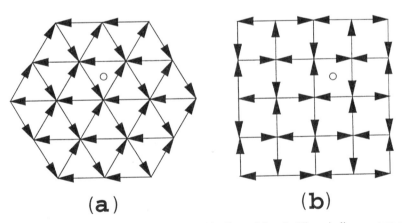

Figure 4.7 Triangular and square lattices with directed bonds. The winding centers are indicated by a circle ($\bigcirc$).

place on a regular triangular lattice, but now the walker has a finite probability of $3 \times 2/3^3 = 2/9$ of staying at the same site. If this site is one of the three points next to the winding center, the winding angle is increased by 2π in one of the six possible configurations that return to the site after three steps. In other words, the walker has a finite probability of having its winding angle increased in the proximity of the center. The amount of this biased increase in angle depends on the structure of the lattice and will be different for other directed lattices. An equivalent physical situation occurs for Brownian motion around a rotating winding center, e.g., a rotating reflecting disc that does not set the surrounding gas or liquid into motion (see Fig. 4.8).

Since this angular symmetry breaking is already present in the above simple example of a directed walk, it is quite likely to occur in more realistic physical systems, such as with screw dislocations in the underlying medium. The winding angle distribution for two *chiral* polymers [5] should also show an angular asymmetry. We thus use the term *chiral winding center* to indicate that each time the

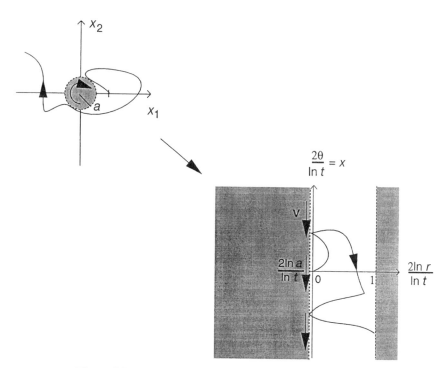

Figure 4.8 Brownian motion around a rotating winding center

polymer comes close to the center, it finds it easier to wind around in one direction as opposed to the other. (Of course, to respect the reflecting boundary conditions, there must be no additional interaction with the winding center. The case of additional attractive or repulsive interaction with the winding center will be discussed later.)

After mapping to the rescaled ζ-plane introduced in Section 4.2, the above situations can be modeled by a downward-moving, reflecting wall on the vertical axis. Each time a random walker hits this wall, its vertical position $x = 2\theta/\ln t$ is changed by a small amount, $2\Delta\theta/\ln t$. Let us now determine the net shift Δx in x due to the motion of the wall for a walker that survives for a time $\tilde{t}$ in the rescaled ζ-plane, before it is absorbed at the right-hand wall. (Recall that t is the time in the original system, while $\tilde{t}$ refers to the time in the rescaled ζ-plane, after the conformal mapping.)

To obtain the full solution, it is necessary to solve the two-dimensional diffusion equation with moving boundary conditions. Since we are mainly interested in the exponential tails of the winding angle distribution, we restrict our analysis to the limit of large times $\tilde{t}$ and determine the shift in x due to encounters with the reflecting moving wall in this limit. A Brownian walker that has survived for a sufficiently long time $\tilde{t}$ forgets its initial horizontal position. The mean number of encounters with the reflecting wall, and consequently the shift Δx in x due to the motion of the wall, is then expected to be simply proportional to the considered time interval. Applying the central limit theorem in the limit $\tilde{t} \to \infty$, the probability distribution of Δx is given by

$$p_\Delta(\Delta x) = \frac{1}{\sqrt{2\pi\beta^2\tilde{t}}} \exp\left[-\frac{(\Delta x - \alpha\tilde{t})^2}{2\beta^2\tilde{t}}\right]. \tag{18}$$

The parameters α and β are related to the velocity v of the wall (chirality of the defect) by $\alpha \propto \beta \propto v$. Presumably, Eq. (18) can be obtained directly from properties of random walks, providing the exact coefficients for these proportionalities.

The tail of the winding angle distribution is then given by

$$p_R^c(x) \propto \int_0^\infty d\tilde{t} \int_{-\infty}^\infty d(\Delta x) \frac{\partial P_{\alpha,1}(0,\tilde{t})}{\partial \tilde{t}} \frac{1}{\sqrt{2\pi\tilde{t}}} \exp\left[-\frac{(x - \Delta x)^2}{2\tilde{t}}\right]$$

$$\propto \int_0^\infty d\tilde{t} \exp[-(\pi^2/4)\tilde{t}/2] \frac{1}{\sqrt{2\pi\tilde{t}(1+\beta^2)}} \exp\left[-\frac{(x - \alpha\tilde{t})^2}{2\tilde{t}(1+\beta^2)}\right]$$

$$= \exp\left[\frac{\alpha x}{1+\beta^2} - \frac{|x|}{1+\beta^2}\sqrt{\alpha^2 + \frac{\pi^2}{4}(1+\beta^2)}\right], \tag{19}$$

which is valid for large $|x|$. The second factor on the right-hand side of the first line of Eq. (19) is the probability that the random walker is absorbed at the right-hand wall at time $\tilde{t}$ (see Eq. (16)) and for large $\tilde{t}$ is dominated by the slowest mode ($v = 0$). The second factor is the probability that the random walk would have the vertical coordinate x after time $\tilde{t}$ if there were no motion of the wall. The effect of the moving wall on the winding angle distribution is thus a systematic shift in the slopes of the exponential tails. For small values of chirality, the slopes on the two sides are changed to $\pi/2 \pm \alpha$. Due to this explicit velocity dependence, these asymmetric distributions are clearly nonuniversal. At large chiralities, the slopes vanish as α/β^2, resulting in quite wide distributions. Apparently, strong chirality of a defect increases the probability of entanglements. Figure 4.9 shows our simulation results for the winding angle distribution for a walk on the above-mentioned directed triangular lattice. The asymmetry due to the shift is clearly visible, and the winding angle distribution is wider than for a stationary wall. This case thus exemplifies the strong chirality limit discussed in the previous paragraph. We also simulated a square lattice with directed bonds, as indicated in Fig. 4.7(b). The corresponding winding angle distribution is shown in Fig. 4.10. The distribution is again asymmetric, but

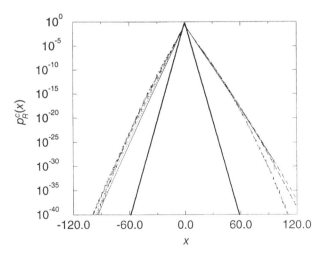

Figure 4.9 Winding angle distribution around the preferred direction for a random walk on a directed triangular lattice for $t = 243$ (dot-dashed), 729 (long dashed), 2187 (dashed), 6561 (dotted), and 19,684 (solid). The scaling variable is $x = 2\theta/\ln(2t)$. The thick solid line is the distribution given in Eq. (17).

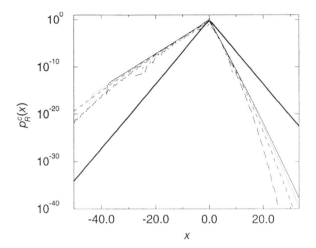

Figure 4.10 Winding angle distribution for a random walk on a directed square lattice for $t = 38$ (dot-dashed), 152 (long dashed), 608 (dashed), 2432 (dotted), and 9728 (solid). The thick solid line is the distribution given in Eq. (17).

not as wide as in the previous case and more similar to that expected in the weak chirality regime.

So far, we have assumed that the winding angle changes by the same amount each time the walker returns to the winding center. It is more realistic to assume that the change in winding angle has a certain probability distribution. A possible example is provided by polymers with randomly changing chirality [19]. In the limit $t \rightarrow \infty$, the total change in winding angle due to the chirality is on an average

$$\langle \Delta\theta \rangle = \int_{-\infty}^{\infty} d(\Delta\theta)\, p(\Delta\theta)\Delta\theta,$$

where $p(\Delta\theta)$ is the probability distribution for $\Delta\theta$. The variance for the total change in the scaling variable x is

$$(\delta\Delta x)^2 \simeq \ln t \left(\frac{2\delta\Delta\theta}{\ln t} \right)^2$$

and vanishes in the limit $t \rightarrow \infty$. The effect of random chirality on the winding angle distribution is identical to that of uniform chirality and is zero when segments of positive and negative chirality occur equally often.

Finally, we want to emphasize that the results of this section are based on the assumption that there is no interaction between the polymer and the winding center besides the chirality. When there is an additional repulsive interaction, we have to choose absorbing boundary conditions, in which case the random walk never hits the moving wall and the motion of the wall (the chirality) has no effect at all on the winding angle distribution. On the other hand, when the polymer is bound to the winding center due to an attractive interaction, the number of returns to the winding center, and consequently the systematic shift in the winding angle due to chirality, are proportional to the length t (see Section 4.3). In the presence of both an attractive interaction and a hard-core repulsion (probably the most realistic case [20]), the polymer performs a phase transition from a bound to a free state depending on the temperature, and we expect the results of this subsection to apply near the transition temperature.

4.3 The Winding Angle Distribution of Confined Polymers

Up to now we have considered only cases where the polymer could wander infinitely far away from the winding center. However, there are many physical situations where polymers indeed are confined to some region around the winding center. When the diameter of the container is small compared to the length of a DP, or when the polymer density is so large that for each of them just a small cylinder is available, the winding angle distribution will be fundamentally different from the previous section. Random walk segments of time Δt that are small compared to the total length of the walk but large compared to the time it takes to make a winding, have identical winding angle distributions. Applying the central limit theorem, we can therefore predict that the total winding angle distribution of a confined polymer has the form

$$p_{\mathrm{con}}(\theta) \propto \exp[-a\theta^2/2t], \tag{20}$$

where $1/a$ is the variance in the winding angle per unit time.

When the winding center is chirally asymmetric, there is an additional shift in the mean winding angle. In the following three subsections, we will determine the winding angle distribution for polymers confined between two cylinders, polymers bound to an attractive winding center, and bound polymers winding around chiral centers.

4.3.1 POLYMERS CONFINED BETWEEN TWO CYLINDERS

We start with the simple model of a polymer confined between two concentric cylinders (see Fig. 4.11). The inner cylinder is the winding center. The outer one is the wall of the container or represents the repulsion of the neighboring poly-

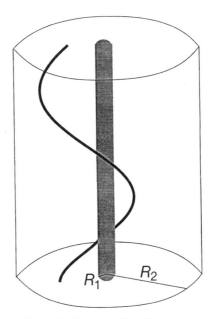

Figure 4.11 A directed polymer confined between two cylinders

mers. This situation is equivalent to a random walk confined between two concentric rings of radii R_1 and R_2. After a long time, the probability $p(r)$ to find the
walker at a given radius r is independent of time and of the angle. For reflecting
boundary conditions at the outer and the inner ring, the walker is then with equal
probability at any site between the two rings, leading to

$$p(r) = \frac{1}{r \ln(R_2/R_1)}.$$

The variance of the winding angle per unit time is $1/r^2$ when the walker is at radius r. Since the variances of different segments of the random walk are independent, we can add them up, leading in the limit of large t to

$$\frac{1}{a} = \int_{R_1}^{R_2} \frac{1}{r^2} p(r) dr = \int_{R_1}^{R_2} \frac{dr}{r^3 \ln(R_2/R_1)} = \frac{t}{2 \ln(R_2/R_1)} \left(\frac{1}{R_1^2} - \frac{1}{R_2^2} \right).$$

It is more physically relevant to use absorbing boundary conditions with $p(r)$
$= 0$ for $r = R_1, R_2$. We solve the diffusion equation

$$\frac{\partial P(r, \phi, t)}{\partial t} = \frac{1}{2} \left(\frac{\partial^2 P}{\partial r^2} + \frac{1}{r} \frac{\partial P}{\partial r} + \frac{1}{r^2} \frac{\partial P}{\phi^2} \right)$$

with the ansatz

$$P(r, \phi, t) = \sum_{n=0}^{\infty} p(r) \cos(n\phi) \exp[-\lambda_n t].$$

For long times, the angular dependence vanishes and the mode with the slowest decay (the smallest eigenvalue λ_0) dominates. (Note that ϕ is not the winding angle but the azimuthal angle, which takes only values between 0 and 2π.) Since we normalize the winding angle distribution with respect to the walkers that do not get absorbed, the factor $\exp[-\lambda_0 t]$ drops out, and the probability to find the walker after time t at radius r is given by the solution of

$$\frac{\partial^2 p(r)}{\partial r^2} + \frac{1}{r} \frac{\partial p(r)}{\partial r} + 2\lambda_0 p(r) = 0,$$

with the boundary conditions given above, and with the normalization condition $\int_{R_1}^{R_2} p(r)dr = 1$. The general solution of this (Bessel) differential equation can be written in form of an integral:

$$p(r) = \int_0^{\pi} [C_1 \cos(\sqrt{2\lambda_0} r \sin \zeta) + C_2 \cos(\sqrt{2\lambda_0} r \cos \zeta) \ln(\sqrt{2\lambda_0} r \sin^2 \zeta)]d\zeta.$$

The values of λ_0, C_1, and C_2 are obtained by matching two consecutive zeros of this function to $r = R_1$ and $r = R_2$, and by normalizing properly. In general, the solution cannot be written down in a closed form and has to be found numerically. In the case where $R_1 \ll R_2$, the values of λ_0 and C_1/C_2 can be found analytically since $C_1 \gg C_2$ in this limit and the first two zeros of $p(r)$ are given by the conditions $\ln(\sqrt{2\lambda_0} R_1) = C_1/C_2$ and $\sqrt{2\lambda_0} R_1 \simeq 2.4$ (the first zero of the Bessel function J_0). In the limit $R_2 - R_1 \ll 1$, we find $p(r) = C \sin(\pi(r-R_1)/(R_2-R_1))$, where C is the normalization constant.

Exercise: Set $R_1 = 1$ and determine numerically the variance

$$\frac{1}{a} = \int_{R_1}^{R_2} \frac{p(r)dr}{r^2}$$

as a function of R_2.

4.3.2 *POLYMERS BOUND TO AN ATTRACTIVE WINDING CENTER*

A polymer can also be confined by an attractive winding center (see Fig. 4.12): A DP subject to an attractive potential of radius b_0 and binding energy U_0 per unit length is bound to that winding center. For temperatures above the crossover value of $T^* \propto b_0 \sqrt{U_0}$, the polymer is only weakly bound and wanders horizontally over a large localization length, $l_{\perp}(T) \simeq b_0 \exp[(T/T^*)^2]$ [8]. The mean vertical

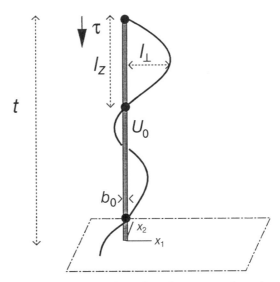

Figure 4.12 A directed polymer bound to an attractive rod

distance l_z between consecutive intersections of the polymer with the defect is consequently proportional to $l_\perp^2$. Over this distance, the polymer can be approximated by a directed walk that returns to its starting point (the winding center) after a time l_z.

Using the result in Eq. (14), we can derive the winding angle distribution $p_A^0(\tilde{x})$ for such confined random walks. Each walk that returns to its starting point after time t (in the z-plane) is composed of two walks of length $t/2$ going from the starting point to $z(t/2)$. As we saw in Subsection 4.2.1, almost all walks of length $t/2$, when mapped on the plane $2\zeta/\ln(t/2)$, have their endpoint on a vertical wall at distance 1 from the origin. The winding angle distribution for these walks is given by Eq. (14), with t replaced by $t/2$. The probability that a walk that returns to its starting point has a winding angle θ is therefore obtained by adding the probabilities of all combinations of two walks of length $t/2$ whose winding angles add up to θ. That is,

$$p_A^0(\tilde{x}) = \int_{-\infty}^{\infty} dy \, \frac{\pi}{4\cosh^2(\pi y)} \frac{\pi}{4\cosh^2(\pi(\tilde{x}-y))}$$

$$= \frac{\pi}{2\sinh^2(\pi\tilde{x}/2)}\left(\frac{\pi\tilde{x}}{2}\coth\left(\frac{\pi\tilde{x}}{2}\right)-1\right), \tag{21}$$

where $\tilde{x} = 2\theta/\ln(t/2R^2)$.

For large $\tilde{x} \approx x$, the previous expression decays as $x \exp[-\pi x]$. A polymer of length t is roughly broken up into t/l_z segments between contacts with the attractive columnar defect. We can assume that the winding angle of each segment is independently taken from the probability distribution in Eq. (21) with $t \approx l_z$. Adding the winding angle distributions of all segments leads to a Gaussian distribution centered around $\theta = 0$, with a variance proportional to $L \ln(l_z)/l_z$.

4.3.3 CONFINED POLYMERS WINDING AROUND CHIRAL CENTERS

When the winding center has a chiral asymmetry, the mean winding angle is increased by some finite amount $\Delta\theta$ per unit time. The winding angle distribution is consequently modified to

$$p_c(\theta) = \exp[-\tilde{a}(\theta - t\Delta\theta)^2/t],$$

with a mean winding angle proportional to the length of the polymer (see also ref. [20]). $1/\tilde{a}$ is larger than $1/a$, since the variances in the number of returns to the winding center and in $\Delta\theta$ both contribute to the variance of the winding angle. For weak chirality, $1/\tilde{a}$ is close to $1/a$, and the main effect of the chirality is just a shift of winding angle distribution. For strong chirality, we expect $1/\tilde{a} \gg 1/a$, and the winding angle distribution becomes very broad (similar to the situation discussed in Section 4.2.4).

When the polymer is confined not by a container but by neighboring polymers, it will not just wind around one of these neighbors. Kamien and Nelson [22] have shown that when chirality is strong, screw dislocations proliferate throughout the polymer crystal.

4.4 Winding Angles in Random Media and for Self-avoiding Polymers

So far we have considered only winding topologies that can be mapped onto ideal random walks. However, when the medium in which the polymer is embedded is nonhomogeneous, the energies of different polymer configurations are different. Examples are polymers in gels and porous media [23] and magnetic flux lines in high-T_c superconductors, pinned by oxygen impurities [21]. We consider the case of *quenched* randomness, where one end of the polymer is fixed [24].

The behavior of a DP in the presence of short-range correlated randomness is modeled by a directed path on a lattice with random bond energies [25]. In three or less dimensions, the polymer is always pinned at sufficiently long length scales. An important consequence of the pinning is that the path wanders away from the origin much more than a random walk, with its transverse fluctuations scaling as t^ν, where $\nu \approx 0.62$ in three dimensions and $\nu = 2/3$ in two dimensions [26, 27]. The probability of such paths returning to the winding center are thus greatly reduced, and the winding probability distribution is expected to change.

We examined numerically the windings of a directed path along the diagonal of a cubic lattice (see Fig. 4.5). An energy randomly chosen between 0 and 1 was assigned to each bond of this lattice. Since the statistical properties of the pinned path are the same at finite and zero temperatures, we determined the winding angle of the path of minimal energy by a transfer matrix method. For each realization of randomness, this method [25] finds the minimum energy of all paths terminating at different points, with different winding numbers. This information is then updated from one time step to the next. From each realization we thus extract an optimal angle as a function of t. The probability distribution is then constructed by examining 2700 different realizations of randomness. To improve the statistics, we averaged over positive and negative winding angles.

The resulting distribution is shown in Figure 4.13, with a scaling variable $x = \theta/2\sqrt{\ln t}$. This scaling form is motivated by that of self-avoiding walks, which in two dimensions follow a Gaussian distribution

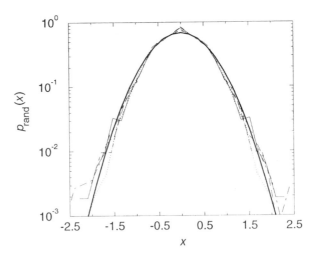

Figure 4.13 Winding angle distribution for a directed path in a random three-dimensional system for $t = 120$ (dotted), 240 (dashed), 480 (long dashed), 960 (dot-dashed), and 1920 (solid). The thick solid line is the Gaussian distribution in Eq. (23).

$$p_{SA}\left(x = \frac{\theta}{2\sqrt{\ln t}}\right) = \frac{1}{\sqrt{\pi}}\exp(-x^2). \tag{22}$$

The result of the data collapse in Fig. 4.13 agrees well with the Gaussian distribution

$$p_{rand}\left(x = \frac{\theta}{2\sqrt{\ln t}}\right) = \sqrt{\frac{1.5}{\pi}}\exp(-1.5x^2). \tag{23}$$

Directed paths in random media and self-avoiding walks share a number of features that make the similarity in their winding angle distributions plausible. Both walks meander away with an exponent larger than the random walk value of 1/2. (The exponent of 3/4 for self-avoiding walks is larger than $\nu \approx 0.62$ for polymers in three dimensions.) As a result, the probability of returns to the origin is vanishingly small in the limit $t \to \infty$ for both types of paths, and the properties of the winding center are expected to be irrelevant. (A simple scaling argument suggests that the number of returns to the origin scales as $N(t) \propto 1/t^{1-2\nu}$.) The conformal mapping of Section 4.2 cannot be applied in either case: The density and size of impurities in a random medium become coordinate dependent under this mapping, as does the excluded volume effect. The winding angle distribution for self-avoiding walks in Eq. (22) has been calculated using a more sophisticated mapping [9, 13]. Because a similar exact solution is not currently available for polymers in random media, we resort to the scaling argument presented next.

Let us divide the self-avoiding walk, or the directed path, in segments going from $t/2$ to t, from $t/4$ to $t/2$, etc., down to some cutoff length of the order of the lattice spacing, resulting in a total number of segments of the order of $\ln t$ (see Fig. 4.14). The statistical self-similarity of the walks suggests that a segment

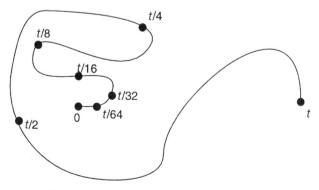

Figure 4.14 Division of the self-avoiding walk into self-similar segments

length $t/2^n$ can be mapped onto a segment of length $t/2^{n+1}$ after rescaling by a factor of $1/2^{\nu}$. Under this rescaling, the winding angle is (statistically speaking) conserved, and consequently all segments have the same winding angle distribution. Convoluting the winding angle distributions of all segments, and assuming that the correlations between segments do not invalidate the applicability of the central limit theorem, leads to a Gaussian distribution with a width proportional to $\ln t$. This argument does not work for the random walks considered in Section 4.2 since the finite radius of the winding center is a relevant parameter. Different segments of the walk are therefore not statistically equivalent, as they see a winding center of different radius after rescaling.

In the pure system, we had to distinguish between repulsive, neutral, and chiral winding centers. Since in the presence of point impurities the polymer does not return to the winding center as often, these differences are now irrelevant. In fact, it can be shown that even an attractive force between the winding center and the polymer cannot bind the polymer to the winding center, as long as it does not exceed some finite threshold [28]. Perhaps not surprisingly, the main conclusion of this section is that the pinning to point randomness decreases entanglement.

4.5 Discussion and Conclusions

Topological entanglements present strong challenges to our understanding of the dynamics of polymers and flux lines. In this chapter, we examined the windings of a single directed polymer around a columnar winding center as well as the winding of two DPs around each other. By focusing on even this simple physical situation, we were able to uncover a variety of interesting properties: The probability distributions for the winding angles can be classified into a number of universality classes characterized by the presence or absence of underlying symmetries or boundary conditions.

For free DPs in a homogeneous medium, we find a number of exponentially decaying distributions: If there is no interaction at all between the polymer and the winding center (corresponding to reflecting boundaries for random walks) we obtain the distribution in Eq. (17). Removing this conservation (absorbing boundaries or repulsive interaction between the polymer and the winding center) leads to the distribution in Eq. (14), whose tails decay twice as fast.

A completely new set of distributions is obtained for chirally asymmetric situations, where the polymer is preferentially twisted in one direction at the winding center. These distributions have asymmetric exponential tails, with decay constants that depend on the degree of chirality. Strong chirality appears to lead to quite broad distributions. A remaining challenge is to find the complete form of this probability distribution by solving the two-dimensional diffusion equation with moving boundary conditions.

When the polymer is confined to a finite volume around the winding center, the winding angle distribution becomes Gaussian, with a width proportional to the length of the polymer. In the presence of chiral asymmetry, the mean winding angle is proportional to the length of the polymer, and we have to distinguish again between the limits of weak and strong chirality.

For nonideal walks, with a vanishing probability to return to the origin, the properties of the winding center are expected to be irrelevant. Both self-avoiding walks in $d = 2$ dimensions and polymers pinned by point impurities in $d = 3$ have wandering exponents ν larger than 1/2 and fall into this category. We presented a scaling argument (supported by numerical data) that in this case the probability distribution has a Gaussian form in the variable $\theta/\sqrt{\ln t}$. Not surprisingly, wandering away from the center reduces entanglement. The characteristic width of the Gaussian form is presumably a universal constant that has been calculated exactly for self-avoiding walks in $d = 2$. It would be interesting to see if this constant (only estimated numerically for the impurity pinned polymers in $d = 3$) can be related to other universal properties of the walk. Changing the correlations of impurities (and hence the exponent ν) may provide a way of exploring such dependence.

There are certainly other universality classes not explored in this chapter. For example, we did not consider the case of a long-range interaction between the polymer and the winding center. Also, the mapping of a DP in a nematic solvent on a random walk is correct only to first approximation [4, 5]. Due to long-range correlations within the nematic solvent, the number of returns of the polymer to the winding center does not increase with $\ln t$, but with $\ln(\ln t)$ for very large t.

The results of this chapter also provide conjectures for the winding of nondirected polymers around a rod. If the self-interaction of the polymer can be neglected for some reason, the results of Section 4.2 can be applied since the parameter τ is now the internal coordinate of the polymer. To a first approximation, this might be correct for a polymer close to the θ-point, but three-point interactions that ultimately swell the θ-polymer will eventually invalidate the result [2].

When the polymer swells to give $\nu > 1/2$, as is the case for a self-avoiding random walk in three dimensions, the winding center is no longer important. The projection into the plane perpendicular to a rod is a walk that wanders away from the winding center faster than an ideal random walk. We can therefore apply the results of Section 4.4 and conclude that the winding angle distribution is Gaussian, with a variance proportional to ln t. When the polymer is in the collapsed state, its winding angle distribution is again different and has yet to be found. Since collapsed polymers are relatively compact, they can be approximated by Hamiltonian walks that visit each site within a volume of the size of the polymer exactly once [29, 30]. We thus conclude this chapter with the open problem of determining the probability distribution for windings in the collapsed state by examining the behavior of Hamiltonian walks.

Acknowledgments

We thank M. E. Fisher, A. Grosberg, Y. Kantor, P. LeDoussal, and S. Redner for helpful discussions. B. D. is supported by the Deutsche Forschungsgemeinschaft (DFG) under Contract No. Dr 300/1-1. M. K. acknowledges support from NSF grant number DMR-93-03667.

References

[1] P.-G. de Gennes, *J. Chem. Phys.* **55:** 572 (1971).

[2] M. Doi and S. F. Edwards. *The Theory of Polymer Dynamics.* Clarendon Press, Oxford, 1986.

[3]. A. Grosberg and S. Nechaev, *Adv. Polym. Sci.* **106:** 1–29 (1993).

[4]. D. R. Nelson, in *Observation, Prediction and Simulation of Phase Transitions in Complex Fluids,* M. Baus, L. F. Rull, and J. P. Ryckaert, Eds., Kluwer, Netherlands, 1995.

[5]. P.-G. de Gennes, in *Polymer Liquid Crystals,* A. Cafieri, W. R. Kringbaum, and R. B. Meyer, Eds., Academic Press, New York, 1982.

[6] D. R. Nelson, *Phys. Rev. Lett.* **60:** 1973 (1988).

[7] M. Rubinstein and S. P. Obukhov, *Phys. Rev. Lett.* **65:** 1279 (1990).

[8] D. R. Nelson, in *Phase Transitions and Relaxation in Systems with Competing Energy Scales,* T. Riste and D. C. Sherrington, Eds., Kluwer, Dordrecht, Boston, 1993.

[9] B. Duplantier and H. Saleur, *Phys. Rev. Lett.* **60:** 2343 (1988).

[10] B. Drossel and Mehran Kardar, *Phys. Rev. E* **53:** 5861 (1996).

[11] J. Rudnick and Y. Hu, *J. Phys. A: Math. Gen.* **20:** 4421 (1987).

[12] F. B. Knight, *Essentials of Brownian Motion and Diffusion,* American Mathematical Society, Providence, RI, 1981.

[13] H. Saleur, *Phys. Rev. E* **50:** 1123 (1994).

[14] R. Durett, *Brownian Motion and Martingales in Analysis,* Wadsworth, Belmont, CA, 1984.

[15] F. Spitzer, *Trans. Am. Math. Soc.* **87**: 187 (1958).

[16] C. Bélisle, *Ann. Prob.* **17**: 1377 (1989).

[17] J. W. Pitman and M. Yor, *Ann. Prob.* **14**: 733 (1986).

[18] A. Comtet, J. Desbois, and C. Monthus, *J. Stat. Phys.* **73**: 433 (1993).

[19] J. V. Selinger and R. L. B. Selinger, *Phys. Rev. Lett.* **76**: 58 (1996).

[20] B. Houchmandzadeh, J. Lajzerowicz, and M. Vallade, *J. de Phys. I* **2**: 1881 (1992).

[21] G. Blatter, M. V. Feigel'man, V. B. Geshkenbein, A. I. Larkin, and V. M. Vinokur, *Rev. Mod. Phys.* **66**: 1125 (1994).

[22] R. D. Kamien and D. R. Nelson, *Phys. Rev. E* **53**: 650 (1996).

[23] G. M. Foo, R. B. Pandey, and D. Stauffer, *Phys. Rev. E* **53**: 3717 (1996).

[24] P. Le Doussal and J. Machta, *J. Stat. Phys.* **64**: 541 (1991).

[25] M. Kardar, *Directed Paths in Random Media,* Fluctuating Geometries in Statistical Mechanics and Field Theory, F. David, P. Ginzparg, and J. Zinn-Justin, Eds., Pg. 1, Elsevier, Amsterdam, 1996.

[26] J. G. Amar and F. Family, *Phys. Rev. A* **41**: 3399 (1990).

[27] J. M. Kim, A. J. Bray, and M. A. Moore, *Phys. Rev. A* **44**: 2345 (1991).

[28] L. Balents and M. Kardar, *Phys. Rev. E* **49**: 13,030 (1994).

[29] H. Orland, C. Itzykson, and C. de Dominicis, *J. Physique Lett.* **46**: L353 (1985).

[30] V. S. Pande, C. Joerg, A. Y. Grosberg, and T. Tanaka, *J. Phys. A: Math. Gen.* **27**: 6231 (1994).

Chapter 5 | Bulk and Interfacial Polymer Reaction Kinetics

Ben O'Shaughnessy

Department of Chemical Engineering
Columbia University, New York, New York

Abstract

We describe theoretical approaches to reacting polymer systems where chains bearing reactive groups can react in a bulk or at a polymer–polymer interface. Methods are introduced through four detailed examples. Beginning with exact formulations, we introduce Wilemski and Fixman's "closure approximation" that unreacted polymers have equilibrium internal chain configurations. Universally employed with the exception of renormalization group studies, it allows closure in terms of the reactive group coordinates only. Closure in terms of two-body correlation functions is still complicated by higher-order correlations. We deal systematically with these. For interfacial problems they are responsible for the long-time first-order kinetics.

Bulk reaction theory is developed for end-functionalized polymers, each of N chain units, dispersed at small initial density n_0 in an unentangled melt. For large Q, the local functional group reactivity, kinetics are diffusion-controlled (DC) before the longest polymer relaxation time τ: The number of reactions per unit volume grows as $\mathcal{R}_t \approx x^3(t)n_0^2 \sim t^{3/4}$ (de Gennes' result). This determines the long time rate constant, $k_\infty \approx R^3/\tau \sim 1/N^{1/2}$ (Doi's result). Here R is the polymer coil size and x_t the rms displacement of a chain end. As a second bulk problem, we treat polymers in dilute solutions (good solvent). Reactions are so weakened by fast hydrodynamical relaxation and excluded volume repulsions between functional groups that the DC limit does not exist, no matter how large Q is. Kinetics are always mean field (MF); i.e., the reaction rate is proportional to the equilibrium reactive group contact probability, giving $k_\infty \sim 1/N^{.15}$ for reactive end groups.

Theory is also developed for reaction kinetics at melt A–melt B interfaces. Three distinct regions are identified in the Q-n_B^∞ plane, each with a unique sequence of kinetics for the number of reacted chains per unit area, $\mathcal{R}_t$. Here n_A^∞ and n_B^∞ are the number densities of end-functionalized chains in either bulk, with the convention $n_A^\infty \leq n_B^\infty$. If Q exceeds a density-dependent threshold, second-order DC kinetics onset for $t < \tau$, with $\mathcal{R}_t \approx x_t^4 n_A^\infty n_B^\infty$. Logarithmic corrections arise in marginal cases. This leads to $\mathcal{R}_t \sim t/\ln t$ for unentangled chains, whereas for entangled melts consecutive regimes $\mathcal{R}_t \sim t/\ln t$, $\mathcal{R}_t \sim t^{1/2}$, and $\mathcal{R}_t \sim t + \ln t$ exist. Which regimes are realized depends on Q and n_B^∞. For very small Q, only simple MF second-order kinetics pertain with time-independent second-order rate constant, $\mathcal{R}_t \sim n_A^\infty n_B^\infty t$.

A unique feature of interfacial reactions is a long-time first-order DC regime with $\mathcal{R}_t$

THEORETICAL AND MATHEMATICAL
MODELS IN POLYMER RESEARCH

Copyright © 1998 by Academic Press.
All rights of reproduction in any form reserved.
ISBN 0-12-304140-6/$25.00.

$\approx x_t n_A^\infty$. The reaction rate is determined by the more dilute A side. If the reactive chains in the denser B bulk are overlapping, $n_B^\infty R^3 > 1$, and if Q is large enough, these kinetics onset before τ. Then $\mathcal{R}_t \sim t^{1/4}$ for unentangled melts, whilst for entangled cases consecutive regimes $\mathcal{R}_t \sim t^{1/4}$, $\mathcal{R}_t \sim t^{1/8}$ and $\mathcal{R}_t \sim t^{1/4}$ arise, some or all of which may be realized depending on Q and n_B^∞. The final first-order regime is always governed by center of gravity diffusion, $\mathcal{R}_t \sim t^{1/2}$.

At a certain timescale the interface saturates with copolymer product and, essentially, reactions cease. This may prevent the onset of the final first-order DC regime if the reactivity is very small—$Q < Q^\dagger$ with $Q^\dagger \sim 1/N^{1/2}$ (unentangled melts) or $Q^\dagger \sim 1/N^{3/2}$ (entangled).

5.1 Introduction

This chapter describes recent theoretical progress in the field of polymer reaction kinetics. The challenge is to adapt standard theoretical approaches, which deal with inert polymer chains, to situations (such as those depicted in Figs. 5.1 and 5.2) where some or all of the chains in a polymer solution or melt carry one or more chemically reactive groups. Theoretical methods to analyze *irreversible* interpolymeric reactions are described. Most commonly, these reactions produce permanent covalent bonds, but also of interest are systems where the "reaction" event entails an irreversible electronic transition (see below) without bond formation. The principal aim is to determine reaction rates as a function of properties such as polymer molecular weight, concentration of reactive chains, and reactivity of functional groups.

Several examples will be treated in detail. These fall into two classes: bulk reactions and interfacial reactions. In the former, reactive polymers within a single bulk phase can meet and react anywhere. In the latter, two immiscible polymer phases are present, each containing chains that may react with chains in the other phase only; reactions are then restricted to the thin interfacial region separating the phases. Not surprisingly, we will see that reaction kinetics are profoundly modified by the presence of an interface; however, interesting parallels exist between the two cases, and by first undertaking the bulk problem we are greatly assisted in the interfacial analysis.

Practically important systems can belong to any of several classes. Various types of bulk reaction systems are illustrated in Fig. 5.1. The simplest case, end-functionalized chains dilutely dispersed within a polymer solution or melt of unreactive but otherwise identical chains, is shown in Fig. 5.1(a). Linear free radical polymerization (FRP), probably the most technologically important application of all, belongs to this class [1–8]. In FRP "living" chains, bearing

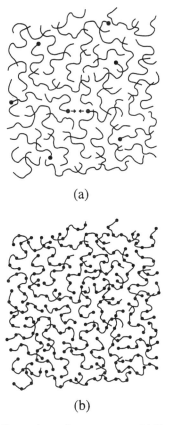

(a)

(b)

Figure 5.1   Examples of bulk reacting polymer systems. (a) End-functionalized polymer
chains dilutely dispersed within a polymer solution or melt of inert but otherwise identical chains. Linear free radical polymerizations belong to this
class. (b) Solution or melt of reactive chains, *all* of which carry a certain
number of reactive groups along their backbone. This situation is realized in
the cross-linking of rubbers or gels.

highly reactive radical end groups, react interpolymerically with one another to
produce "dead" chains, the final polymer product. (However, significant compli-
cations include the fact that the living chains are growing and their distribution is
extremely polydisperse [7].) Initially, these reactions occur in a medium that is a
dilute dead polymer solution. The concentration of this solution gradually increases as more dead chains are generated, frequently attaining levels close to the
melt during the latter stages. Consequently, the essential ingredients in any sys-

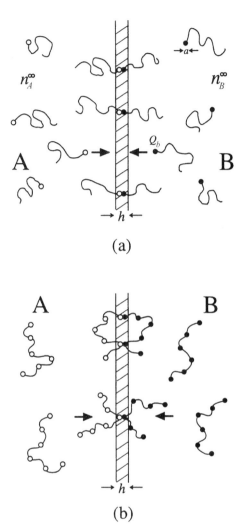

(a)

(b)

Figure 5.2 Examples of polymer reactions occurring at an interface. (a) The interface of
width h separates bulk melts A and B in which some of the polymers are reactive. Initially, reactive end groups, with reactivity Q_b and of size a, are pre-
sent at uniform densities n_A^∞ and n_B^∞. We adopt the convention $n_A^\infty \leq n_B^\infty$; for
unequal densities, the smaller A density governs long-time reaction rates
whereas the higher B density determines characteristic timescales. A–B reactions are confined to the interface, each reaction producing an AB diblock
copolymer. (b) As (a), but now each reactive chain has many reactive sites
along its backbone.

tematic model of FRP are theories of irreversible interpolymeric reaction kinetics for arbitrary concentrations of the inert background medium.

Figure 5.1(b) depicts a similar situation to that in 5.1(a), but every chain in the system now carries many groups. This is realized in the cross-linking of rubbers or gels [1, 9] (e.g., vulcanization of rubbers [1, 10]). Important issues include how the kinetics of cross-link formation vary with reactive group density, which may be controlled, for instance, by radiation dosage.

Much recent research has focused on the reactive reinforcement of polymer–polymer interfaces [11–13]. Here reactions conducted under melt conditions generate bridging copolymers (Fig. 5.2), which enhance the interfacial fracture toughness and yield stress after cooling. Commercial applications entail the simultaneous mechanical mixing of two thermodynamically immiscible polymer species ("reactive processing"), reactions occurring at interfaces separating droplets from the continuous phase. In addition to their direct reinforcing effect, the interfacial copolymer products also promote the mixing itself, apparently both by suppression of droplet coalescence rates and through surface tension reduction [14–17].

Any of the situations depicted in Figs. 5.1 and 5.2 are also in principle realizable in model studies where the "reactive groups" may be optically excited species [18]. As a probe of interpolymeric kinetics, phosphorescence quenching [19, 20] has been the most successful to date. Here the phosphorescence of an excited chromophore group (e.g., benzil) is quenched if it encounters a quencher (e.g., anthryl) belonging to another polymer: This is the "reaction" event. By monitoring the phosphorescence intensity, the experimentalist can infer the reaction rate. Another approach is to follow the recombination of radical groups whose creation by laser photolysis can be carefully controlled. The typical aim of such model studies is to determine molecular-weight dependence of rate constants under carefully controlled conditions, which may be difficult to realize in the applications cited above. Measuring reaction rates in this way can reveal fundamental information on polymer statics and dynamics both in the bulk and at surfaces [10, 21–23], where many issues remain open.

Finally, we should mention that a great variety of macromolecular reactions occur in living organisms [24, 25]. One can view the relatively simple synthetic polymers we discuss here as models for the more complex reacting macromolecules found in biology.

Given these physical systems, successful theories must aim to calculate reaction rates as a function of time t; polymer molecular weight or, equivalently, number of monomer units, N; initial density of reactive groups, $n(0)$; concentration of the inert polymer solution medium; fraction of polymers that are reactive; number of reactive groups per reactive chain; and local reactivity of the reactive groups, Q.

A more basic issue is whether or not the universality familiar in high polymer statics and dynamics applies also to reaction kinetics. Can we expect universal scaling laws regardless of local details? For example, it is natural to ask to what extent the reaction rate is determined by the local chemistry, as embodied by Q. We define Q as the reaction probability per unit time given that a pair of reactive groups are in "contact," i.e., are within a distance a equal to the size of one chain unit (see Fig. 5.3). Presumably reaction rates will also reflect overall static and dynamical polymer properties (the encounter of two reactive groups necessitates the encounter of their host polymer coils) such as how rapidly a polymer chain unit bearing a reactive group diffuses in space. The basic quantity describing this diffusion is x_t, the root mean square (rms) displacement after time t,

$$x_t = a\left(\frac{t}{t_a}\right)^{1/z},$$ (1)

(a)

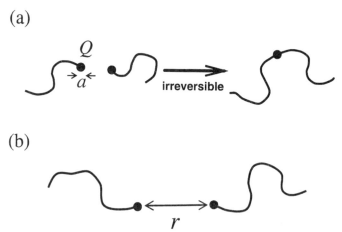

(b)

Figure 5.3 (a) If two functional end groups of size a contact one another, irreversible reaction occurs with probability Q per unit time (the notation Q_b is used for interfacial systems). (b) In the absence of reactions, the probability distribution for the separation r between two chain ends is the equilibrium one, $P_{eq}(\mathbf{r})$. For dilute solutions of end-functionalized polymers in good solvents, the small r behavior, $P_{eq} \sim r^g$, so diminishes the ability of reactive ends to meet that mean field kinetics apply even for highly reactive end groups: there is no diffusion-controlled limit.

where t_a is the diffusion time corresponding to a and z is the dynamical exponent. In addition to dealing with $z = 2$ for small molecules and for polymers at long times (simple Fickian diffusion), we will need to consider various values of z to describe short-time polymer dynamics ($z = 3, 4, 8$).

We will verify later that reaction rates do in general depend on dynamical polymer properties such as the structure of x_t. In fact, in certain cases there is no dependence whatsoever on the local reactive group chemistry as described by Q. In such cases the theoretical limit $Q \to \infty$, for example, is well defined and in this limit there is no dependence on "capture radius" i.e., the size a of the reactive groups. This universality was first found by Wilemski and Fixman [26] and then physically explained by Doi [27], in the context of cyclization (intramolecular reactions between reactive end groups) in unentangled melts (where Rouse dynamics [10, 28] apply; see below). First principles renormalization group (RG) studies of intramolecular reaction kinetics confirm this universality [29–33]. The same universality was demonstrated by Doi [34] and de Gennes [35] for intermolecular reactions in melts, also confirmed by RG treatments [36].

Unlike melts, in dilute and semidilute solutions static interpolymeric correlations are also important (see Fig. 5.3(b)). The ability of a pair of reactive groups to meet—each group belonging to a different polymer—is diminished by interpolymeric repulsive short-range interactions between chain units belonging to the host coils ("excluded volume" interactions [28, 10]). Except in melts, where these interactions are screened out, it was shown in refs. [36] and [37] that they exert a crucial influence on reaction kinetics (see Section 5.3). Excluded volume interactions diminish the equilibrium contact probability P_{eq}^{cont} between two reactive groups belonging to different chains (see Fig. 5.3(b)):

$$P_{eq}^{cont} \equiv \int_{|\mathbf{r}| \leq a} P_{eq}(\mathbf{r}) d^d r, \qquad P_{eq}(\mathbf{r}) \sim r^g. \qquad (2)$$

Here $P_{eq}(\mathbf{r})$ is the equilibrium probability distribution for the separation between two chain ends, $\mathbf{r}$, and d is the dimension of space (in much of the following it will prove conceptually helpful to keep d general, setting $d = 3$ in our final results). The "correlation hole" exponent [38–41] g measures how rapidly $P_{eq}(\mathbf{r})$ vanishes as $\mathbf{r} \to 0$ and reflects the strength of excluded volume repulsions. For ideal, noninteracting chains (realized in melts) $g = 0$ and $P_{eq}(0)$ is finite. But for dilute solutions in good solvents, g is nonzero and its value depends on the positioning of the reactive groups on the polymer backbones: for $d = 3$, $g \approx 0.27$ for end groups, $g \approx 0.46$ for one end and one internally positioned group, and $g \approx 0.71$ when both groups are internal.

5.1.1 SCALING ARGUMENTS: BULK REACTIONS

Before detailed calculation, let us develop some physical intuition with the aid of simple scaling arguments. These arguments are loose and merely designed to motivate the detailed calculations of Sections 5.2–5.5. There we will establish two classes of reaction kinetics: mean field (MF) and diffusion-controlled (DC). Let us first consider reactions in the bulk.

1. In the MF picture, interparticle spatial correlations are unaffected by reactions. Hence, the reaction rate equals the *equilibrium* probability that a pair is in contact, P_{eq}^{cont}, multiplied by the number of pairs in the system, multiplied by Q. This leads to second-order reaction kinetics, $\dot{n} = -kn^2(t)$, with rate constant $k = QVP_{eq}^{cont}$ where $n(t)$ is the density of reactive groups that have not reacted after time t and V is the volume of the reaction vessel. For the ideal case ($g = 0$) Eq. (2) gives $P_{eq}^{cont} \approx a^d/V$ whence $k = Qa^d$.

2. If reactions are limited by diffusion, then after time t those groups that have reacted will be those initially belonging to pairs whose separation was less than x_t; i.e., those within diffusive range of one another. Consider the ideal case. Then the number of such pairs per unit volume is $n^2(0)x_t^d$, which gives a time-dependent second-order rate constant, $k(t) \approx dx_t^d/dt \sim t^{d/z-1}$.

When do reacting systems obey MF kinetics and when do they obey DC kinetics [42]? Roughly, MF kinetics are favored by high dimensions (where reaction-induced correlations can more easily dissipate away) and strong intergroup repulsions, or low local reactivities Q (when reactions are a weak perturbation of equilibrium [34, 43]). To quantify this, consider two reactive groups, attached to two different polymer chains, initially close enough such that diffusion could have brought them together by time t, i.e., initially separated by less than x_t. Now to estimate the probability of reaction after time t for this pair, P_2^{react}, let us use equilibrium correlations. Since P_2^{react} equals Qt times the fraction of the time for which the pair was in contact, we have

$$P_2^{react}(t) = Qt \int_{|\mathbf{r}| \leq a} d^d r \, P_{eq}(\mathbf{r}|r < x_t) \approx Qt_a \left(\frac{t}{t_a}\right)^{1-(d+g)/z} \tag{3}$$

where

$$P_{eq}(\mathbf{r}|r < x) \equiv \frac{P_{eq}(\mathbf{r})}{\int_{|\mathbf{r}| \leq x} d^d r \, P_{eq}(\mathbf{r})} \approx \frac{1}{x^d} \left(\frac{r}{x}\right)^g \tag{4}$$

is the equilibrium conditional probability that two groups are separated by $\mathbf{r}$, given that they are within x of each other. We used the small argument scaling form of $P_{eq}(\mathbf{r})$ shown in Eq. (2). From this result we see that if $d + g > z$ then no matter how large Q may be, as time increases the fraction of pairs of this type that actually succeed in reacting is becoming ever smaller. Pair correlations are thus weakly perturbed from equilibrium and we expect MF kinetics to pertain, $k = QVP_{eq}^{cont}$, even for highly reactive end groups.

On the other hand, if $d + g < z$ then $P_2^{react}(t)$ is an increasing function of time. This defines a characteristic timescale t_2^* such that $P_2^{react}(t_2^*) = 1$:

$$t_2^*/t_a = 1/(Qt_a)^{z/(z-d-g)}, \quad (d + g < z). \tag{5}$$

For times greater than t_2^*, any pair initially within diffusive range will definitely have reacted: These are DC kinetics. Note that the earliest regime is always MF because equilibrium conditions prevail before reactions are "switched on" at $t = 0$. In summary,

$$\underline{\text{BULK}} \quad \begin{array}{l} (i) \quad d + g > z \rightarrow \text{MF kinetics} \\[1em] (ii) \quad d + g < z \rightarrow \left\{ \begin{array}{ll} \text{MF kinetics,} & (t < t_2^*) \\ \text{DC kinetics,} & (t > t_2^*) \end{array} \right. \end{array} \tag{6}$$

According to Eq. (6) there are two classes of reaction systems. For one class, kinetics are always MF. For the other, kinetics are asymptotically always DC. The determining criterion [36] is whether $d + g > z$ or $d + g < z$. This is a generalization of the criterion for ideal systems ($g = 0$) where the two classes reduce to $d > z$ and $d < z$. For ideal systems, when $d > z$, space is explored in a dilute fashion since the volume explored, x_t^d, grows more rapidly than t. On the other hand, if $d < z$ then the exploration is dense, $x_t^d < t$. De Gennes [35] called these two cases "noncompact" and "compact" exploration, respectively. Equation (6) tells us that, more generally, to determine the nature of reaction kinetics one must consider also the strength of static repulsions between reactive species as measured by g. We will see in Section 5.3 that for dilute polymer solutions these correlation effects are the decisive factor.

5.1.2   SCALING ARGUMENTS: INTERFACIAL REACTIONS

In Section 5.4 it will be shown that similar general principles apply also to reactions at polymer–polymer interfaces [44–50]. For simplicity, we restrict the following interfacial discussion to ideal cases ($g = 0$). It turns out that second-order MF kinetics are associated with high spatial dimensions and second-order DC kinetics with low dimensions, as in the bulk. However, there are two fundamental differences: (1) The kinetics are now determined by whether $d + 1 > z$ or $d + 1 < z$. (2) At the longest times a crossover to *first-order* DC kinetics occurs in all cases [48–50]. There is no bulk analog to (2).

To understand these new features, consider chain types A and B as in Fig. 5.4, on either side of the interfacial region of width h where the reactive groups can meet and react. Let us define the local group reactivity to be Q_b, with $Q \equiv (h/a)Q_b$. It turns out that Q emerges as a natural renormalized group reactivity, coarse-grained over the scale h. We take the ideal case, $g = 0$, and we assume A and B have identical dynamics. Consider those pairs of A–B reactive groups that are initially within x_t of one another (see Fig. 5.4). Then the reaction probability after time t for a typical such pair (cf. Eq. (3) for the bulk) is

$$P_2^{\text{react}}(t) \approx Q_b t \left(\frac{a}{x_t}\right)^d \frac{h}{x_t} \approx (Q t_a)\left(\frac{t}{t_a}\right)^{1-(d+1)/z} \tag{7}$$

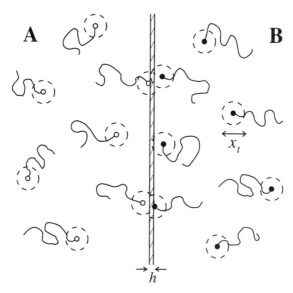

Figure 5.4   Interfacial reactions. The timescale $t_{\frac{1}{2}}^*$ is the time at which the reaction probability reaches unity for a pair that was initially close enough to have diffused and met within time t. For times greater than $t_{\frac{1}{2}}^*$, reactions are confined to those pairs whose exploration volumes (indicated here by dashed lines) overlap at time t at the interface. Thus, $\mathcal{R}_t$ is just the number of such pairs per unit area, $\mathcal{R}_t \approx n_A^\infty n_B^\infty x_t^4$ (second-order DC kinetics).

Here the h/x_t factor is the fraction of the time t for which the A group was at the interface, and $(a/x_t)^d$ is the A–B contact probability, given that A is at the interface. Now if $d + 1 < z$, this defines a timescale t_2^* analogously to the bulk case, Eq. (5):

$$\frac{t_2^*}{t_a} = \frac{1}{(Qt_a)^{z/(z-d-1)}}, \quad (d+1<z). \tag{8}$$

For times greater than t_2^* it is inevitable that the A–B pair will have reacted. Hence, the reaction rate will become limited by the rate at which diffusion can bring A–B pairs together at the interface and a hole of size x_t will grow in the two-body correlation function. On the other hand, if $d + 1 > z$ then P_2^{react} is a decreasing function of time and kinetics are always MF. Corresponding to Eq. (6), we thus have

INTERFACE ($g = 0$, short times)   (i) $d + 1 > z \rightarrow$ MF kinetics

$$(ii)\ d + 1 < z \rightarrow \begin{cases} \text{MF kinetics,} & (t < t_2^*) \\ \text{DC kinetics,} & (t > t_2^*) \end{cases} \tag{9}$$

Evidently, the short-time kinetics are formally like those of a bulk problem in $d + 1$ dimensions. In some respects, the effect of an interface is to increase the dimensionality from $d \rightarrow d + 1$. Indeed, it will be shown in Sections 5.4 and 5.5 that during the DC regime above the reaction rate per unit area, $\dot{\mathcal{R}}_t$, obeys second-order kinetics, $\dot{\mathcal{R}}_t = k^{(2)} n_A^\infty n_B^\infty$ with rate constant $k^{(2)} \approx dx_t^{d+1}/dt$, just as if this were a bulk system with one extra dimension. Here n_A^∞, n_B^∞ are the far-field reactive group densities on the A and B sides of the interface (see Fig. 5.2(a)). An important special case, which we will first meet in Section 5.4, is the marginal one, $z = d + 1$. Section 5.4 deals with Rouse chains for which $z = 4$, which in real three-dimensional space is marginal. Essentially, this is like the $d + 1 < z$ case, with a corresponding crossover at t_2^*. The distinctive features are in the form of t_2^* and the logarithmic corrections in the reaction rate.

The longest time behavior is quite different. At long enough times any reactive group of type A initially within x_t of the interface will have reacted by time t. Then the interface becomes like a homogeneous reacting surface and a one-dimensional DC situation develops. To see this, consider $P_m(t)$, the probability that an A group has reacted with *any* B group:

$$P_m(t) \approx Q_b t \frac{h}{x_t} n_B^\infty a^d \tag{10}$$

where we adopt the convention that the B side is the more dense, $n_B^\infty \geq n_A^\infty$. In Eq. (10), h/x_t is the fraction of the time t for which A is at the interface, and $n_B^\infty a^d$ is the fraction of the time when A is at the interface for which it is also in contact with any B. Defining $P_m(t_m^*) = 1$, we see that reaction is inevitable for $t > t_m^*$, where

$$t_m^*/t_a = 1/(Q t_a n_B^\infty a^d)^{z/(z-1)}. \tag{11}$$

Thus, for $t > t_m^*$ all A particles within x_t of the interface will have reacted; i.e., we have first-order kinetics, $\mathcal{R}_t \approx x_t n_A^\infty$.

In fact, the situation is a little more complex than this discussion suggests. First, it may be that $t_{\frac{1}{2}}^* > t_m^*$ in which case the DC kinetics of Eq. (9) will never be realized. Second, in the case when $t_{\frac{1}{2}}^* < t_m^*$, the transition to first-order DC kinetics in fact occurs at a timescale different to t_m^*. Finally, in polymer systems there are in reality several regimes each with different z. We postpone detailed discussion of these issues until Section 5.4.

5.1.3   EXPERIMENTAL SYSTEMS: TYPICAL GROUP REACTIVITIES Q

We end this section with a few comments on the magnitude of local reactivities Q in real reactive polymer systems. A natural dimensionless measure is the parameter $Q t_a$, which can be written

$$Q t_a \approx \frac{k^{\text{small}}}{k_{\text{rad}}^{\text{small}}}. \tag{12}$$

Here $k^{\text{small}} = Q a^3$ denotes the MF expression for the small molecule analog reaction in the bulk, i.e., that describing reactions between free reactive groups removed from their host polymer chains. The quantity $k_{\text{rad}}^{\text{small}}$ denotes the same for radical groups for which $k_{\text{rad}}^{\text{small}} \approx a^3/t_a$ reaches its maximum value, controlled by the local diffusion rate $1/t_a$. With $a \approx 3$ Å and $t_a \approx 10^{-10}$ sec, one has $k_{\text{rad}}^{\text{small}} \approx 10^8$ liters mol^{-1} sec^{-1}, which roughly coincides with reported radical rate constants [1]. Consistent with the picture of local diffusion supplying an upper limit, this order of magnitude, $k^{\text{small}} = 10^8 - 10^9$ liters mol^{-1} sec^{-1}, is never exceeded.

Now the important point is that for virtually all nonradical species, measured rate constants k^{small} are at least six orders of magnitude less reactive than this upper limit. This tells us that

$$Q t_a \lesssim 10^{-6} \quad \text{(nonradicals)}. \tag{13}$$

In fact, the great majority are orders of magnitude smaller than even this level. Thus, in practice, $Q t_a$ is almost always tiny: DC regimes are frequently inaccessible and simple MF kinetics apply. The exceptions are relatively exotic species such as radicals (or electronically excited species as in model experiments).

5.2   Bulk Reactions: Unentangled Melts

To introduce the basic theoretical approaches we begin with a detailed treatment of perhaps the simplest physically realized system: identical end-functionalized reactive polymers, as in Fig. 5.3, dilutely dispersed within an unentangled melt of unreactive but otherwise identical chains. This is an example of the general situation depicted in Fig. 5.1(a). Chain statistics in melts are ideal [28, 10, 51] due to screening effects; thus chain ends are uncorrelated with each other, $g = 0$, and the general criterion determining the class of reaction kinetics, Eq. (6), reduces to $d > z$ (noncompact) or $d < z$ (compact).

Anticipating second-order rate kinetics (reaction rate proportional to the number of unreacted chain pairs), we aim to calculate the rate constant, $k(t)$, which governs the decay of the number density of unreacted chains, $n(t)$:

$$\frac{dn}{dt} = -k(t)\, n^2(t). \tag{14}$$

Reactions commence at $t = 0$. Experimentally, this corresponds to the instant of radical-producing laser flash photolysis or, for example, the moment at which chromophores are excited to a triplet state in a phosphorescence-quenching study. By "dilute reactive chains" we mean $n(0)R^d \ll 1$ where R is the rms coil size; i.e., the reactive chains are (on average) nonoverlapping. Generally, $k(t)$ turns out to be time dependent, saturating to its asymptotic value $k_\infty \equiv k(\infty)$ after times large compared to the longest polymer relaxation time, τ.

5.2.1   ROUSE DYNAMICS

It is well established [28, 10, 51] that single chain dynamics are "ideal" in a melt of chains that are sufficiently short so that entanglement effects are unimportant. Hydrodynamic and excluded volume interactions are irrelevant, and the simple "Rouse" or "phantom chain" model is applicable: Neither intrachain nor interchain interactions are present, other than interactions between nearest-neighbor chain units holding a given chain together (connectivity). The Rouse model views a single chain as N units at locations $[\mathbf{R}(n)]|_{n=0}^{N}$ connected by harmonic springs with spring constant k such that the energy of a configuration is

$$H_0([\mathbf{R}(n)]) = \frac{k}{2} \sum_{n=1}^{N} (\mathbf{R}_n - \mathbf{R}_{n-1})^2 \to \frac{k}{2} \int_0^N dn \left(\frac{d\mathbf{R}}{dn} \right)^2 \tag{15}$$

after taking the continuous limit. This implies a Gaussian weighting, proportional to $\exp -H_0/k_B T$, for each chain configuration, which in turn implies ideal random walk statistics. Using this weighting to determine the mean square size of one unit, a^2, we identify $k = 3k_B T/a^2$, where k_B and T are Boltzmann's constant and temperature. The form of H_0 leads to the following dynamics [28] (in the absence of reactions)

$$\nu \frac{\partial \mathbf{R}}{\partial t} = -\frac{\delta H_0}{\delta \mathbf{R}(n)} + \mathbf{f}(n, t) = \frac{3k_B T}{a^2} \frac{\partial^2 \mathbf{R}}{\partial n^2} + \mathbf{f}(n, t) \tag{16}$$

where ν is the monomer friction coefficient and the random Gaussian force $\mathbf{f}$, with zero mean and correlations $\langle f_i(n, t) f_j(n', t') \rangle = 2\nu k_B T \delta(n - n') \delta(t - t') \delta_{ij}$, represents collisions with surrounding monomers. Eq. (16) balances the random force, the drag force, and the force due to the springs attached to the nth monomer. Inertial effects are irrelevant for the timescales of interest. In Appendix 5.A we present simple physical arguments to motivate how Eqs. (15) and (16) lead to the following well-known forms for R and τ and for the rms displacement of a chain end, x_t:

$$x_t \approx \begin{cases} a(t/t_a)^{1/4}, & (t \ll \tau) \\ R(t/\tau)^{1/2}, & (t \gg \tau) \end{cases} \qquad \frac{\tau}{t_a} = N^2, \qquad \frac{R}{a} = N^{1/2} \quad \text{(Rouse)}. \tag{17}$$

Thus, Rouse dynamics are characterized by two dynamical exponents: $z = 4$ for $t < \tau$ and $z = 2$ for $t > \tau$; in three-dimensional space we have an early compact regime ($d < z$) followed by a noncompact one ($d > z$). This last regime is simple Fickian diffusion.

5.2.2 ADDING REACTIVE SINK TERMS TO ROUSE DYNAMICS

How can we introduce reactive groups into the dynamics described by Eq. (16)? Reactions are a peculiar "interaction" between chain ends whose effect is to reduce the normalization of different chain configurations by different amounts. A natural framework [26, 52] is the Fokker-Planck equation describing the evolution of $\Phi_t[\mathbf{R}_A, \mathbf{R}_B]$, namely, the number density of pairs of reactive chains whose configurations at time t are near $[\mathbf{R}_A(n)]|_{n=0}^N$ and $[\mathbf{R}_B(n)]|_{n=0}^N$, respectively. We choose the time-dependent normalization of Φ_t to be such that $V^{-2} \int d[\mathbf{R}_A] d[\mathbf{R}_B] \Phi_t = n^2(t)$. Suppose that the Nth unit of every reactive chain is a reactive group. Throughout, we will use lowercase variables to denote the positions of chain ends: $\mathbf{r}_A \equiv \mathbf{R}_A(N)$, for example. Thus, $\int d[\mathbf{R}_A]$ denotes integration over all values of $[\mathbf{R}_A(n)]|_{n=0}^N$ or, equivalently, over all values of $[\mathbf{R}_A(n)]|_{n=0}^{N-1}$ and $\mathbf{r}_A$. Reaction terms are easily introduced into the following exact Fokker-Planck equation (we consider general dimensionality and set $d = 3$ in our final results):

$$\frac{\partial \Phi_t}{\partial t} = F\Phi_t - Qa^d \delta(\mathbf{r}_A - \mathbf{r}_B)\Phi_t$$

$$- Qa^d \int d[\mathbf{R}_C] \Phi_t^{(3)}[\mathbf{R}_A, \mathbf{R}_B, \mathbf{R}_C]\{\delta(\mathbf{r}_C - \mathbf{r}_A) + \delta(\mathbf{r}_C - \mathbf{r}_B)\} \qquad (18)$$

where

$$F = \sum_{i=A,B} \int_0^N dn \frac{\delta}{\delta \mathbf{R}_i(n)} \cdot \left[\frac{k_B T}{\nu} \frac{\delta}{\delta \mathbf{R}_i(n)} + \frac{1}{\nu} \frac{\delta H_0[\mathbf{R}_i]}{\delta \mathbf{R}_i(n)} \right] \qquad (19)$$

is the evolution operator for Φ_t in the absence of reactions and $\Phi_t^{(3)}[\mathbf{R}_A, \mathbf{R}_B, \mathbf{R}_C]$ is the number density of chain triplets with configurations close to $[\mathbf{R}_A]$, $[\mathbf{R}_B]$, and $[\mathbf{R}_C]$. Note that the relation $\dot{\Phi}_t = F\Phi_t$ is in the present case an exact description of the two-chain density distribution Φ_t because different chains are statistically independent in the Rouse model. The sink terms on the right-hand side of Eq. (18) describe the three ways in which reactions can diminish Φ_t. (1) The first *two-body* sink term represents reactions between A–B pairs whose reactive ends are located at $\mathbf{r}_A$, $\mathbf{r}_B$. The delta functions restrict reactions to $\mathbf{r}_A$ and $\mathbf{r}_B$ values such that the ends are in contact (i.e., within a of one another). (2), (3) The remaining two sink terms describe reactions involving just *one* reactive chain end of an A–B pair whose ends are at $\mathbf{r}_A$, $\mathbf{r}_B$. Such a reaction involves a third chain, weighted by the appropriate three-chain correlation function. These are *many-chain* terms; were they absent, one would have a closed two-chain system. (Note that the A, B labels here do not denote chains with chemically different functional groups: All reactive polymers are identical.)

If we can calculate Φ_t, then we can obtain the reaction rate $\dot{n}$ according to

$$\dot{n} = -Qa^d \rho_t^s, \qquad \rho_t^s \equiv \rho_t(0, 0) \equiv \int d[\mathbf{R}_A] d[\mathbf{R}_B] \Phi_t[\mathbf{R}_A, \mathbf{R}_B]\delta(\mathbf{r}_A)\delta(\mathbf{r}_B) \qquad (20)$$

where $\rho_t(\mathbf{r}_A, \mathbf{r}_B)$ is the density of chain end pairs at $\mathbf{r}_A$, $\mathbf{r}_B$ and is normalized to the total number of pairs in the system. This exact relation equates the reaction rate to Q times the number of pairs of reactive chain ends that are in contact.

A more natural object than Φ_t is the same quantity but with its normalization $n^2(t)$ factored out, $q_t \equiv \Phi_t/n^2(t)$. From Eqs. (14) and (20), this object is directly related to the rate constant:

$$k(t) = Qa^d \int d[\mathbf{R}_A] d[\mathbf{R}_B] q_t \delta(\mathbf{r}_A)\delta(\mathbf{r}_B), \qquad q_t \equiv \frac{\Phi_t}{n^2(t)}. \qquad (21)$$

The dynamics of q_t are immediately obtained from Eq. (18) as

$$
\frac{\partial q_t}{\partial t} = Fq_t - Qa^d\delta(\mathbf{r}_A - \mathbf{r}_B)q_t
$$

$$
- \frac{Qa^d}{n^2(t)} \int d[\mathbf{R}_C] \Phi_t^{(3)}[\mathbf{R}_A, \mathbf{R}_B, \mathbf{R}_C]\{\delta(\mathbf{r}_C - \mathbf{r}_A) + \delta(\mathbf{r}_C - \mathbf{r}_B)\} + 2\frac{k(t)}{n(t)} \Phi_t \tag{22}
$$

after using Eq. (14) to eliminate $\dot{n}$.

We can now obtain a self-consistent expression for q_t in terms of the Green's function, $G_t[\mathbf{R}_A, \mathbf{R}_B, \mathbf{R}'_A, \mathbf{R}'_B]$, describing *nonreactive* Rouse chains, i.e., with Q and $k(t)$ set to zero in Eq. (22). G_t is the probability, in the *absence* of reactions, that two Rouse chains have configurations $[\mathbf{R}_A, \mathbf{R}_B]$ at time t, given initial configurations $[\mathbf{R}'_A, \mathbf{R}'_B]$. "Solving" for q_t in terms of G_t, multiplying both sides of the "solution" by $Qa^d\delta(\mathbf{r}_A)\delta(\mathbf{r}_B)$, and integrating over all $[\mathbf{R}_A, \mathbf{R}_B]$, we obtain the basic relation for $k(t)$.

$$
k(t) = Qa^d - Qa^d \int_0^t dt' \int d[\mathbf{R}_A]d[\mathbf{R}_B]d[\mathbf{R}'_A]d[\mathbf{R}'_B]\delta(\mathbf{r}_A)\delta(\mathbf{r}_B)
$$

$$
\times G_{t-t'}[\mathbf{R}_A, \mathbf{R}_B, \mathbf{R}'_A, \mathbf{R}'_B]\delta(\mathbf{r}'_A - \mathbf{r}'_B)\frac{Qa^d}{n^2(t')} \Phi_{t'}[\mathbf{R}'_A, \mathbf{R}'_B]
$$

$$
+ 2Qa^d \int_0^t dt' \int d[\mathbf{R}_A]d[\mathbf{R}_B]d[\mathbf{R}'_A]d[\mathbf{R}'_B]\delta(\mathbf{r}_A)\delta(\mathbf{r}_B)G_{t-t'}[\mathbf{R}_A, \mathbf{R}_B, \mathbf{R}'_A, \mathbf{R}'_B]
$$

$$
\times \left\{ \frac{k(t')}{n(t')} \Phi_{t'}[\mathbf{R}'_A, \mathbf{R}'_B] - \frac{Qa^d}{n^2(t')} \int d[\mathbf{R}'_C]\delta(\mathbf{r}'_A - \mathbf{r}'_C)\Phi_{t'}^{(3)}[\mathbf{R}'_A, \mathbf{R}'_B, \mathbf{R}'_C] \right\} \tag{23}
$$

after using Eq. (21). Note that the two delta functions multiplying Φ_t^3 in Eq. (22) give the same contribution after integration. The origin of the first term on the right-hand side of Eq. (23) is the initial condition $q_0[\mathbf{R}_A, \mathbf{R}_B] = V^2 P_{eq}[\mathbf{R}_A, \mathbf{R}_B]$. Here, P_{eq} is the equilibrium two-chain distribution function, normalized to unity, which gives $1/V^2$ after multiplication by the delta functions and integration, $\int d[\mathbf{R}_A]d[\mathbf{R}_B]\delta(\mathbf{r}_A)\delta(\mathbf{r}_B)P_{eq} = 1/V^2$.

5.2.3   *THE CLOSURE APPROXIMATION: EQUILIBRIUM CHAIN CONFIGURATIONS*

Unfortunately, it is not simple to extract $k(t)$ from Eq. (23) because we do not know how reactions have distorted Φ_t and $\Phi_t^{(3)}$ away from equilibrium. To quantify this, let us define $P_t^{int}([\mathbf{R}_A, \mathbf{R}_B]|\mathbf{r}_A, \mathbf{r}_B)$ to be the normalized probability of chain configurations $[\mathbf{R}_A]$, $[\mathbf{R}_B]$ *given* that the two chain ends are located at $\mathbf{r}_A$, $\mathbf{r}_B$. That is, P_t^{int} is the weighting for internal chain configurations. Similarly, we define $P_t^{(3),int}$ to be the three-chain internal configuration probability distribution. Then Φ_t and $\Phi_t^{(3)}$ can be written

$$\Phi_t[\mathbf{R}_A, \mathbf{R}_B] = P_t^{int}([\mathbf{R}_A, \mathbf{R}_B]|\mathbf{r}_A, \mathbf{r}_B)\rho_t(\mathbf{r}_A, \mathbf{r}_B),$$

$$\Phi_t^{(3)}[\mathbf{R}_A, \mathbf{R}_B, \mathbf{R}_C] = P_t^{(3),int}([\mathbf{R}_A, \mathbf{R}_B, \mathbf{R}_C]|\mathbf{r}_A, \mathbf{r}_B, \mathbf{r}_C)\rho_t^{(3)}(\mathbf{r}_A, \mathbf{r}_B, \mathbf{r}_C) \quad (24)$$

where $\rho_t(\mathbf{r}_A, \mathbf{r}_B)$ is the density of chain end pairs at $\mathbf{r}_A$, $\mathbf{r}_B$ and $\rho_t^{(3)}$ the density of chain end triplets.

The complexity of Eq. (23), then, is that P_t^{int} and $P_t^{(3),int}$ are not equal to the equilibrium distributions: Reactions induce nonequilibrium correlations not only between chain ends but also for all the other degrees of freedom. Thus, the problem does not close in terms of the chain ends. This is a many-body difficulty that requires a near-exact RG approach [53, 54, 29, 36, 32, 33]. Apart from RG studies, virtually all other works in this field [26, 55, 34, 27, 35, 56, 37] have dealt with this difficulty by employing the somewhat ad hoc but highly successful "closure approximation," first used by Wilemski and Fixman. This amounts to assuming that unreacted chain configurations are *unchanged from equilibrium*:

$$P_t^{int} \approx P_{eq}^{int}, \qquad P_t^{(3),int} \approx P_{eq}^{(3),int} \qquad \text{("closure approximation")} \quad (25)$$

where P_{eq}^{int} and $P_{eq}^{(3),int}$ are the equilibrium two-chain and three-chain conditional probability distributions. This closes the problem in terms of chain ends. In every case where first principles RG analyses have been performed [29, 36, 32], the corresponding closure approximation analysis has captured the correct scaling laws (in t, N, etc.). Only prefactors are incorrect. Specifically, Doi has used the above scheme to calculate both cyclization rates [27] and intermolecular reaction rates [34] assuming Rouse dynamics. RG treatments of the same two problems [29, 36, 32], lead to the identical molecular weight exponents. The only differences arise in the constant prefactor. The results obtained by de Gennes, using the same self-consistent framework, for the short-time kinetics of Rouse chains reacting intermolecularly are similarly supported by RG treatments [31], which give the same powers of time t but slightly different prefactors. The RG analyses show that the success of this approximation scheme lies in the fact that a reaction sink is a weak perturbation, in the sense that it does not modify static and dynamic exponents. For the remainder of this chapter we will employ this approximation and, given their unreliability, prefactors of order unity will be discarded.

Let us first define $\rho_t(\mathbf{r}|0)$ and $\rho_t(\mathbf{r}|0, 0)$ to be the conditional chain end densities at $\mathbf{r}$, given one and two chain ends, respectively, at the origin:

$$\rho_t(\mathbf{r}|0) \equiv \frac{\rho_t(\mathbf{r}, 0)}{n(t)}, \qquad \rho_t^{(3)}(\mathbf{r}|0, 0) \equiv \frac{\rho_t(\mathbf{r}, 0, 0)}{\rho_t^s}. \tag{26}$$

In Appendix 5.B it is shown that making the approximation of Eq. (25) in Eq. (23) leads to a much simpler self-consistent expression for $k(t)$:

$$k(t) = Qa^d - Qa^d \int_0^t dt'\, S^{(d)}(t - t')k(t')$$

$$+ 2Qa^d \int_0^t dt' \int d\mathbf{r}' d\mathbf{r}_B\, G_{t-t'}(0, 0, \mathbf{r}' + \mathbf{r}_B, \mathbf{r}_B)k(t')\{\rho_t(\mathbf{r}'|0) - \rho_{t'}^{(3)}(\mathbf{r}'|0, 0)\}. \tag{27}$$

Here, $G_t(\mathbf{r}_A, \mathbf{r}_B, \mathbf{r}'_A, \mathbf{r}'_B)$ is the *equilibrium* chain end propagator; it gives the net weighting for two ends to arrive at $\mathbf{r}_A$, $\mathbf{r}_B$ given that the initial chain configurations were distributed as in equilibrium. This is a well-known object. The return probability, $S^{(d)}(t)$, is the probability that two ends are in contact at t given initial contact:

$$S^{(d)}(t) \equiv \int d^d\mathbf{r}'\, G_t(0, 0, \mathbf{r}', \mathbf{r}') \approx \frac{1}{x_t^d}. \tag{28}$$

The relation $S^d \approx 1/x_t^d$, which will be proved in the next section (Eq. (51) with $g = 0$), follows from the well-known properties of G_t.

5.2.4   *ROLE OF THREE-BODY CORRELATION FUNCTIONS*

Although Eq. (27) is a great simplification relative to Eq. (23), there remains a fundamental many-body difficulty: The undetermined three-body function $\rho^{(3)}$ prevents the closure of Eqs. (27) and (20) in terms of the one-body and two-body functions $n(t)$ and $\rho_t(\mathbf{r}, 0)$, respectively. It will now be demonstrated [57] that if one makes a simple physically motivated assumption then one may show that for small densities, $n_0 R^3 \ll 1$, the terms involving the two- and three-body conditional densities profiles, $\rho^{(3)}(\mathbf{r}|0, 0)$ and $\rho(\mathbf{r}|0)$, exactly cancel one another. Thus, the role of these higher-order correlations is to annihilate the third term on the right-hand side of Eq. (27).

Our assumption is that the more reactive groups placed at the origin, the lower the conditional density: Chemical reactivity can only induce anticorrelations. Thus,

$$\rho_t^{(3)}(\mathbf{r}|00) \leq \rho_t(\mathbf{r}|0) \leq n(t) \qquad \text{(assumption)}. \tag{29}$$

For the remainder of this section we specialize always to $d = 3$. To explore the constraint imposed by (29), let us consider $\bar{q}(\mathbf{r}, 0) \equiv \rho(\mathbf{r}, 0)/n^2(t)$. This quantity is a two-body version of $q_t[\mathbf{R}_A, \mathbf{R}_B]$, introduced in Eq. (21). Note that $k(t) = \bar{q}(0, 0)/Qa^3$. We now return to Eq. (22) and repeat the operations described in the paragraph following it, except now instead of multiplying by $Qa^3\delta(\mathbf{r}_A)\delta(\mathbf{r}_B)$ we multiply by $\delta(\mathbf{r} - \mathbf{r}_A)\delta(\mathbf{r}_B)$ before integrating over $[\mathbf{R}_A, \mathbf{R}_B]$. Implementing the closure approximation we obtain

$$\bar{q}(\mathbf{r}, 0) = 1 - \int_0^t dt' \int d\mathbf{r}' G_{t-t'}(\mathbf{r}, 0, \mathbf{r}', \mathbf{r}')k(t')$$

$$+ \int_0^t dt' \int d\mathbf{r}' d\mathbf{r}_B \, G_{t-t'}(\mathbf{r}, 0, \mathbf{r}' + \mathbf{r}_B, \mathbf{r}_B)k(t')\{\rho_{t'}(\mathbf{r}'|0) - \rho_{t'}^{(3)}(\mathbf{r}'|0, 0)\}, \quad (30)$$

which is the $\mathbf{r}$-dependent version of Eq. (27). This equation is conveniently expressed in terms of the propagator $G_t^{\text{sep}}(\mathbf{r}, \mathbf{r}')$ (see Eq. (46) below, with $g = 0$), namely, the nonreactive probability density two chain ends are separated by $\mathbf{r}$ at time t given initial separation $\mathbf{r}'$:

$$\bar{q}(\mathbf{r}, 0) = 1 - \int_0^t dt' G_{t-t'}^{\text{sep}}(\mathbf{r}, 0)k(t') + \varphi_t(\mathbf{r}) \tag{31}$$

where

$$\varphi_t(\mathbf{r}) \equiv \int_0^t dt' \int d\mathbf{r}' G_{t-t'}^{\text{sep}}(\mathbf{r}, \mathbf{r}')k(t')\mu_t(\mathbf{r}'), \quad \mu_t(r) \equiv 2\{\rho_t(\mathbf{r}|0) - \rho_t^{(3)}(\mathbf{r}|0, 0)\}. \quad (32)$$

Now, from Eq. (29) and Eqs. (31) and (32), we have

$$0 \le \mu_t(\mathbf{r}) \le 2n(t), \quad \int_0^t dt' k(t') \ge \int_0^t dt' \int d\mathbf{r}' k(t')\mu_t(\mathbf{r}'). \tag{33}$$

We have used the fact that the assumption, Eq. (29), implies that $\bar{q} < 1$. Thus, the magnitude of the second term on the right-hand side of Eq. (31) must exceed that of the third term, $\int_0^t dt' G_{t-t'}^{\text{sep}}(\mathbf{r}, 0)k(t') \ge \varphi_t(\mathbf{r})$. This is the full inequality that must be obeyed by the function $\mu_t(\mathbf{r}')$ involving the unknown $\rho^{(3)}$. Integrating over $\mathbf{r}$ yields a less demanding inequality, the last of the three listed above.

Our aim is to show that the third term on the right-hand side of Eq. (27) is much less than the first term Qa^3. This is equivalent to showing that $\varphi_t(0) \ll 1$, as may be seen by noting that if one sets $r = 0$ in Eq. (31) and multiplies by Qa^3, one recovers Eq. (27). We will show that $\varphi_t(0)$ is indeed small for the true (unknown) $\mu_t(\mathbf{r}')$ by determining the function $\mu_{t'}^{\text{max}}(\mathbf{r}')$, which maximizes $\varphi_t(0)$ (for a given time t) subject to the constraints in (33). Then we will show that even in this case $\varphi_t(0) \ll 1$. Note that this function, μ^{max}, can only produce an equal or bigger $\varphi_t(0)$ value than the $\mu^{\max}$ function that would be produced by respecting the full unintegrated constraint, $\overline{q} \leq 1$.

The function $\mu^{\max}$ is calculated in Appendix 5.C. Substituting $\mu = \mu^{\max}$ into the expression for $\varphi_t(\mathbf{r})$ in Eq. (32), we find $\varphi_t(0) \ll 1$ (see Eqs. (C5) and (C6)). This is the worst possible case. Thus, $\varphi_t(0) \ll 1$ must be true for the actual μ and we conclude that in the dilute limit we can delete the third term on the right-hand side of Eq. (27) to obtain a simple self-consistent expression for $k(t)$:

$$k(t) = Qa^d - Qa^d \int_0^t dt' S^{(d)}(t - t')k(t'). \tag{34}$$

5.2.5   *REACTION RATES*

Equation (34) is simple to solve for $k(t)$. Laplace transforming $t \to E$ and specializing to $d = 3$, we obtain

$$k(E) = \frac{Qa^3}{E[1 + Qa^3 S^{(3)}(E)]}. \tag{35}$$

Using Eqs. (17) and (28) we can say the following about $S^{(3)}(E)$: For $E \ll 1/\tau$, $S^{(3)}(E)$ approaches its $E = 0$ limit, $\int_0^\infty dt S^{(3)}(t)$; for $E \gg 1/\tau$, $S^{(3)}(E) \sim E^{-1/4}$. Thus,

$$k(E \gg 1/\tau) = \frac{Qa^3/E}{1 + (Et_2^*)^{-1/4}}, \qquad k(E \ll 1/\tau) = \frac{Qa^3/E}{1 + Qa^3 \int_0^\infty dt S^{(3)}(t)}, \tag{36}$$

where

$$t_2^* = t_a/(Qt_a)^4 \tag{37}$$

is the timescale introduced in the introduction, Eq. (5). This timescale defines a threshold local group reactivity $\hat{Q}$ such that $t_2^* = \tau$ when $Q = \hat{Q}$:

$$\hat{Q}t_a = 1/N^{1/2}. \tag{38}$$

It follows from Eq. (36) that for high reactivities, $Q > \hat{Q}$, k exhibits a short-time ($t < \tau$) time-dependent DC regime with $k \sim t^{-1/4}$:

$$k(t < t_2^*) = Qa^3,$$
$$\qquad\qquad\qquad\qquad (Q > \hat{Q}) \tag{39}$$
$$k(t_2^* < t < \tau) \approx \frac{dx_t^3}{dt} \approx \frac{R^3}{\tau}\left(\frac{\tau}{t}\right)^{1/4}.$$

The time-dependent DC result, first obtained by de Gennes [35], follows from the scaling form $S^{(3)} \approx 1/x_t^3 \sim t^{-3/4}$.

At long times $k(t)$ approaches a constant, k_∞, whose form depends on reactivity Q. This is the result first obtained by Doi [34]:

$$k(t \gg \tau) = k_\infty = \begin{cases} Qa^3, & (Q < \hat{Q}) \\ R^3/\tau \sim 1/N^{1/2}, & (Q > \hat{Q}) \end{cases}. \qquad (40)$$

For weakly reactive systems, $Q < \hat{Q}$, the MF result $k = Qa^3$ holds for all times. Only for $Q > \hat{Q}$ does the long-time rate constant exhibit molecular weight dependence according to the universal scaling law, $k_\infty \sim 1/N^{1/2}$. In this limit, the universality is reflected by the fact that local details such as Q do not appear in k.

The threshold reactivity for this universal behavior is in fact almost never realized for real chemically reacting systems, with the exception of radicals. For Rouse dynamics to apply, one must have $N \lesssim 300$ (to avoid entanglements [51]) and thus $\hat{Q}t_a$ is always greater than a value of order 10^{-1}. Thus, even for the largest N values the reacting species would require reactivities roughly within an order of magnitude of the maximum possible radical reactivities. In practice, only radicals (or optically excitable species) achieve these levels.

5.3 Bulk Reactions: Dilute Solutions

In the previous section we saw (Eqs. (39) and (40)) that reacting chains in the melt exhibit DC kinetics at sufficiently high reactivities ($Q > \hat{Q}$) and sufficiently long times ($t > t_2^*$). By DC kinetics we mean that any pair of reactive chain ends within diffusive range will react. This short-time diffusion-control leaves its mark on the long-time rate constant, which can be written $k_\infty \approx DR$ where $D \approx R^2/\tau$ is the long-time polymer center of gravity diffusivity. This has the same form as the Smoluchowski result [58] for small molecules, $k = Da$, but with the capture radius a being replaced by the coil size R. That is, the small-time compact exploration of space ensures reaction between any pair of groups should their host coil volumes overlap.

What happens in a dilute solution of reactive polymers in a good swelling solvent, where hydrodynamical and excluded volume interactions are no longer screened out? For very small group reactivities, $Qt_a \ll 1$, MF kinetics must apply, as analyzed by Khokhlov [43]. This is always true, no matter what the concentration regime. But what happens for highly reactive groups? This limit is frequently and somewhat misleadingly referred to as the diffusion-controlled (DC) limit.

We consider now the same dilute end-functionalized chains of Section 5.2 but with the background melt replaced by good solvent. As implied by the introductory discussion, the kinetics are completely changed. We now show that no matter how large Q or t, kinetics are always of MF form: the DC limit does not exist [36, 37]. The results that follow have been derived by nearly rigorous RG methods in ref. [36]; however we adopt below the most flexible approach, using the closure approximation (Eq. (25)) as in ref. [37].

5.3.1   SELF-CONSISTENT EXPRESSION FOR RATE CONSTANT $k(t)$

Much of the analysis for dilute solutions is similar to that for the simple Rouse dynamics of the previous section. There are several important differences, however, and we will find entirely different reaction kinetics. First, if we were to define ρ_t^s as we did for Rouse dynamics, Eq. (20), we would obtain $\rho_t^s = 0$ since the equilibrium probability of zero separation between two chain ends vanishes in the presence of excluded volume repulsions. Thus, we must broaden one of the delta functions in the integrand, $\delta(\mathbf{r}_B) \rightarrow \Delta(\mathbf{r}_B)$, where

$$\Delta(\mathbf{r}) \equiv \frac{1}{a^d} \Theta(a - r) \qquad (41)$$

and $\Theta(r)$ is the step function. The function Δ now has width equal to the chain unit size a. With this replacement, Eq. (20) remains correct: That is, it equates the reaction rate to Q times the number of chain end pairs in contact, now correctly defined.

The starting point is again the Fokker-Planck equation, Eq. (18), except that F, describing the system in the absence of reactions, is now the appropriate two-chain operator incorporating intrachain and interchain excluded volume and hydrodynamical effects [36]. Note that this is only a correct description of the chain pair distribution function Φ_t in the present limit of dilute reactive chains where the probability of three-chain interactions is very small. The reaction sinks must also be broadened in Eq. (18), $\delta \rightarrow \Delta$ in every case. That is, we must explicitly allow a possible reaction whenever two reactive groups are within a of each other.

Proceeding as for the Rouse case, the dynamics of q_t are unchanged from Eq. (22) except that all sink δ-functions are replaced by Δ-functions. Replacing $\delta(\mathbf{r}_B) \rightarrow \Delta(\mathbf{r}_B)$ in Eqs. (20) and (21) and solving for q_t in terms of G_t as before, we now multiply both sides of this solution by $Qa^d \delta(\mathbf{r}_A)\Delta(\mathbf{r}_B)$ instead of $Qa^d \delta(\mathbf{r}_A)\delta(\mathbf{r}_B)$, and again integrate over all $[\mathbf{R}_A, \mathbf{R}_B]$. This leads again to Eq. (23), but with all $\delta \rightarrow \Delta$ with the exception of the $\delta(\mathbf{r}_A)$ factors. In addition, the first term on the right-hand side, formerly Qa^d for the Rouse case, is now replaced as

$$Qa^d \rightarrow Qa^d \int d[\mathbf{R}_A]d[\mathbf{R}_B]\delta(\mathbf{r}_A)\Delta(\mathbf{r}_B)q_0 = QVP_{eq}^{cont}. \qquad (42)$$

Here the initial condition for q_t is still $q_0 = V^2 P_{eq}$ (but with a different P_{eq}) and the equilibrium pair contact probability P_{eq}^{cont} of Eq. (2) is

$$P_{eq}^{cont} = a^d V \int d[\mathbf{R}_A] d[\mathbf{R}_B] \delta(\mathbf{r}_A) \Delta(\mathbf{r}_B) P_{eq}[\mathbf{R}_A, \mathbf{R}_B]. \tag{43}$$

After making the closure approximation of Eq. (25), i.e., assuming equilibrium internal chain configurations, we obtain the analog of Eq. (27). This equation again includes a term involving $\rho_t^{(3)}$, analogous to the third term on the right-hand side of Eq. (27). This term may be discarded using similar arguments to those described in Section 5.2.4 for the Rouse case. We then arrive at the analog of Eq. (34):

$$k(t) = QV P_{eq}^{cont} - Q a^d \int_0^t dt' S^{(d)}(t-t') k(t'), \tag{44}$$

where the definition of the return probability $S^{(d)}(t)$ (replacing Eq. (28)) is now

$$S^{(d)}(t) \equiv \int d^d r \, d^d r'_a \, d^d r'_b \Delta(\mathbf{r}) G_t(0, \mathbf{r}, \mathbf{r}'_A, \mathbf{r}'_B) \Delta(\mathbf{r}'_A - \mathbf{r}'_B)$$

$$\approx \int d^d r \, d^d r' \Delta(\mathbf{r}) G_t(0, \mathbf{r}, \mathbf{r}', \mathbf{r}'). \tag{45}$$

We were able to replace $\Delta(\mathbf{r}'_A - \mathbf{r}'_B) \rightarrow \delta(\mathbf{r}'_A - \mathbf{r}'_B)$ with negligible error for times $t \gg t_a$ because G_t is insensitive to changes in initial positions $\mathbf{r}'$, which are small compared to x_t. From Eq. (45) we can express $S^{(d)}(t)$ in terms of G^{sep}, the propagator for the separation between two chain ends (in the absence of reactions) as follows:

$$S^{(d)}(t) = \frac{1}{a^d} \int_{|\mathbf{r}| \leq a} d^d r \, G_t^{sep}(\mathbf{r}, 0), \quad G_t^{sep}(\mathbf{r}, \mathbf{r}') \equiv \int d^d R \, G_t(\mathbf{R}, \mathbf{R} + \mathbf{r}, \mathbf{R}', \mathbf{R}' + \mathbf{r}'). \tag{46}$$

$G_t^{sep}(\mathbf{r}, \mathbf{r}')$, the probability that a pair of chain ends are separated by $\mathbf{r}$ at t given initial separation $\mathbf{r}'$, is the full propagator G_t summed over all possible initial locations of the ends' center of gravity.

Before examining the return probability closely, let us set down the important features of equilibrium static correlations for dilute polymers. These correlations are quantified by $P_{eq}(\mathbf{r})$, the equilibrium probability distribution for the separation $\mathbf{r}$ between a pair of chain ends, Eq. (2). Now repulsion is felt when two coils try to interpenetrate, i.e., for r smaller than the coil size R: This is the nonideal regime with finite g as discussed in the introduction. When two coils are far apart they do not feel one another; i.e., ideal behavior is recovered ($g = 0$). Thus, [10, 39]

$$P_{eq}(\mathbf{r}) = \begin{cases} 1/V \, (r/R)^g, & (r < R) \\ 1/V, & (r > R) \end{cases} \quad R = aN^\nu \tag{47}$$

The rms size of a single polymer coil, R, is determined by ν, the Flory exponent [10]. For $d = 3$ its value is $\nu \approx 3/5$. Note that excluded volume repulsions swell a chain relative to the noninteracting case, $R \sim N^{1/2}$.

5.3.2   STRUCTURE OF RETURN PROBABILITY $S^{(d)}(t)$

To obtain the rate constant $k(t)$ from Eq. (44), we need the return probability. For ideal chains this quantity had the simple structure $S^{(d)} \sim 1/x_t^d$ (Eq. (28)). What is its form in nonideal cases where interchain equilibrium correlations are present? To answer this question [37] we will invoke three principles, the first being detailed balance, which dictates that

$$P_{eq}(\mathbf{r}')G_t^{sep}(\mathbf{r}, \mathbf{r}') = P_{eq}(\mathbf{r})G_t^{sep}(\mathbf{r}', \mathbf{r}). \tag{48}$$

Second, we assert that $G_t^{sep}(\mathbf{r}, \mathbf{r}')$ is approximately independent of $\mathbf{r}'$ for values of $\mathbf{r}'$ much less than x_t. This is simply the principle of finite memory: The initial condition $\mathbf{r}'$ will be forgotten after a time scale much greater than the diffusion time corresponding to $\mathbf{r}'$; i.e., when the width of G^{sep} as a function of r greatly exceeds r', one may simply replace r' with zero. Then, specializing Eq. (48) to small values of $\mathbf{r}$ and $\mathbf{r}'$, we obtain

$$\frac{G_t^{sep}(\mathbf{r}, 0)}{P_{eq}(\mathbf{r})} = \frac{G_t^{sep}(\mathbf{r}', 0)}{P_{eq}(\mathbf{r}')}, \quad (r, r' \ll x_t). \tag{49}$$

Since this must be true for all small values of r and r', we conclude that $G_t^{sep}(\mathbf{r}, 0)/P_{eq}(\mathbf{r})$ is *independent* of r for $r \ll x_t$, i.e., the small r behavior of G^{sep} must match that of $P_{eq}(\mathbf{r})$ displayed in Eq. (2). That is, $G_t^{sep}(\mathbf{r}, 0) = r^g f(t)$ for some function $f(t)$. Finally, let us assume that the only scales in $G_t^{sep}(\mathbf{r}, 0)$ are r and x_t. Then the dimensions of G^{sep} force the following structure:

$$G_t^{sep}(\mathbf{r}, 0) \approx \frac{1}{x_t^d}\left(\frac{r}{x_t}\right)^g, \quad (r \ll x_t). \tag{50}$$

This result is physically reasonable; $G_t^{sep}(\mathbf{r}, 0)$ is essentially equal to the conditional equilibrium probability that the groups are separated by r, given that they lie within x_t of each other [36]. That is, the dynamics of the relative position of the reactive groups are ergodic in a region of volume x_t^d only, so within that portion of $\mathbf{r}$-space relative probabilities are *as in equilibrium*. Inserting the above expression for G^{sep} into the definition of $S^{(d)}(t)$, Eq. (46), we arrive at the following general form [37].

$$S^{(d)}(t) \approx \frac{1}{a^d}\int_{|\mathbf{r}| \leq a} d^d r \frac{1}{x_t^d}\left(\frac{r}{x_t}\right)^g \approx \frac{1}{x_t^d}\left(\frac{a}{x_t}\right)^g \sim \frac{1}{t^{(d+g)/z}}. \tag{51}$$

Note that, as one would expect, excluded volume repulsions make it harder for groups to find one another. Accordingly the return probability is *diminished* by the factor $(a/x_t)^g$. Note also that if we set $g = 0$, recovering the ideal case dealt with in Section 5.2, this proves Eq. (28), which was central in determining the form of $k(t)$ for Rouse chains.

5.3.3   REACTION RATES: ABSENCE OF DIFFUSION-CONTROLLED LIMIT

Let us now specialize the general behavior of $S^{(d)}(t)$, Eq. (51), to the present problem for which there are two regimes. For times much less than the longest polymer relaxation time τ, it is well established [10, 59] that dilute solution dynamics are dominated by hydrodynamical interactions, leading to $z = d$. On the corresponding spatial scales, $x_t \ll R$, excluded volume repulsions are strong (Eq. (47)) with the correlation hole exponent $g \approx 0.27$ for chain ends. The long-time regime, $t > \tau$, is characterized by simple Fickian diffusion ($z = 2$) and ideal statistics ($g = 0$). Setting $d = 3$, Eq. (51) therefore implies

$$S^{(3)}(t) \sim \begin{cases} 1/t^{1+g/3}, & (t < \tau) \\ 1/t^{3/2}, & (t > \tau). \end{cases} \tag{52}$$

The crucial point is that $S^{(3)}(t)$ decays more rapidly than $1/t$ for *all* timescales. In consequence the Laplace transform is dominated by the small-time cutoff at t_a:

$$S^{(3)}(E) \approx \int_{t_a}^{\infty} dt \, e^{-Et} \frac{1}{x_t^3} \left(\frac{a}{x_t} \right)^g \approx \frac{t_a}{a^3}, \quad (Et_a \ll 1). \tag{53}$$

The cutoff corresponds to the fact that $S^{(3)}(t) \approx 1/a^3$ for $t < t_a$.

Returning to Eq. (44), the solution for $Et_a \ll 1$ is

$$k(E) = \frac{QVP_{\mathrm{eq}}^{\mathrm{cont}}/E}{1 + Qa^3 S^{(3)}(E)} = \frac{QVP_{\mathrm{eq}}^{\mathrm{cont}}/E}{1 + Qt_a}. \tag{54}$$

We conclude that reaction kinetics are *always* mean field. Using the explicit form for $P_{\mathrm{eq}}^{\mathrm{cont}}$, obtained from Eq. (2) after using Eq. (47), we obtain the result of refs. [36]:

$$k(t) = k_\infty = \tilde{Q}VP_{\mathrm{eq}}^{\mathrm{cont}} \approx \tilde{Q}a^3 \left(\frac{a}{R} \right)^g = \tilde{Q}a^3 N^{-\nu g}, \qquad \tilde{Q} \equiv Q/(1 + Qt_a). \tag{55}$$

The only remnant of diffusion-controlled effects is in the renormalized local reactivity $\tilde{Q}$, which saturates at $1/t_a$ for $Q > 1/t_a$ as it must do.

The rather surprising conclusion [36, 37] is that there is no diffusion-controlled reaction limit in dilute polymer solutions in good solvents. No matter how large Q may be, k is never time dependent (apart from weak logarithmic corrections [36]) and no long-time result of the form $k \approx R^3/\tau$ is ever valid. The important issue [36, 37] is the value of the "reaction exponent" θ describing the decay of the return probability, $S^{(3)}(t) \sim t^{-\theta}$. We have found $\theta = (d + g)/z$ for general dimensionality. Whether MF or DC kinetics apply depends on whether $S^{(d)}(t)$ decays more or less rapidly than $1/t$ during the short-time regime $t < \tau$. Thus, the general criterion is $d + g > z$ or $d + g < z$, as anticipated in the introduction. Now, if one were to naively apply the compact versus noncompact criterion, then $z = d$ would suggest marginal kinetics. In fact, excluded volume repulsions "tip the balance" in favor of MF kinetics: Because $g > 0$ (for $d < 4$) it is always true that $d + g > z$.

The above analysis is virtually identical for reactions between reactive groups with various other locations along the chain backbones [36]. For the limiting cases, we simply substitute the appropriate value of g into Eq. (55). Using $\nu = 3/5$, the following molecular weight dependencies are predicted:

$$k \sim \begin{cases} N^{-0.16}, & \text{(end–end)} \\ N^{-0.28}, & \text{(end–interior)} \\ N^{-0.43}, & \text{(interior–interior)} \end{cases} \tag{56}$$

Phosphoresence-quenching experiments [19] are consistent with a very low power law for end groups. Interesting future experiments may probe reactions between chains bearing internally positioned reactive groups. Here the power law is rather strong, $k \sim 1/N^{0.43}$, and should be considerably easier to measure accurately than the very weak decay for end groups.

5.4   Interfacial Reactions: Unentangled Melts

How are bulk reaction kinetics modified by the introduction of an interface? In this section we consider the same type of reacting system as investigated in Section 5.2, but now two immiscible bulk phases are present and reactions are restricted to the interfacial region, as shown schematically in Fig. 5.2(a). Theoretically, our problem is one of Rouse chains reacting at an interface [45–50, 60]. Only the specific case $d = 3$ will be treated.

Referring to Fig. 5.2(a), we have two three-dimensional bulk polymer unentangled melt phases A and B, each containing a certain fraction of reactive end-functionalized chains. The initial number densities of reactive groups (of size a) are n_A^∞ and n_B^∞. These may have any values in the range $0 < n_A^\infty \leq n_B^\infty \leq 1/(Na^3)$, where the maximum allowed density corresponds to every chain end in the bulk being reactive. Generally, the densities in either bulk may be unequal, and our convention is always that the B side is the denser. The local reactivity of functional groups is denoted Q_b (rather than Q, see below). Typically the width h of the thin two-dimensional interfacial region, which is the locus of all reaction events, is somewhat larger than a but of the same order of magnitude. We assume A and B chains have identical diffusion dynamics and possess the same number of chain units, N.

In the following, $\mathbf{r}_A$ and $\mathbf{r}_B$ will always denote the spatial coordinates of A or B reactive end groups, respectively. We define $\rho_{AB}(\mathbf{r}_A, \mathbf{r}_B)$ to be the density of reactive chain end A–B pairs at $\mathbf{r}_A, \mathbf{r}_B$, normalized to the total number of reactive pairs in the system. The three-body correlation function $\rho_{ABB}(\mathbf{r}_A, \mathbf{r}_B, \mathbf{r}'_B; t)$ denotes the probability density to find an A–B–B triplet at locations $\mathbf{r}_A, \mathbf{r}_B, \mathbf{r}'_B$. A similar definition applies to $\rho_{ABA}(\mathbf{r}_A, \mathbf{r}_B, \mathbf{r}'_A; t)$. Of particular importance is the interfacial pair density, $\rho_{AB}^s(t)$. This is the two-body density correlation function evaluated at the interface,

$$\rho_{AB}^s(t) \equiv \rho_{AB}(0, 0; t). \tag{57}$$

Translational invariance parallel to the interface is assumed, implying that $\rho_{AB}^s(t)$ is spatially uniform. The next few subsections will assume a bare interface. In the final subsection we consider the influence on reaction kinetics of the accumulation of reaction products.

5.4.1   *INTERFACIAL DENSITY CORRELATION FUNCTION $\rho_{AB}^s(t)$*

The reaction rate is closely related to $\rho_{AB}^s(t)$. Reactions are "switched on" at $t = 0$, and we seek the reaction rate per unit area, $\dot{\mathcal{R}}_t$. This equals the number of A–B reactive group pairs per unit area that are in contact at the interface, $ha^3\rho_{AB}^s(t)$, multiplied by the local reactivity Q_b:

$$\dot{\mathcal{R}}_t = \lambda \rho_{AB}^s(t), \qquad \lambda \equiv Q_b ha^3 \equiv Qa^4, \tag{58}$$

where λ is a naturally occurring measure of reactive strength. Clearly, Eq. (58) has implicitly taken the limit $h \to 0$; that is, our model can only be interpreted on scales large compared to h and $t_h \equiv (h/a)^4 t_a$. On these scales, the quantity $Q \equiv Q_b h/a$ emerges as an effective coarse-grained local reactivity.

A self-consistent expression for $\rho_{AB}^s(t)$, essentially analogous to Eq. (27) for the bulk problem, can be obtained using similar reasoning. This includes making the closure approximation (Eq. (25)), i.e., assuming that internal chain configurations are as in equilibrium and so the only reaction-induced nonequilibrium correlations are those between the ends of different chains. The A and B bulk phases occupy $x > 0$ and $x < 0$, respectively, where the x direction lies orthogonal to the interface, and the superscript T will always denote a three-dimensional vector parallel to the interface. Thus by definition the x-component of $\mathbf{r}^T$ vanishes. In refs. [49] and [50], by applying Doi's [61] second-quantization representation for many-particle reacting systems, it was shown that

$$\rho_{AB}^s(t) = n_A^\infty n_B^\infty - \lambda \int_0^t dt' S^{(4)}(t - t')\rho_{AB}^s(t')$$

$$- \lambda \int_0^t dt' \int d\mathbf{r}_A^{T'} d\mathbf{r}_B' G_{t-t'}(0, 0, \mathbf{r}_A^{T'}, \mathbf{r}_B')\rho_{ABB}(\mathbf{r}_A^{T'}, \mathbf{r}_B', \mathbf{r}_A^{T'}; t')$$

$$- \lambda \int_0^t dt' \int d\mathbf{r}_A' d\mathbf{r}_B^{T'} G_{t-t'}(0, 0, \mathbf{r}_A', \mathbf{r}_B^{T'})\rho_{ABA}(\mathbf{r}_A', \mathbf{r}_B^{T'}, \mathbf{r}_B^{T'}; t'), \qquad (59)$$

where the equilibrium chain end propagator, $G_t(\mathbf{r}_A, \mathbf{r}_B, \mathbf{r}_A', \mathbf{r}_B')$, is the probability density an A–B reactive end group pair is at $\mathbf{r}_A$, $\mathbf{r}_B$ at time t given initial locations $\mathbf{r}_A'$, $\mathbf{r}_B'$, in the *absence* of reactions. Here

$$S^{(4)}(t) \equiv \int d\mathbf{r}^{T'} G_t(0, 0, \mathbf{r}^{T'}, \mathbf{r}^{T'}) \approx \frac{1}{x_t^4} \qquad (60)$$

is the two-body "return probability," namely, the probability density an A–B pair of chain ends is in contact at time t at the interface, given it was in contact somewhere within the interface at $t = 0$. Note that $S^{(4)}(t) \approx 1/x_t^4$, a relation derived in Appendix 5.D, has the same form as the return probability in a four-dimensional bulk problem (compare to Eq. (28)).

The events corresponding to the three reaction sink-derived terms in Eq. (59) are depicted in Fig. 5.5. Each reaction integral subtracts off interfacial pairs that are incorrectly counted by the first term, $n_A^\infty n_B^\infty$. These wrongly counted pairs involve either one or two reactive end groups that have reacted at some previous time. (1) The two-body integral involving $S^{(4)}(t)$ represents depletion in $\rho_{AB}^s(t)$ due to pairs of reactive A–B ends that failed to reach the origin because their members reacted with *one another* at some earlier time t' (2), (3). The last two terms in Eq. (59) measure depletion in ρ_{AB}^s due to many-body effects; they subtract off any A–B reactive pair that would have arrived at the origin at time t but failed to because *only one* member of the pair reacted with any other reactant in the system at the interface at some earlier time t'. Such an event involves a third end group, weighted by the appropriate three-body correlation function.

The major difficulty here is that Eq. (59) does not close in terms of two-body correlation functions, with higher-order correlations being involved in the last

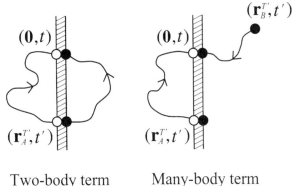

Two-body term   Many-body term

Figure 5.5 The reduction in the number density of reactive A–B groups in contact at the origin from the value it would have in the absence of reactions originates from three terms, Eq. (59). The two-body term counts those A–B pairs that would have been at the origin at time t, but failed to arrive because *both* members reacted at an earlier time t' at point $\mathbf{r}_A^{T'}$. The many-body terms count A–B pairs that would have been at the origin had there been no reactions, but failed to arrive because *one* member of the pair (the A member here) reacted at an earlier time.

two reaction terms. In ref. [49] it was shown that after making certain simple assumptions on the boundedness of the three-body correlation functions, these integral terms when relevant (at long times) can in fact be expressed in terms of ρ_{AB}^s. Consider $\rho_{ABB}(\mathbf{r}_A^{T'}, \mathbf{r}_B', \mathbf{r}_A^{T'}; t')$ appearing in Eq. (59). Now at time t', values of $\mathbf{r}_B'$ with sufficiently large x-coordinate, $x'_B \gg x_{t'}$, are beyond diffusional range of the interface and therefore uncorrelated with it. This implies

$$\rho_{ABB}(\mathbf{r}_A^{T'}, \mathbf{r}_B', \mathbf{r}_A^{T'}; t') \approx \rho_{AB}^s(t') n_B^\infty, \quad \text{when } x'_B \gg x_{t'}. \tag{61}$$

The essential point of the argument in ref. [49] was that *a fraction of order unity* of the value of the integral term containing ρ_{ABB} derives from integration over points in which ρ_{ABB} can be written as in Eq. (61). An analogous statement holds for the other many-body term. Now, suppose we make the approximation of Eq. (61) in the ρ_{ABB} many-body term of Eq. (59). Then, since the two-chain propagator G_t can be expressed as a product of one-chain propagators $G_t^{(1)}$ (see Appendix 5.D, Eq. (D1)), the value of this term would be $\lambda n_B^\infty \int_0^t dt' S^{(1)}(t-t') \rho_{AB}^s(t')$, where

$$S^{(1)}(t) \equiv \int d\mathbf{r}^T G_t^{(1)}(0, \mathbf{r}^T) \approx \frac{1}{x_t} \tag{62}$$

is the one-dimensional return probability, namely, the probability that a reactive end group (either A or B) initially within the interface returns to an interfacial site after time t. It is defined in terms of the single chain end propagator $G_t^{(1)}(\mathbf{r}, \mathbf{r}')$. The relation $S^{(1)} \approx 1/x_t$ follows immediately from the structure of $G_t^{(1)}$ (see Appendix 5.D).

Accepting the above, it follows that Eq. (59) can be written from ref. [49]

$$\rho_{AB}^s(t) = n_A^\infty n_B^\infty - \lambda \int_0^t dt' S^{(4)}(t-t') \rho_{AB}^s(t') - \lambda n(t) \int_0^t dt' S^{(1)}(t-t') \rho_{AB}^s(t'), \quad (63)$$

where $n(t) = A(t)(n_A^\infty + n_B^\infty)$, with $A(t)$ being an unknown time-dependent function of order unity. In ref. [49] it was argued that the interfacial density on the more dilute A side vanishes at long times; it was then shown that this forces the asymptotic value of $n(t)$ to be exactly the far-field density in the more dense phase:

$$n(t \to \infty) = n_B^\infty. \quad (64)$$

Because the many-body term is actually dominant only at times long enough that $n(t)$ is close to its asymptotic value, for simplicity we will replace $n(t) \to n_B^\infty$ in the following. The relation of Eq. (63) is now a closed one for $\rho_{AB}^s(t)$.

5.4.2   CHARACTERISTIC TIMESCALES

Setting $n(t) \to n_B^\infty$ in Eq. (63), we can now solve for ρ_{AB}^s and obtain the reaction rate using (58). This gives the following expression for its Laplace transform, $\dot{\mathcal{R}}_t(E)$:

$$\dot{\mathcal{R}}_t(E) = \frac{\lambda n_A^\infty n_B^\infty}{E[1 + \lambda S^{(4)}(E) + \lambda n_B^\infty S^{(1)}(E)]}. \quad (65)$$

This is quite general, and the return probabilities $S^{(1)}(E)$ and $S^{(4)}(E)$ must now be specialized to Rouse dynamics. A number of different kinetic regimes will emerge for different values of Q (or λ) and n_A^∞, n_B^∞.

Now, as discussed in the introduction, for ideal situations such as in melts, interfacial reaction kinetics can be termed "compact" or "noncompact," depending on $d + 1 < z$ or $d + 1 > z$, by analogy to a bulk problem of $d + 1$ dimensions. With $d = 3$, this means the present Rouse case is marginal at short times, $z = 4$, and we expect logarithmic corrections to DC kinetics at these times. The long-time Fickian behavior is noncompact, $z = 2$, suggesting MF kinetics. These must yield at some time to one-dimensional DC kinetics. In fact, we will see that for very reactive and dense systems these longer-time MF kinetics are entirely eclipsed by the final first-order DC regime. These various features are all contained within the specific Rouse form of the return probabilities $S^{(4)}(E)$ and $S^{(1)}(E)$. These are determined in Appendix 5.E, Eqs. (E2) and (E3).

For a given E value, the reaction rate $\dot{R}_t(E)$ in Eq. (65) is determined by which of the three terms in the denominator is the greatest. For short times $t < \tau$, corresponding to large E values, $E > 1/\tau$, it is natural to express $\dot{R}_t(E)$ in terms of the timescales t_2^*, t_m^*, and T_l defined as follows:

$$\frac{t_2^*}{t_h} \equiv e^{1/Qta}, \qquad \frac{t_m^*}{t_a} \equiv \frac{1}{(Qt_a n_B^\infty a^3)^{4/3}}, \qquad T_l \equiv t_l \ln(T_l/t_h), \qquad (66)$$

where

$$t_l \equiv t_a (n_B^\infty a^3)^{-4/3}. \qquad (67)$$

If $t_l < \tau$, then t_l is identified as the diffusion time corresponding to the mean distance between reactive groups on the dense B side. Similarly, if t_2^* and t_m^* are less than τ, they correspond to the timescales discussed in the introduction. The t_2^* expression above is then the marginal version of Eq. (8) ($z = d + 1$, see discussion following Eq. (9)), and the t_m^* expression is identified as that of Eq. (11) with $z = 4$, $d = 3$. If any of these timescales exceed τ, they become physically irrelevant but remain useful as a way of expressing terms in Eq. (65).

Similarly, the long-time ($t > \tau$, $E < 1/\tau$) form of $\dot{R}_t(E)$ can conveniently be written in terms of $t_{m,R}^*$ and $t_{m,W}^*$, where

$$\frac{t_{m,W}^*}{\tau} \equiv \frac{1}{(Q\tau n_B^\infty a^4/R)^2}, \qquad \frac{t_{m,R}^*}{\tau} \equiv \left(\frac{\int_0^\infty dt S^{(4)}(t)}{n_B^\infty \tau/R}\right)^2 = \left(\frac{\ln(\tau/t_h)}{n_B^\infty R^3}\right)^2. \qquad (68)$$

With these definitions, we can write for large E

$$\dot{R}_t(E) = \frac{\lambda n_A^\infty n_B^\infty}{E}\left\{1 + \frac{\ln(1/Et_h)}{\ln(t_2^*/t_h)} + (Et_m^*)^{-3/4}\right\}^{-1} \qquad (A)$$

$$= \frac{\lambda n_A^\infty n_B^\infty}{E}\left\{1 + \frac{\ln(1/Et_h)}{\ln(t_2^*/t_h)}\left[1 + \frac{(Et_l)^{-3/4}}{\ln(1/Et_h)}\right]\right\}^{-1} \qquad (B)$$

$$(E \gg 1/\tau) \qquad (69)$$

where T_l has been defined such that the two terms in the square brackets are equal at $E = 1/T_l$,

$$\left[\frac{(Et_l)^{-3/4}}{\ln(1/Et_h)}\right]_{E=1/T_l} = 1. \qquad (70)$$

Similarly, for small E

$$\dot{\mathcal{R}}_t(E) = \frac{\lambda n_A^\infty n_B^\infty}{E} \left\{ 1 + \lambda \int_0^\infty dt\, S^{(4)}(t) + (Et_{m,W}^*)^{-1/2} \right\}^{-1} \quad \textbf{(C)}$$

$$= \frac{\lambda n_A^\infty n_B^\infty}{E} \left\{ 1 + \lambda \int_0^\infty dt\, S^{(4)}(t)[1 + (Et_{m,R}^*)^{-1/2}] \right\}^{-1} \quad \textbf{(D)}$$

$$(E \ll 1/\tau). \tag{71}$$

The two relevant physical parameters are Q and n_B^∞. For different densities and functional group reactivities, the reaction kinetics will be different. Consider short times, $E > 1/\tau$. From Eqs. (69) and (70) it is clear that the ordering of the timescales t_2^*, t_m^*, T_l, and τ completely determines the sequence of kinetic regimes. This is because these timescales determine the values of E at which one term will start to dominate over another term. (For example, for $E > 1/t_m^*$ the 1 term in the denominator of Eq. (65) exceeds the $\lambda n_B^\infty S^{(1)}(E)$ term, and vice versa for $E < 1/t_m^*$.) Now, as one changes Q and n_B^∞, so the timescales also will change; if, as a result, their ordering changes, then the sequence of regimes will change. We note that from their definitions the magnitude of t_m^* always lies between those of t_2^* and T_l. Thus, there are only two orderings we need consider:

$$T_l < t_m^* < t_2^*, \quad (Q < Q^*); \qquad t_2^* < t_m^* < T_l, \quad (Q > Q^*). \tag{72}$$

The timescales are degenerate at a certain critical reactivity Q^*:

$$t_2^* = t_m^* = T_l, \quad (Q = Q^*); \qquad Q^* t_a \equiv \frac{1}{\ln(T_l/t_h)}. \tag{73}$$

5.4.3   KINETIC SEQUENCE: FOUR REGIONS IN THE Q-n_B^∞ PLANE

In Fig. 5.6 we depict the Q-n_B^∞ plane, in which we have drawn the lines $t_2^* = \tau$, $t_m^* = \tau$, $T_l = \tau$, and $t_2^* = t_m^* = T_l$. This last line may also be called $Q = Q^*$. (In those regions where they are irrelevant, parts of these lines are not shown; this will become clear below.) This defines four regions, each having a unique sequence of reaction kinetic behaviors. Let us consider these regions one by one, referring to Fig. 5.6 throughout. In each region, and for different E values, we will identify which of the three terms in the denominator of Eq. (65) are dominant: 1, $\lambda S^{(4)}(E)$, or $\lambda n_B^\infty S^{(1)}(E)$. We will "scan" from large E (small times) to small E (large times).

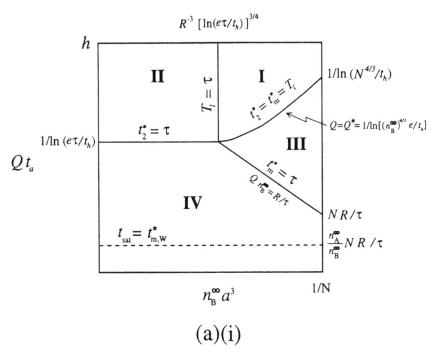

$$\frac{(a)(i)}{}$$

(a)(i)

Figure 5.6   Interfacial reactions, unentangled melts, one reactive group per reactive polymer. "Phase diagram" of reaction kinetics as a function of renormalized reactivity $Q \equiv Q_b h/a$ and density of the denser phase n_B^∞. Axes are logarithmic and units are chosen such that $t_a = a = 1$. Maximum possible density is $n_B^\infty a^3 = 1/N$ (all chains functionalized). $\tau = N^2 t_a$ is the Rouse polymer relaxation time; all other timescales are defined in Tables 5.1 and 5.2. (a)(i) There are four regions in the phase diagram. The line $t_{sat} = t_{m,W}^*$ is shown for the special case of constant n_A^∞ / n_B^∞ (note the convention, $n_A^\infty \le n_B^\infty$). (a)(ii) Three types of kinetic behavior occur, in the three regions labeled S^{conc}, S^{dil}, and W. These symbols denote, respectively, "Strong Concentrated," "Strong Dilute" and "Weak." In each region the sequence of kinetics is indicated. T_m^* denotes t_m^* in region III and $t_{m,W}^*$ in region IV (III and IV are subregions of W and have the same kinetic sequences). Below the $t_{sat} = t_{m,W}^*$ line, interface saturation occurs before the onset of the DC1 regime: kinetics are then always MF until saturation. (b) As (a)(i), but reactivities and densities are expressed in terms of molecular weight N.

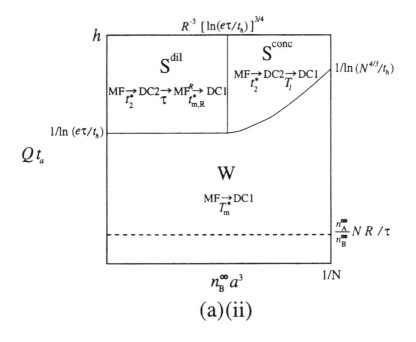

(a)(ii)

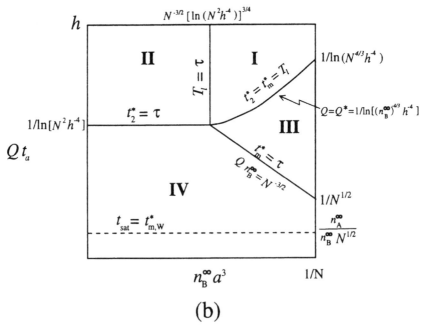

(b)

Figure 5.6   Continued

Region I, "Strong Concentrated": $Q > Q^*$, $T_I < \tau$

From Eq. (72) we have $t_2^* < T_I < \tau$ in this region. Thus, by inspection of Eq. (69)(B) we see that for $E^{-1} < t_2^*$ the 1 term dominates; for $t_2^* < E^{-1} < T_I$ the $S^{(4)}$ term dominates; and for $T_I < E^{-1} < \tau$ the $S^{(1)}$ term dominates. In fact the $S^{(1)}$ term continues to dominate for all $E^{-1} > \tau$ since it is an increasing function of E^{-1} and since the $S^{(4)}$ term saturates at $E^{-1} = \tau$ (see Eq. (E3)). Thus,

$$\dot{\mathcal{R}}_t(E) \approx \begin{cases} \lambda n_A^\infty n_B^\infty/E, & (E^{-1} < t_2^*) \\ n_A^\infty n_B^\infty/ES^{(4)}(E), & (t_2^* < E^{-1} < T_I) \\ n_A^\infty/ES^{(1)}(E), & (E^{-1} > T_I) \end{cases} \tag{74}$$

Region II, "Strong Dilute": $t_2^* < \tau < T_I$

Region II is similar to region I, but now the densities are so low that T_I exceeds τ and becomes irrelevant. From the ordering of timescales and Eq. (69)(B), the 1 term dominates for $E^{-1} < t_2^*$ whereas for $t_2^* < E^{-1} < \tau$ the $S^{(4)}$ term dominates. Looking at Eq. (71)(D), one sees that at larger $E^{-1} > \tau$ the $S^{(4)}$ term, despite saturating to its $E = 0$ value, remains dominant until $E^{-1} = t_{m,R}^*$, at which point the $S^{(1)}$ term becomes the greatest. This term remains dominant thereafter. Thus,

$$\dot{\mathcal{R}}_t(E) \approx \begin{cases} \lambda n_A^\infty n_B^\infty/E, & (E^{-1} < t_2^*) \\ n_A^\infty n_B^\infty/ES^{(4)}(E), & (t_2^* < E^{-1} < \tau) \\ n_A^\infty n_B^\infty/E\int_\infty^0 dt\, S^{(4)}(t), & (\tau < E^{-1} < t_{m,R}^*) \\ n_A^\infty/ES^{(1)}(E), & (E^{-1} > t_{m,R}^*) \end{cases} \tag{75}$$

Region III, "Weak": $Q < Q^*$, $t_m^* < \tau$

Now Eq. (72) implies $t_m^* < t_2^*$. Thus from Eq. (69)(A), the 1 term dominates for $E^{-1} < t_m^*$ after which point the $S^{(1)}$ term takes over. This term remains dominant thereafter since it increases more rapidly than the $S^{(4)}$ term. Thus,

$$\dot{\mathcal{R}}_t(E) \approx \begin{cases} \lambda n_A^\infty n_B^\infty/E, & (E^{-1} < t_m^*) \\ n_A^\infty n_B^\infty/ES^{(1)}(E), & (E^{-1} > t_m^*) \end{cases} \tag{76}$$

Region IV, "Weak": $t_2^* > \tau$, $t_m^* > \tau$

In this region both $S^{(1)}$ and $S^{(4)}$ terms are less than 1 for all $E^{-1} < \tau$, as is clear from Eq. (69)(A). For $E^{-1} > \tau$, the $S^{(4)}$ term saturates but we see from Eq. (71)(C) that the $S^{(1)}$ keeps growing and catches up with the 1 term at $E^{-1} = t^*_{m,W}$. Thus,

$$\dot{\mathcal{R}}_t(E) \approx \begin{cases} \lambda n_A^\infty n_B^\infty / E, & (E^{-1} < t^*_{m,W}) \\ n_A^\infty n_B^\infty / E S^{(1)}(E), & (E^{-1} > t^*_{m,W}) \end{cases} \tag{77}$$

Since the kinetic sequences are the same in regions III and IV, we have given them the same label, "weak." The only difference is in the timescale, either t^*_m or $t^*_{m,W}$, separating the two kinetic regimes.

5.4.4   *FOUR TYPES OF REACTION KINETICS: MF, MF^R, DC1, DC2*

Looking at Eqs. (74)–(77) we observe that four classes of reaction kinetics are realized. When $\dot{\mathcal{R}}_t = \lambda n_A^\infty n_B^\infty / E$, we have second-order mean field (MF) kinetics, where the reaction rate is proportional to the equilibrium value of the number of reactive pairs in contact at the interface:

$$\dot{\mathcal{R}}_t = k^{(2)} n_A^\infty n_B^\infty, \qquad k^{(2)} \approx \lambda \equiv Q_b h a^3 \qquad \text{(MF kinetics).} \tag{78}$$

The form $\dot{\mathcal{R}}_t(E) = n_A^\infty n_B^\infty / E S^{(4)}(E)$ for $E^{-1} < \tau$, with $S^{(4)}(E)$ given by Eq. (E3), leads again to second-order kinetics but this time of diffusion-controlled (DC) form:

$$\dot{\mathcal{R}}_t = k^{(2)} n_A^\infty n_B^\infty, \qquad k^{(2)} \approx \frac{a^4}{t_a \ln(t/t_h)} \qquad \text{(DC2 kinetics).} \tag{79}$$

(For details on the inverse Laplace transformation in this case, see ref. [49].) Note that if the dynamics were compact (rather than marginal as here) then since $S^{(4)}(t) \sim 1/x_t^4$ we would obtain $\dot{\mathcal{R}}_t \sim (dx_t^4/dt) n_A^\infty n_B^\infty$. This is the second-order DC kinetics behavior discussed following Eq. (9). We label this class DC2. The present case is an instance of this, but since it is marginal a logarithmic correction arises. However, the term DC remains appropriate since a depletion region of size x_t grows in the two-body correlation function [49] and the reaction rate is independent of λ.

In one case, region II, the form $\dot{\mathcal{R}}_t(E) = n_A^\infty n_B^\infty / E \int_0^\infty S^{(4)}$ arises, leading again to second-order kinetics with a constant $k^{(2)}$:

$$\dot{\mathcal{R}}_t = k^{(2)} n_A^\infty n_B^\infty, \qquad k^{(2)} \approx \frac{1}{\int_0^\infty dt\, S^{(4)}(t)} \approx \frac{R^4}{\tau \ln(\tau/t_h)} \qquad (MF^R \text{ kinetics).} \tag{80}$$

We call these kinetics MF^R (for "renormalized" MF) because, coarse-grained on scales R, τ, they are of MF form with a renormalized local reactivity.

Finally, the behavior $\dot{\mathcal{R}}_t(E) = n_A^\infty/ES^{(1)}(E)$ is very different from the other forms above. These are *first-order* DC kinetics with first-order rate constant $k^{(1)}$:

$$\dot{\mathcal{R}}_t = k^{(1)}n_A^\infty, \qquad k^{(1)} \approx \frac{dx_t}{dt} \sim \begin{cases} t^{-3/4}, & (t < \tau) \\ t^{-1/2}, & (t > \tau) \end{cases} \qquad \text{(DC1 kinetics)}. \qquad (81)$$

We label these kinetics DC1. In every case, the final kinetic regime is of this type; at long enough times, every A reactive group arriving at the interface will definitely react, and a depletion hole grows on the dilute A side. Hence the reaction rate itself involves the A density, n_A^∞, whereas the density-dependent timescale t_m^* involved the density on the more concentrated B side.

5.4.5   *PHASE DIAGRAM IN Q-n_B^∞ PLANE*

We have seen that in each region there is a different sequence of reaction kinetic regimes in time. As remarked, regions III and IV exhibit the same sequence, with only the definition of the relevant timescale being different. Each regime of reaction kinetics is one of the four types described in the previous subsection. Following Eqs. (74)–(77), we have indicated the full behavior in each region in Fig. 5.6. This figure may be used with Tables 5.1 and 5.2, which list the definitions of all relevant timescales. We postpone discussion of the "saturation line" (labeled $t_{\text{sat}} = t_{m,W}^*$) appearing in region IV until the next subsection.

To relate this phase diagram to a given experimental system, one needs $\epsilon \equiv Qt_a \approx (h/a)k^{\text{small}}/k_{\text{rad}}^{\text{small}}$ (cf. discussion surrounding Eqs. (12) and (13), with $Q \to Q_b$) and the volume fraction $\phi_B \equiv n_B^\infty a^3$ of reactive groups on the denser side of the interface.

Referring to Fig. 5.6, the number of reactions per unit area is predicted to exhibit the following behaviors in each region. Below, $D \approx R^2/\tau$ denotes the long-time self-diffusivity of a polymer chain. For very large Q and n_B^∞, we have

$$\mathcal{R}_t \approx \begin{cases} \lambda n_A^\infty n_B^\infty t, & (t < t_2^*, \text{MF}) \\ n_A^\infty n_B^\infty a^4 t/[t_a \ln(t/t_h)], & (t_2^* < t < T_l, \text{DC2}) \\ n_A^\infty a(t/t_a)^{1/4}, & (T_l < t < \tau, \text{DC1}) \\ n_A^\infty(Dt)^{1/2}, & (\tau < t, \text{DC1}) \end{cases} \qquad \text{(Strong Concentrated)} \quad (82)$$

The behavior at high reactivities, $(Qt_a > 1/\ln(\tau/t_h))$, but at lower densities, such that the reactive polymers are well below their overlap threshold, is similar except now the $t^{1/4}$ regime is replaced by a MF^R regime:

Table 5.1 Unentangled chains: Formulas for characteristic timescales and Q^*. The volume fraction of reactive end groups in the B and A bulks are denoted $\phi_B \equiv n_B^\infty a^3$ and $\phi_A \equiv n_A^\infty a^3$, respectively. Factors of e have been introduced to ensure continuity of reaction rates.

	Rouse $t^{1/4}$ $t_h < t < \tau$	Fickian $t^{1/2}$ $t > \tau$
t_m^*	$t_a(Qt_a\phi_B)^{-4/3}$	
t_2^*	$(t_h/e)e^{1/Qt_a}$	
T_l	$t_a\phi_B^{-4/3}[\ln(e\phi_B^{-4/3}t_a/t_h)]^{4/3}$	
$t_{m,W}^*$		$\tau[Q\tau(a/R)\phi_B]^{-2}$
$t_{m,R}^*$		$\tau\left[\dfrac{\ln(e\tau/t_h)}{\phi_B(R/a)^3}\right]^2$
t_{sat}		$1/(Q\phi_A\phi_B R/a)$ or $\tau/(N\phi_A)^2$
Q^*	$\dfrac{1}{t_a \ln[e\phi_B^{-4/3}t_a/t_h]}$	

Table 5.2 Unentangled chains: Identical to Table 5.1, but now all quantities are expressed in terms of $\epsilon \equiv Qt_a$ and molecular weight N.

	Rouse $t^{1/4}$ $t_h < t < \tau$	Fickian $t^{1/2}$ $t > \tau$
t_m^*/t_a	$(\epsilon\phi_B)^{-4/3}$	
t_2^*/t_a	$(h/a)^4 e^{1/\epsilon}$	
T_l/t_a	$\phi_B^{-4/3}[\ln(\phi_B^{-4/3}a/h)]^{4/3}$	
$t_{m,W}^*/t_a$		$N^{-1}(\epsilon\phi_B)^{-2}$
$t_{m,R}^*/t_a$		$N^{-1}\phi_B^{-2}[\ln(N^2 h/a)]^2$
t_{sat}/t_a		$1/(\epsilon\phi_A\phi_B N^{1/2})$ or $1/\phi_A^2$
Q^*t_a	$1/\ln(\phi_B^{-4/3}a/h)$	

$$\mathcal{R}_t \approx \begin{cases} \lambda n_A^\infty n_B^\infty t, & (t < t_2^*, \text{MF}) \\ n_A^\infty n_B^\infty [a^4/[t_a \ln(t/t_h)]]t, & (t_2^* < t < \tau, \text{DC2}) \\ n_A^\infty n_B^\infty [a^4/t_a \ln(\tau/t_h)]t, & (\tau < t < t_{m,R}^*, \text{MF}^R) \\ n_A^\infty (Dt)^{1/2}, & (t > t_{m,R}^*, \text{DC1}) \end{cases} \quad \text{(Strong Dilute).} \quad (83)$$

For less reactive functional groups the behavior is simple: The initial linear MF regime crosses over directly into the long-time DC1 regime. The actual power laws in time are different for subregions III and IV:

$$\mathcal{R}_t \approx \begin{cases} \lambda n_A^\infty n_B^\infty t, & (t < t_m^*, \text{MF}) \\ n_A^\infty a(t/t_a)^{1/4}, & (t_m^* < t < \tau, \text{DC}) \\ n_A^\infty (Dt)^{1/2}, & (t > \tau, \text{DC1}) \end{cases} \quad \text{(Weak, region III)} \quad (84)$$

and

$$\mathcal{R}_t \approx \begin{cases} \lambda n_A^\infty n_B^\infty t, & (t < t_{m,W}^*, \text{MF}) \\ n_A^\infty (Dt)^{1/2}, & (t > t_{m,W}^*, \text{DC1}) \end{cases} \quad \text{(Weak, region IV).} \quad (85)$$

5.4.6 *SATURATION OF THE INTERFACE*

The entire analysis of this section has assumed that the interface is unaltered by reaction products. This is valid at sufficiently short times. However, for end-functionalized chains, reactions generate an increasingly dense layer of A–B diblock copolymer product, which beyond a certain timescale must inhibit further reactions (see Fig. 5.7). Roughly speaking, at sufficiently high surface density,

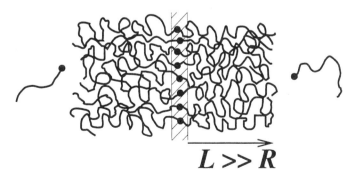

$$L \gg R$$

Figure 5.7 Reaction rates are exponentially suppressed when the interface becomes crowded with diblock product at values of surface density above $1/N^{1/2}a^2$. This corresponds to the diblock brush being stretched, i.e., its size L being much bigger than the unperturbed chain dimension R. (Reprinted with permission. See ref. (46). © 1996 American Chemical Society.)

the diblock brush formed at the A–B interface becomes extended relative to the unperturbed coil size $R = N^{1/2}a$; when this happens the density of mobile bulk chains near the interface will drop significantly below the bulk values n_A^∞ and n_B^∞ with consequent [46] reduction in k (see Fig. 5.7).

It can been shown [62–65] that the surface density at which such diblocks start to feel one another laterally—and therefore become stretched—is one diblock per interface area $N^{1/2}a^2$. This is simple to understand if one notes that an imaginary plane slicing through an unperturbed bulk polymer melt is intersected by one chain per area $N^{1/2}a^2$ since there are R/Na^3 chains per unit area of such a surface that are close enough (i.e., within distance $R = N^{1/2}a$) to intersect it (the volume of one chain is Na^3). Hence, any attempt to load a surface beyond this areal density must perturb the chains relative to their Gaussian dimensions in a free melt: The chains cannot fit into the space unless they stretch. According to a detailed analysis in ref. [46], the rate constant is exponentially suppressed for diblock surface densities ρ_{surf} above this critical level,

$$k \sim \exp(-9Na^4\rho_{\text{surf}}^2), \qquad \rho_{\text{surf}} > 1/N^{1/2}a^2. \tag{86}$$

Essentially, therefore, reactions stop at a time t_{sat} where $\mathcal{R}_{t_{\text{sat}}} = 1/N^{1/2}a^2$.

To determine the value of t_{sat}, note first that it must exceed τ; even at the maximum density, $n_A^\infty = 1/Na^3$, and even if reactions were diffusion-controlled, a time τ would be required to react all those chains with ends within R of the interface and thereby saturate it. Note also that under MF^R kinetics, to within a logarithmic prefactor, $\mathcal{R}_t \approx (n_A^\infty/n_B^\infty)/R^2$ at $t = t^*_{m,R}$ (see Eqs. (68) and (80)). Since this surface density is less than $1/N^{1/2}a^2$, hence, $t_{\text{sat}} > t^*_{m,R}$. It follows that at the time t_{sat}, reaction kinetics are either MF or DC1 (see Fig. 5.6). Using the $t > \tau$ expressions for $\mathcal{R}_t$ implied by Eqs. (78) and (81) and equating these to $\mathcal{R}_{t_{\text{sat}}}$, one obtains

$$t_{\text{sat}} = \begin{cases} 1/[Qa^5Rn_A^\infty n_B^\infty], & (t_{\text{sat}} < t^*_{m,W}, Q < Q^\dagger) \\ \tau/(Nn_A^\infty a^3)^2, & (t_{\text{sat}} > t^*_{m,W}, Q > Q^\dagger) \end{cases} \tag{87}$$

Note that $t^*_{m,W}$ is a decreasing function of Q at fixed n_B^∞ (see Eq. (68)). At the boundary between the two cases in Eq. (87), Q has the value $Q^\dagger$ given below.

$$t_{\text{sat}} = t^*_{m,W}, \qquad Qt_a = Q^\dagger t_a \equiv \frac{n_A^\infty}{n_B^\infty} N \frac{R}{a} \frac{t_a}{\tau}. \tag{88}$$

The line $Q = Q^\dagger$ is indicated in the phase diagrams of Fig. 5.6. For simplicity, we have assumed that the ratio n_A^∞/n_B^∞ is fixed as n_B^∞ varies. Below this line, saturation of the interface at time t_{sat} truncates the $t > \tau$ MF regime before the final DC1 kinetics can be realized. Above it, saturation occurs after the onset of DC1 kinetics and the full "weak" sequence is realized.

5.5   Interfacial Reactions: Entangled Melts

Above a certain species-dependent molecular weight dynamical properties of polymer melts are strongly affected by entanglements [51, 10, 28]. Diminished center of gravity diffusivities and raised viscosities are the result of entanglements suppressing the mobility of individual chains [51]. Correspondingly, polymer reactions occurring at an interface separating two entangled bulk melts must be slowed down relative to the unentangled situation studied in the previous section. Moreover, one expects different reaction kinetics since entanglements modify power laws characterizing small-time monomer diffusion.

In this section we briefly discuss the influence of entanglements on interfacial reaction kinetics. We consider the identical situation to that of the previous section, but now both A and B chains are entangled; That is, $N > N_e$, where N_e is the entanglement threshold [51, 10, 28] (assumed for simplicity to be the same for A and B). For brevity's sake, results in this section will simply be presented without detailed derivation. The basic principles are the same as those that emerged from the unentangled analysis. (For detailed derivations, see ref. [50].)

5.5.1   REPTATION DYNAMICS

Our theoretical framework is the "reptation" model, which has successfully explained many entangled phenomena, although significant anomalies remain [28, 10]. This theory models entanglements as inhibiting lateral chain motion on a certain entanglement scale r_e corresponding to a portion of chain equal in length to the threshold N_e: In effect, each chain is confined to a "tube" of diameter r_e. Note that the random walk statistics of Eq. (15) are unchanged by entanglements; thus the rms coil size remains $R = N^{1/2}a$ and the tube diameter is $r_e = N_e^{1/2}a$. We assume the tube diameter exceeds the interface width h.

In Appendix 5.F it is shown that the reptation model leads to the following behavior for the rms displacement of a chain end:

$$x_t = \begin{cases} a(t/t_a)^{1/4}, & (t < t_e \equiv N_e^2 t_a, \text{ "Rouse"}) \\ r_e(t/t_e)^{1/8}, & (t_e < t < t_b \equiv N^2 t_a, \text{ "breathing"}) \\ r_b(t/t_b)^{1/4}, & (t_b < t < \tau \equiv (R/r_b)^4 = N^3 t_a/N_e, \text{ "tube"}) \\ R(t/\tau)^{1/2}, & (\tau < t, \text{ "Fickian"}). \end{cases} \quad (89)$$

Here, $r_b \equiv r_e (N/N_e)^{1/4}$ and the longest polymer relaxation or reptation time is $\tau = (N^3/N_e)t_a$.

Thus, there are four regimes with the following sequence of dynamical exponents: $z = 4, 8, 4, 2$. Recalling that $z = d + 1 = 4$ is the marginal boundary, we see that two of these regimes are marginal (occurring at small times $t < \tau$) and one noncompact (long times, $t > \tau$). These regimes are familiar from the unentangled Rouse case, and correspondingly $S^{(4)}(E)$ exhibits logarithmic behavior when $z = 4$ and saturates for $E \ll 1/\tau$ where $z = 2$. In addition, there is one short-time $z = 8$ regime; being compact, for some values of Q and n_B^∞ we anticipate algebraic short-time DC2 kinetics and long-time DC1 kinetics with new exponents.

5.5.2   PHASE DIAGRAM IN Q-n_B^∞ PLANE

The calculational procedure is similar to the Rouse case but somewhat more complex. The starting point is again the general expression for $\dot{\mathcal{R}}_t(E)$ of Eq. (65) into which one inserts appropriate forms of $S^{(4)}$ and $S^{(1)}$ using Eqs. (60) and (62). The value of $\dot{\mathcal{R}}_t(E)$ depends on which is the dominant term, and this competition generates the Q-n_B^∞ phase diagram shown in Fig. 5.8(a)(i) and (b). This has many

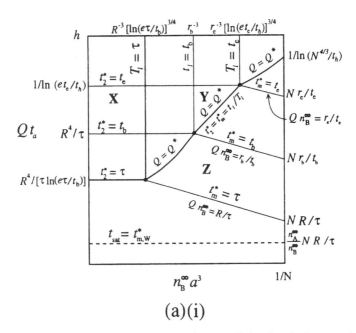

(a)(i)

Figure 5.8   Phase diagram, as Fig. 5.6 but now for entangled melts. The longest relaxation time is now the reptation time $\tau = (N^3/N_e)t_a$; all other timescales are defined in Tables 5.3 and 5.4. (a)(i) There are many more regions than in Fig. 5.6(a)(i) for Rouse chains, since relevant timescales can now occur within any of four reptation regimes (see Tables 5.3 and 5.4). The $Q = Q^*$ line (defined by $t_2^* = t_m^* = t_l/T_l$) has three segments. Each segment has different dependencies on n_B^∞, as shown in Table 5.3. The line $t_{sat} = t_{m,W}^*$ is shown for constant n_A^∞/n_B^∞.

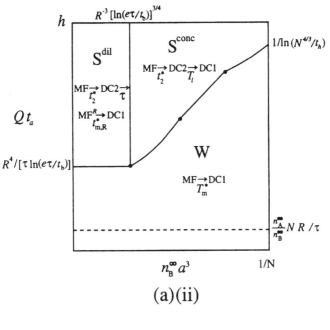

$$(a)(ii)$$

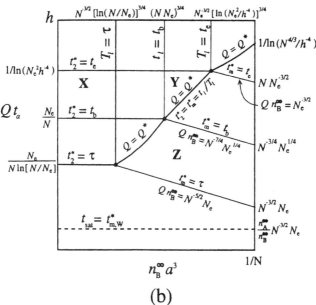

$$(b)$$

Figure 5.8 (a)(ii) The same three types of kinetic behavior occur as for unentangled systems, Fig. 5.6(a)(ii). Again, T_m^* denotes either t_m^* or $t_{m,W}^*$, depending on whether it is less than or exceeds τ. Below the $t_{sat} = t_{m,W}^*$ line, there is again no DC1 regime. (b) As (a)(i), but reactivities and densities are expressed in terms of N and N_e.

261

Table 5.3 Entangled chains: Characteristic timescales and Q^*. Only one of the displayed expressions for every quantity is relevant, depending on the dynamical regime in which it occurs. ϕ_B and ϕ_A have the same meaning as in Table 5.1. The notation "t_l/T_l" denotes whichever is the relevant of the two timescales; t_l is relevant for the compact reptation regime, T_l for marginal ones.

	Unentangled Rouse $t^{1/4}$ $t_h < t < t_e$	*Breathing* $t^{1/8}$ $t_e < t < t_b$	*Tube* $t^{1/4}$ $t_b < t < \tau$	*Fickian* $t^{1/2}$ $t > \tau$
t_m^*	$t_a(Qt_a\phi_B)^{-4/3}$	$t_e[Qt_e(a/r_e)\phi_B]^{-8/7}$	$t_b[Qt_b(a/r_b)\phi_B]^{-4/3}$	
t_2^*	$(t_h/e)e^{1/Qt_a}$	$t_e(Qt_a)^{-2}$	$(t_b/e)e^{1/[Q(a/r_b)4t_b]}$	
t_l/T_l	$t_a\phi_B^{-4/3} \times$ $[\ln(e\phi_B^{-4/3}t_a/t_h)]^{4/3}$	$t_e[\phi_B(r_e/a)^3]^{-8/3}$	$t_b[\phi_B(r_b/a)^3]^{-4/3} \times$ $\{\ln[e(\phi_B r_b^3/a^3)^{-4/3}]\}^{4/3}$	
$t_{m,W}^*$				$\tau[Q\tau(a/R)\phi_B]^{-2}$
$t_{m,R}^*$				$\tau\left[\dfrac{\ln(e\tau/t_b)}{\phi_B(R/a)^3}\right]^2$
t_{sat}				$1/(Q\phi_A\phi_B R/a)$ or $\tau/(N\phi_A)^2$
Q^*	$\dfrac{1}{t_a\ln[e\phi_B^{-4/3}t_a/t_h]}$	$(1/t_a)[\phi_B(r_e/a)^3]^{4/3}$	$\dfrac{(r_b/a)^4}{t_b\ln[e(\phi_B r_b^3/a^3)^{-4/3}]}$	

similarities with the Rouse phase diagram (Fig. 5.6(a)(i), (b)). However, there are many more subregions: although t_2^*, t_m^*, t_l, and T_l are again defined in terms of crossovers between the different terms in the denominator in Eq. (65), each timescale may occur in any of the four different reptation regimes. Shown in Tables 5.3 and 5.4 is the full list of possible formulas for these timescales, depending on which regime they happen to fall in (which in turn depends on the values of Q and n_B^∞).

The discussion of saturation effects is identical to that for unentangled systems (see Subsection 5.3.3). The saturation time t_{sat} is again given by Eq. (87), with τ understood as the reptation time. Similar remarks apply to $Q^\dagger$ of Eq. (88).

The most important feature that is unchanged from the unentangled Rouse case (see Fig. 5.6(a)(ii)) is that there remain only three distinct regions in terms of the sequence of reaction kinetics. The same four classes of reaction kinetics

Table 5.4 Entangled chains: Identical to Table 5.3, but now all quantities are expressed in terms of $\epsilon \equiv Qt_a$, entanglement threshold N_e, and molecular weight N.

	Unentangled Rouse $t^{1/4}$ $t_h < t < t_e$	*Breathing* $t^{1/8}$ $t_e < t < t_b$	*Tube* $t^{1/4}$ $t_b < t < \tau$	*Fickian* $t^{1/2}$ $t > \tau$
t_m^*/t_a	$(\epsilon\phi_B)^{-4/3}$	$N_e^{2/7}(\epsilon\phi_B)^{-8/7}$	$N^{-1/3}N_e^{1/3}(\epsilon\phi_B)^{-4/3}$	
t_2^*/t_a	$(h/a)^4 e^{1/\epsilon}$	N_e^2/ϵ^2	$N^2 e^{N_e/(\epsilon N)}$	
t_l/t_a or T_l/t_a	$\phi_B^{-4/3} \times$ $[\ln(\phi_B^{-4/3}a/h)]^{4/3}$	$\phi_B^{-8/3}N_e^{-2}$	$\phi_B^{-4/3}(NN_e)^{-1} \times$ $[\ln(N/N_e^3)]^{4/3}$	
$t_{m,W}^*/t_a$				$N_e(N\epsilon\phi_B)^{-2}$
$t_{m,R}^*/t_a$				$N_e^{-1}\left[\dfrac{\ln(N/N_e)}{\phi_B}\right]^2$
t_{sat}/t_a				$1/(\epsilon\phi_A\phi_B N^{1/2})$ or $N/(N_e\phi_A^2)$
Q^*t_a	$1/\ln(\phi_B^{-4/3}a/h)$	$\phi_B^{4/3}N_e^2$	$N_e/[N\ln(\phi_B^{-4/3}(NN_e)^{-1})]$	

are involved: MF, DC2, MF^R, and DC1. The two MF regimes are second-order with time-independent rate constant; DC2 denotes second-order kinetics with $k^{(2)} \approx dx_t^4/dt$; DC1 denotes first-order kinetics with $k^{(1)} \approx dx_t/dt$. In all marginal cases there are logarithmic corrections. Note that again the line $Q = Q^\dagger$ divides the W region into two parts, as for the Rouse case.

The three regions with their kinetic sequences are illustrated in Fig. 5.8(a)(ii). One interesting qualitatively new feature is that whereas for Rouse dynamics the DC2 regime was always marginal, with $\mathcal{R}_t \sim t/\ln t$, now we have in addition a truly compact DC2 regime for $z = 8$ (the breathing modes) where $\mathcal{R}_t \sim x_t^4 \sim t^{1/2}$ (see discussion following Eq. (9)).

The phase diagram of Fig. 5.8 allows the complete sequence of reaction kinetics to be determined for any experimental system with given Q and n_B^∞ values. These define a point in the plane whose coordinate belongs to either the S^{dil}, S^{conc}, or W regions of Fig. 5.8(a)(ii) where the relevant timescales for each region are indicated. The specific formulas for these timescales depend on where the coordinate lies in the full phase diagram of Fig. 5.8(a)(i) and (b), with the corresponding formulas in Tables 5.3 and 5.4. Finally, the explicit time dependencies for $\mathcal{R}_t$ are determined by the time dependencies of x_t.

5.5.3   EXAMPLES

Consider as examples the three points X, Y, and Z shown in Fig. 5.8(a)(i) and (b). Since the point X belongs to the S^{dil} regime, the relevant timescales are t_2^* and $t_{m,R}^*$. Moreover, its location tells us that $t_e < t_2^* < t_b$. Thus, $\mathcal{R}_t$ is given by

$$
\mathcal{R}_t \approx
\begin{cases}
\lambda n_A^\infty n_B^\infty t, & (t < t_2^*, \text{MF}) \\
n_A^\infty n_B^\infty r_e^4 (t/t_e)^{1/2}, & (t_2^* < t < t_b, \text{DC2}) \\
n_A^\infty n_B^\infty r_b^4 (t/t_b)/\ln(et/t_b), & (t_b < t < \tau, \text{DC2}) \\
n_A^\infty n_B^\infty R^4 (t/\tau)/\ln(e\tau/t_b), & (\tau < t < t_{m,R}^*, \text{MF}^R) \\
n_A^\infty (Dt)^{1/2}, & (t_{m,R}^* < t < t_{\text{sat}}, \text{DC1})
\end{cases}
\qquad (\text{Point X}). \qquad (90)
$$

Now consider point Y, belonging to the S^{conc} regime. Its Q value is the same as that of point X, and hence t_2^* is the same, $t_e < t_2^* < t_b$. However, since n_B^∞ is greater, the crossover time to DC1 kinetics is smaller; in the S^{conc} region this occurs at either t_l or T_l. The location of Y implies $t_e < t_l < t_b$. Thus,

$$
\mathcal{R}_t \approx
\begin{cases}
\lambda n_A^\infty n_B^\infty t, & (t < t_2^*, \text{MF}) \\
n_A^\infty n_B^\infty r_e^{\ 4} (t/t_e)^{1/2}, & (t_2^* < t < t_l, \text{DC2}) \\
n_A^\infty r_e\, (t/t_e)^{1/8}, & (t_l < t < t_b, \text{DC1}) \\
n_A^\infty r_b (t/t_b)^{1/4}, & (t_b < t < \tau, \text{DC1}) \\
n_A^\infty (Dt)^{1/2}, & (\tau < t < t_{\text{sat}}, \text{DC1})
\end{cases}
\qquad (\text{Point Y}). \qquad (91)
$$

In this case there are three phases to the DC1 regime, each with a different power law in time. The breathing modes phase introduces a $t^{1/8}$ law that has no parallel in the unentangled case.

Finally, the point Z has the same n_B^∞ value as Y, but the reactivity is so much lower that $t_2^* > \tau$ and the DC2 regime vanishes: Z now belongs to the W region where the relevant timescale is t_m^*, and its specific location within W tells us that $t_b < t_m^* < \tau$. Thus,

$$
\mathcal{R}_t \approx
\begin{cases}
\lambda n_A^\infty n_B^\infty t, & (t < t_m^*, \text{MF}) \\
n_A^\infty r_b (t/t_b)^{1/4}, & (t_m^* < t < \tau, \text{DC1}) \\
n_A^\infty (Dt)^{1/2}, & (\tau < t < t_{\text{sat}}, \text{DC1})
\end{cases}
\qquad (\text{Point Z}). \qquad (92)
$$

5.6   Discussion

This chapter has highlighted both similarities and differences between bulk and interfacial polymer reaction kinetics. The emphasis has been on techniques and the meaning of approximations that are, implicitly or explicitly, widely employed by theoreticians in the field. The analysis presented in Section 5.2 (Rouse chains in bulk) was especially detailed. We began with a near-exact statement of the problem, eventually reducing this via several approximations to a simple closed integral equation, Eq. (34), essentially the same as that first used by Doi [34]. The "closure approximation," universally employed outside renormalization group analysis, is the most crucial one: the assumption of equilibrium chain configurations. We systematically incorporated higher-order correlations (beyond two-body) and argued that to leading order their contribution cancels with a two-body term (that is, the third term in Eq. (27) may be neglected). Such higher-order correlations also appear in the analysis of interfacial reactions; they are responsible for a crossover at long times to first-order diffusion-controlled (DC) kinetics.

In Sections 5.2 and 5.3 bulk kinetics were studied in the limit where the density of reactive chains is very small. (For discussions of dense reactive chains in the bulk, see refs. [66], and [57].) We saw that reaction kinetics depend critically on the concentration regime. Unentangled melts (Rouse dynamics) were investigated in Section 5.2. For sufficiently reactive groups, $Q > \hat{Q}$, there is a short-time second-order DC regime [35] in which a hole of size x_t develops in the two-body correlation function. Hence, the number of reactions per unit volume increases as $\mathcal{R}_t \approx x^3(t)n^2(0)$. Under Rouse dynamics, this translates to $\mathcal{R}_t \sim t^{3/4}$. Should this short-time DC regime occur, it is remembered in the long-time rate constant [34], $k_\infty \approx DR \sim 1/N^{1/2}$.

Kinetics in dilute solutions, in the presence of a good swelling solvent, are entirely different. The probability that a reaction occurs during one collision between two coils, each bearing a reactive group, is strongly reduced by fast hydrodynamic relaxation and excluded volume repulsions between functional groups. For large N, this probability becomes very small and consequently simple mean field (MF) reaction kinetics *always* pertain. There is no DC limit: Reactions are never controlled by polymer center of gravity diffusion, even for highly reactive radical functional groups. Neither the form $k_\infty \approx DR$ nor early time-dependent behavior arise. MF kinetics mean that k is always time independent and scales as the equilibrium contact probability, giving $k \sim 1/N^\alpha$ with $\alpha \approx 0.16$ for end groups.

In Sections 5.4 and 5.5 melt–melt interfacial reactions were studied for arbitrary values of reactive group densities and local reactivities Q. For large enough Q, there are again second-order small-time DC regimes in which the two-body correlation function develops a depletion region of size x_t, and the number of reactions per unit area increases as the fourth power of the rms monomer displacement, $\mathcal{R}_t \approx x_t^4 n_A^\infty n_B^\infty$. This led to logarithmic corrections, $\mathcal{R}_t \sim t/\ln t$, for unentangled Rouse chains ($x_t \sim t^{1/4}$). For entangled systems, this again gave $\mathcal{R}_t \sim t/\ln t$ during the two short-time $t^{1/4}$ regimes and $\mathcal{R}_t \sim t^{1/2}$ during the $t^{1/8}$ breathing modes regime. As for the bulk, large values of Q are required to realize these DC regimes, and for virtually all chemical reactions save radicals, the early kinetics will almost certainly obey the simple MF law, $\mathcal{R}_t \approx Q a^4 n_A^\infty n_B^\infty t$.

The entirely new feature for interfaces is the appearance of *first-order* kinetics in which a depletion region of size x_t grows in the density field itself. Even for very small values of Q, the long-time kinetics are always first-order and DC, with $\mathcal{R}_t = x_t n_A^\infty$. For unequal reactive group densities on either side of the interface, it is the dilute A side whose density n_A^∞ governs the reaction rate. If the reactive chain density on the denser B side exceeds the overlap threshold, $n_B^\infty R^3 > 1$, and if Q is big enough, then this regime onsets before τ. For unentangled chains, this leads to $\mathcal{R}_t \sim t^{1/4}$. For entangled chains, consecutive short-time regimes $\mathcal{R}_t \sim t^{1/4}$, $\mathcal{R}_t \sim t^{1/8}$, and $\mathcal{R}_t \sim t^{1/4}$ may occur, depending on density. In all cases the final regime is first-order DC and corresponds to the long-time Fickian center of gravity motion, $\mathcal{R}_t \sim t^{1/2}$ (unless saturation effects intervene).

The accumulating copolymer product ultimately saturates the interface after time t_{sat}. Essentially, reactions cease at this point for very long polymers. (We do not discuss here interesting effects such as interface destabilization due to surface tension reduction by copolymer product; in some cases this appears to perpetuate reactions by generating new surface area.) If Q is very small, $Q < Q^\dagger$, then saturation occurs before the final first-order DC kinetics regime has begun. For such cases, kinetics are always MF. Now for "ordinary" chemical species one has $Q_b t_a \lesssim 10^{-6}$ (note that typically $Q \equiv (h/a)Q_b \approx 5Q_b$). Then for unentangled chains of maximum possible length [51], $N \approx 300$, Eq. (88) implies $Q^\dagger t_a \approx 0.06$ which is many orders of magnitude higher than $Q t_a$. Thus, kinetics are always MF until saturation at t_{sat}. Considering maximum reactant densities, we obtain $t_{sat} \gtrsim 6 \times 10^4 \tau \approx 0.5$ sec for $N = 300$. In the case of entangled melts with, say, $N = 10^4$, $Q^\dagger t_a \approx 10^{-4}$. Thus, even here MF kinetics are likely to apply up to t_{sat}. For $N = 10^4$, $N_e \approx 300$ and at maximum densities, the saturation time $t_{sat} > 10^4 \tau \approx 3 \times 10^3$ sec. For radicals, on the other hand, $Q_b t_a \approx 1$. At high densities such systems are located in the S^{conc} or S^{dil} regions of the Q-n_B^∞ phase diagrams of Figs. 5.6 and 5.8, and the full sequence of regimes is realized well before saturation. The value of t_{sat} then depends only on n_A^∞ and ranges from τ at maximum density to arbitrarily large values as density decreases.

We have not attempted in this article to do justice to the huge experimental effort in this field. Here follow a few remarks including suggestions for future experiments. A vast body of data on free radical polymerization exists; these are complex situations, but some measurements can be directly compared to theory of the type discussed here. Particularly interesting are "post-effect" electron spin resonance studies where the radical source is switched off and the subsequent decay of the macroradical population is monitored. Even under (apparently) non-glassy conditions, virtually infinite macroradical lifetimes are observed [67, 68] during the advanced stages when entanglements are important. This appears to be consistent with theory [69, 7]; because [35] $k \sim 1/N^{3/2}$ for entangled bulk polymers, and since reactions between chain pairs of very different lengths are dominated by the short chain [69], it can be shown that macroradicals are unable to terminate their growth if new short ones are not being supplied. Formally this leads to infinite lifetimes.

Unfortunately, data from model experiments involving carefully prepared monodisperse chains are limited. For bulk polymer reactions the strongest support for theory [37] at present comes from studies of concentration dependence of rate constants, from dilute to concentrated polymer solutions [19]. For dilute solutions with good solvent, phosphorescence-quenching measurements of end-functionalized chains [19] are at least consistent with a small exponent $\alpha = 0.16$. However, there is little or nothing for bulk melts. Since the threshold $\hat{Q}$ for DC behavior is actually very high, the only candidates for DC kinetics are radicals or optically excitable species (see the discussion at the end of Section 5.2). But phosphorescence-quenching lifetimes are no match for the long melt relaxation times, rendering the long-time k_∞ inaccessible. The difficulty with radicals is that they are always created in pairs and therefore one expects rapid recombination before polymer dynamics are relevant. Use of radicals or photo-excited groups to examine the short-time $k(t)$ in melts is also problematic; since only a small fraction of the reactive chains contribute (those pairs whose ends happen to be in close proximity), the signal at small times would be very weak.

Interesting experimental approaches that may successfully combat these difficulties include photocleaving end-functionalized polymers to create macroradical–radical pairs, a certain fraction of which are theoretically expected to separate from one another. Since the small radicals have much higher mobility, it can be shown that after a transient they all disappear, leaving behind a fraction (of order unity) of the macroradicals [70]. These are isolated, and their subsequent kinetics should be governed by, for example, the $k_\infty \sim 1/N^{1/2}$ result if the melt is unentangled. Similarly, the short-time $k(t)$ may be accessible by photocleaving internally positioned groups to generate two chains with radical end groups in close proximity. The great advantage is that on the one hand radical pairs are *created* close to one another, thus avoiding the problem of inadequate signals; on the other hand, their immediate recombination (which would also prevent any measurement of $k(t)$) is prevented by spin selection rules [71]. Recombination processes commence after a certain delay time, and one expects the subsequent kinetics to be governed by $k(t)$.

Similar experimental approaches, exploiting the convenience and accuracy of laser-controlled radical production, hold promise for measurements of basic rate laws at interfaces, as well as highly controlled studies of free radical polymerization. A particularly interesting area of future research is the development of accurate laser manipulation of macroradical reactions and polymerizations. Generally, the outlook for badly needed experiments with new levels of accuracy and reproducibility is very encouraging. We hope these will test many of the predictions presented in this chapter, as well as motivate new theoretical work in the field of reacting polymer systems.

Acknowledgments

This work was supported by the National Science Foundation under grant no. DMR-9403566. I am grateful to Dimitris Vavylonis for many valuable discussions.

Appendix 5.A   Rouse Dynamics: rms Displacement x_t of Chain Ends

Here we present simple scaling arguments for the $t^{1/4}$ short-time and $t^{1/2}$ long-time rms displacement of a monomer in the Rouse model, as displayed in Eq. (17). Detailed analysis may be found, for example, in ref. [28].

Since the Hamiltonian H_0 of Eq. (15) (with $k = 3k_BT/a^2$) implies Gaussian statistics, the rms end-to-end distance of the entire polymer coil is $R = N^{1/2}a$, and indeed the rms size of any subcoil consisting of n monomers is $n^{1/2}a$. Under the dynamics of Eq. (16), the relaxation time of this subcoil is $n^2 t_a$, where $t_a \approx \nu a^2/k_BT$ is the single chain unit relaxation time. (One can see this immediately from the Gaussian Green's function for the dynamics with $\mathbf{f}$ deleted.)

Consider a unit at the chain end. After time t the subcoil containing the $n(t) \approx (t/t_a)^{1/2}$ units nearest the chain end will have relaxed. During this time, the end will therefore have suffered a displacement of the order of the rms size of this subcoil, namely, $[n(t)]^{1/2}a$, giving $x_t \approx (t/t_a)^{1/4}a$.

For long times, the center of gravity diffusion obeying Fick's law, $x_t \sim t^{1/2}$, must take over. This happens when the short-time displacement x_t becomes of the order of the coil size, R, at the longest relaxation time τ that is thus determined, $\tau \approx N^2 t_a$. This also determines the prefactor, $x_t \approx R(t/\tau)^{1/2}$ for $t > \tau$.

Appendix 5.B   Implementing the Closure Approximation

In this appendix the "closure approximation" of Eq. (25) is implemented on the exact expression of Eq. (23) for $k(t)$. That is, in Eq. (23) we replace $\Phi_t \to P_{eq}^{int}\rho_t$ and $\Phi_t^{(3)} \to P_{eq}^{(3),int} \rho_t^{(3)}$ (see Eq. (24)).

Firstly, consider the effect this replacement has on the second term on the right-hand side of Eq. (23). In this term the following factor simplifies as

$$\delta(\mathbf{r}_A' - \mathbf{r}_B')\frac{Qa^d}{n^2(t')}\Phi_t[\mathbf{R}_A', \mathbf{R}_B'] \to \delta(\mathbf{r}_A' - \mathbf{r}_B')\frac{Qa^d}{n^2(t')}\rho_{t'}^s P_{eq}^{int}([\mathbf{R}_A', \mathbf{R}_B']|0, 0)$$

$$= \delta(\mathbf{r}_A' - \mathbf{r}_B')k(t')P_{eq}^{int}([\mathbf{R}_A', \mathbf{R}_B']|0, 0) \tag{B1}$$

after using translational invariance and Eqs. (14) and (20).

Now consider the integral in the second term in the curly brackets on the right-hand side of Eq. (23). This becomes

$$\int d[\mathbf{R}_C']\delta(\mathbf{r}_A' - \mathbf{r}_C')\Phi_{t'}^{(3)}[\mathbf{R}_A', \mathbf{R}_B', \mathbf{R}_C']$$

$$\to \rho_{t'}^{(3)}(\mathbf{r}_A', \mathbf{r}_B', \mathbf{r}_C')\int d[\mathbf{R}_C']\delta(\mathbf{r}_A' - \mathbf{r}_C')P_{eq}^{(3),int}([\mathbf{R}_A', \mathbf{R}_B', \mathbf{R}_C']|0, 0, 0)$$

$$= \rho_{t'}^{(3)}(\mathbf{r}_A', \mathbf{r}_B', \mathbf{r}_C')P_{eq}^{int}([\mathbf{R}_A', \mathbf{R}_B']|0, 0) \tag{B2}$$

where we have used the fact that the equilibrium internal chain distributions are independent of the locations of the chain ends: In the Rouse model all chains are independent. The statistical independence of different chains also implies that the two-chain distribution P_{eq}^{int} is obtained after integrating out, in the three-chain internal distribution $P_{eq}^{(3),int}$, the internal coordinates of one chain (namely, $[\mathbf{R}_C(n)]_{n=0}^{n=N-1}$ in the above).

Making the changes of Eqs. (B1) and (B2) in Eq. (23) gives

$$k(t) = Qa^d - Qa^d\int_0^t dt' S^{(d)}(t - t')k(t')$$

$$+ 2Qa^d\int_0^t dt'\int d\mathbf{r}_A' d\mathbf{r}_B' G_{t-t'}(0, 0, \mathbf{r}_A', \mathbf{r}_B')$$

$$\times \left\{\frac{k(t')}{n(t')}\rho_{t'}(\mathbf{r}_A', \mathbf{r}_B') - \frac{Qa^d}{n^2(t')}\rho_{t'}^{(3)}(\mathbf{r}_A', \mathbf{r}_B', \mathbf{r}_A')\right\} \tag{B3}$$

where G_t in the above is the *equilibrium* propagator for chain ends

$$G_t(\tilde{\mathbf{r}}_A, \tilde{\mathbf{r}}_B, \tilde{\mathbf{r}}_A', \tilde{\mathbf{r}}_B') \equiv \int d[\mathbf{R}_A]d[\mathbf{R}_B]d[\mathbf{R}_A']d[\mathbf{R}_B']\delta(\tilde{\mathbf{r}}_A - \mathbf{r}_A)\delta(\tilde{\mathbf{r}}_B - \mathbf{r}_B)$$

$$\times \mathcal{G}_t[\mathbf{R}_A, \mathbf{R}_B, \mathbf{R}_A', \mathbf{R}_B']\delta(\tilde{\mathbf{r}}_A' - \mathbf{r}_A')\delta(\tilde{\mathbf{r}}_B' - \mathbf{r}_B')P_{eq}^{int}([\mathbf{R}_A', \mathbf{R}_B']|0, 0).$$

$$\tag{B4}$$

The quantity $S^{(d)}(t)$, defined in Eq. (28), is the equilibrium chain-end-pair return probability. Substituting into Eq. (B4) the conditional densities $\rho_t(r|0)$ and $\rho_t^{(3)}(r|0, 0)$ as defined in Eq. (26), one obtains the final self-consistent expression for $k(t)$, Eq. (27) in the main text.

Appendix 5.C   Demonstration That $\varphi_t(0) \ll 1$

This appendix shows that $\varphi_t(0) \ll 1$, where $\varphi_t(r)$ is defined in Eq. (32). This is equivalent to showing that one may neglect the third term on the right-hand side of Eq. (27) for $k(t)$ (reactive Rouse chains in bulk).

Our approach will be self-consistent; $k(t')$ in Eq. (31) will be set to the solution that is obtained in Subsection 5.2.5 after neglecting the third term in Eq. (27) (Eqs. (39) and (40)). This $k(t)$ solution asymptotes k_∞ for $t > \tau$ and thus the solution of Eq. (14) is

$$n(t) \approx \frac{n_0}{1 + k_\infty n_0 t} \approx \begin{cases} n_0, & (t < t^*_{m,\text{bulk}}) \\ 1/k_\infty t, & (t > t^*_{m,\text{bulk}}) \end{cases}; \qquad t^*_{m,\text{bulk}} \equiv 1/(k_\infty n_0). \qquad (C1)$$

In fact this expression is valid to first-order for times $t < \tau$ also, since for such times $n(t) \approx n_0$.

Let us now establish the function $\mu^{\max}$. Clearly, from its definition in Eq. (32), $\varphi_t(0)$ is maximized when the integral of $k(t')\mu_{t'}(\mathbf{r}')$ has its maximum value; this corresponds to replacing the last inequality in (33) by an equality. Moreover, $\mu_t(\mathbf{r}')$ should be distributed around the points at which $G^{\text{sep}}_{t-t'}(0, \mathbf{r}')$ in the definition of $\varphi_t(0)$ in Eq. (32) is maximum. But G^{sep} has the well-known scaling form

$$G^{\text{sep}}_t(0, \mathbf{r}) \approx \begin{cases} 1/x_t^3, & (r < x_t) \\ 0, & (r > x_t) \end{cases} \qquad (C2)$$

(see Eq. (50) with $g = 0$). Thus $G^{\text{sep}}_{t-t'}(0, \mathbf{r}')$ is roughly constant (in space) for $r' < x_{t-t'}$ and vanishes for larger r'. For a given r', it achieves maximum values at the maximum available t' values outside the vanishing region. Furthermore, we note that the maximum amplitude of $\mu_{t'}^{\max}$ is of order $n(t')$. Hence, the function maximizing $\varphi_t(0)$ is approximately

$$\mu_{t'}^{\max}(\mathbf{r}') \approx \begin{cases} n(t')\Theta(x_{t-t'} - r'), & (t_c < t' < t) \\ 0, & (t' < t_c) \end{cases}; \qquad \int_0^t k(t')dt' = \int_{t_c}^t dt' \int d\mathbf{r}' k(t')\mu_{t'}^{\max}(\mathbf{r}'), \qquad (C3)$$

where Θ is the step function ($\Theta(x) = 0$ for $x < 0$, $\Theta(x) = 1$ for $x \geq 0$). That is, $\mu_{t'}^{\max}(\mathbf{r}')$ is localized at $(t', \mathbf{r}')$ close to $(t, 0)$, where it has its maximum amplitude. The time t_c defines the lower limit of the support of μ^{max} and is determined by demanding that the total integral of $k\mu^{max}$ yields the required value. Using $n(t)$ from Eq. (C1) and x_t from Eq. (17), its explicit form is

$$t_c \approx \begin{cases} t - \tau[t/(\tau n(t)R^3)]^{2/5}, & (t > t_l) \\ 0, & (t < t_l) \end{cases}; \qquad t_l \equiv \tau(n_0 R^3)^{-2/3}, \qquad (C4)$$

where t_l is the time for a reactive end to diffuse the mean distance reactive groups. For $t < t_l$, $t_c = 0$ corresponds to the maximum value the integral of $k\mu^{max}$ can possibly have, given that it vanishes for $r' > x_{t-t'}$.

Substituting μ^{max} into the expression for $\varphi_t(0)$ in Eq. (32) and integrating over r', one obtains $\varphi_t^{max}(0) \approx \int_{t_c}^t dt' k(t')n(t') \approx \ln[n(t_c)/n(t)]$ after the use of Eq. (14). It is then simple to show that $\varphi_t^{max}(0)$ is $n_0 k(t)t$ for $t < t_l$ and $k_\infty n(t)\tau[t/(\tau n(t)R^3)]^{2/5}$ for $t > t_l$. Thus, using the two possible forms for k_∞ derived in Eq. (40) (depending on whether $Q > \hat{Q}$ or $Q < \hat{Q}$ where $\hat{Q}$ is given by Eq. (38)), we obtain

$$\varphi_t^{max}(0) \approx \begin{cases} n_0 k(t)t, & (t < t_l) \\ (n_0 R^3)^{1/5}(t/t^*_{m,bulk})^{2/5}, & (t_l < t < t^*_{m,bulk}); \\ (\tau/t)^{1/5}, & (t > t^*_{m,bulk}) \end{cases} \qquad (Q > \hat{Q}) \qquad (C5)$$

and

$$\varphi_t^{max}(0) \approx \begin{cases} t/t^*_{m,bulk}, & (t < t_l) \\ (Q/\hat{Q})^{3/5}(n_0R^3)^{1/5}(t/t^*_{m,bulk})^{2/5}, & (t_l < t < t^*_{m,bulk}); \\ (Q/\hat{Q})^{2/5}(\tau/t)^{1/5}, & (t > t^*_{m,bulk}). \end{cases} \qquad (Q < \hat{Q}) \qquad (C6)$$

Noting that $k(t)$ always decreases more slowly than $1/t$ for $t < t_l$ (see Eqs. (39) and (40)), it follows from Eqs. (C5) and (C6) that $\varphi_t^{max}(0) \ll 1$ for $n_0 R^3 \ll 1$.

Appendix 5.D Interfacial Reactions: Proof of $S^{(4)} \sim x_t^{-4}$ and $S^{(1)} \sim x_t^{-1}$

In the absence of reactions, the A and B reactive polymers belong to different bulk phases and diffuse independently; thus the two-particle propagator G_t can be written as a product of single-particle propagators $G_t^{(1)}$:

$$G_t(\mathbf{r}_A, \mathbf{r}_B, \mathbf{r}'_A, \mathbf{r}'_B) = G_t^{(1)}(\mathbf{r}_A, \mathbf{r}'_A)G_t^{(1)}(\mathbf{r}_B, \mathbf{r}'_B). \qquad (D1)$$

For a given regime of dynamical behavior, $G_t^{(1)}$ will have only one characteristic scale, x_t. Dimensional analysis then dictates the scaling form

$$G_t^{(1)}(\mathbf{r}, \mathbf{r}') = \frac{1}{x_t^d} g(\mathbf{r}/x_t, \mathbf{r}'/x_t), \qquad g(\mathbf{u}, \mathbf{v}) \to \begin{cases} f(u_x, v_x), & (|\mathbf{u} - \mathbf{v}| \ll 1) \\ 0, & (|\mathbf{u} - \mathbf{v}| \gg 1) \end{cases} \qquad (D2)$$

where $f(u_x, v_x)$ is a function of order unity for every value of its arguments (u_x and v_x are the x components of $\mathbf{u}$, $\mathbf{v}$, respectively). The fact that f depends on u_x, v_x is a result of the broken translational invariance in the x direction.

Now the scaling form $S^{(4)} \approx 1/x_t^4$ follows immediately from the definition of $S^{(4)}$ in Eq. (60) and Eqs. (D1) and (D2). Similarly, the result $S^{(1)} \approx 1/x_t$ follows from its definition in Eq. (62) and Eq. (D2).

Appendix 5.E   Interfacial Reactions, Rouse Chains: Return Probabilities $S^{(1)}(E)$ and $S^{(4)}(E)$

From Eq. (62) and the form of x_t for Rouse dynamics, Eq. (17), we have

$$\lambda n_B^\infty S^{(1)}(E) \approx \begin{cases} Qt_a n_B^\infty a^3 (Et_a)^{-3/4}, & (E \gg 1/\tau) \\ Q\tau n_B^\infty a^3 (a/R)(E\tau)^{-1/2}, & (E \ll 1/\tau) \end{cases} \tag{E1}$$

whereas Eq. (60) leads to the approximate expression

$$S^{(4)}(E) \approx \int_{t_h}^{\tau} dt' e^{-Et'} \frac{1}{a^4} \frac{t_a}{t'} + \int_{\tau}^{\infty} dt' e^{-Et'} \frac{1}{R^4} \left(\frac{\tau}{t'} \right)^2. \tag{E2}$$

We have introduced a cutoff at t_h. (At shorter times kinetics are as for a standard bulk reaction problem, with $S^{(4)} \approx 1/(hx_t^3)$; one can show that the $t < t_h$ contribution [49] is of the same order as that from the lower limit $t = t_h$ in Eq. (E2).) For $E\tau \gg 1$, the first integral dominates and we can set its upper limit to infinity with small error. Thus,

$$\lambda S^{(4)}(E) \approx \begin{cases} Qt_a \ln(1/Et_h), & E \gg 1/\tau \\ \lambda \int_0^{\infty} dt S^{(4)}(t) \approx Qt_a \ln(\tau/t_h), & E \ll 1/\tau. \end{cases} \tag{E3}$$

Appendix 5.F   Monomer Diffusion in the Reptation Model

This appendix reviews the reptation dynamics [28, 10] that lead to Eq. (89). For short times such that $x_t < r_e$, entanglements are not felt and one recovers Rouse-like unentangled dynamics, $x_t \approx a(t/t_a)^{1/4}$, until $t_e = N_e^2 t_a$. For $t > t_e$, the chain can diffuse up and down the tube only; initially, this curvilinear diffusion is Rouse-like, giving $t^{1/4}$ rms displacement along the tube. This corresponds to $t^{1/8}$ rms displacement in space since the tube itself is a random walk in space with persistence length r_e. Hence, $x_t = r_e(t/t_e)^{1/8}$. These "breathing modes" continue until the

Rouse time $t_b = N^2 t_a$, by which time the chain has relaxed its configuration relative to the tube; for longer times it diffuses coherently along the tube with rms tube displacement $t^{1/2}$, implying $t^{1/4}$ displacement in space. Thus, $x_t = r_b(t/t_b)^{1/4}$ where $r_b = r_e(N/N_e)^{1/4}$. This "tube diffusion" regime persists until the longest polymer relaxation or "reptation" time, $\tau = (R/r_b)^4 t_b = (N^3/N_e)t_a$, by which time the chain has completely diffused out of its initial tube into a new and uncorrelated one. Here $R = N^{1/2}a$ is the rms coil size. The process then repeats itself indefinitely, corresponding to long-time Fickian center of gravity motion, $x_t = R(t/\tau)^{1/2}$.

References

1. P. Flory, *Principles of Polymer Chemistry,* Cornell Univ. Press, Ithaca, NY, 1971.
2. R. G. W. Norrish and R. R. Smith, *Nature (London),* **150:** 336 (1942).
3. E. Trommsdorff, H. Kohle, and P. Lagally, *Makromol. Chem.* **1:** 169 (1948).
4. J. Cardenas and K. F. O'Driscoll, *J. Polym. Sci., Polym. Chem. Ed.* **15:** 1883 (1977).
5. T. J. Tulig and M. Tirrell, *Macromolecules* **14:** 1501 (1981).
6. K. F. O'Driscoll, *Pure and Appl. Chem.* **53:** 617 (1981).
7. B. O'Shaughnessy and J. Yu, *Phys. Rev. Lett.* **73:** 1723 (1994); *Macromolecules* **27:** 5067, 5079 (1994).
8. G. A. Oneil, M. B. Wisnudel, and J. M. Torkelson, *Macromolecules* **29:** 7477 (1996).
9. G. S. Grest, K. Kremer, and E. R. Duering, *Physica A* **194:** 330 (1993).
10. P. G. de Gennes, *Scaling Concepts in Polymer Physics,* Cornell Univ. Press, Ithaca, NY, 1985.
11. C. Creton, E. J. Kramer, C. Y. Hui, and H. R. Brown, *Macromolecules* **25:** 3075 (1992).
12. J. Washiyama, C. Creton, E. J. Kramer, F. Xiao, and C. Y. Hui, *Macromolecules* **26:** 6011 (1993).
13. D. Gersappe, D. Irvine, A. C. Balazs, Y. Liu, J. Sokolov, M. Rafailovich, S. Schwarz, and D. G. Peiffer, *Science* **265:** 1072 (1994).
14. S. Wu, *Polymer* **26:** 1855 (1985).
15. M. Okamoto and T. Inoue, *Polym. Eng. and Sci.* **33:** 175 (1993).
16. U. Sundararaj and C. Macosko, *Macromolecules* **28:** 2647 (1995).
17. C. Scott and C. Macosko, *J. Polym. Sci.: Part B* **32:** 205 (1994).
18. J. Guillet, *Polymer Photophysics and Photochemistry,* Cambridge Univ. Press, Cambridge, 1987.
19. I. Mita and K. Horie, *J. Macromol. Sci., Rev. Macromol. Chem. Phys.* **C27(1):**91 (1987).
20. M. D. Wisnudel and J. M. Torkelson, *J. Polym. Sci., Polym. Phys. Ed.* **34:** 2999 (1996).

21. P.-G. de Gennes, *J. Colloid Interface Sci.* **27**: 189 (1987).
22. X. Zheng, B. B. Sauer B. B., J. G. Van Alsten, S. A. Schwarz, M. H. Rafailovich, J. Sokolov, and M. Rubinstein, *Phys. Rev. Lett.* **74**: 407 (1995).
23. A. Halperin, M. Tirrel, and T. P. Lodge, *Adv. Polm. Sci.* **100**: 31 (1992).
24. B. Alberts, D. Bray, J. Lewis, M. Raff, K. Roberts, and J. D. Watson, *Molecular Biology of the Cell,* third ed., Garland Press, New York, 1994.
25. C. R. Cantor and P. R. Schimmel, *Biophysical Chemistry,* Freeman, New York, 1980.
26. G. Wilemski and M. Fixman, *J. Chem. Phys.* **60**: 866, 878 (1974).
27. M. Doi, *Chem. Phys.* **9**: 455 (1975).
28. M. Doi, and S. F. Edwards, *The Theory of Polymer Dynamics,* Clarendon Press, Oxford, 1986.
29. B. Friedman and B. O'Shaughnessy, *Phys. Rev. Lett.* **60**: 64 (1988).
30. B. Friedman and B. O'Shaughnessy, *Phys. Rev. A* **40**: 5950 (1989).
31. B. Friedman and B. O'Shaughnessy, *J. Phys. II (Paris)* **1**: 471 (1991).
32. B. Friedman and B. O'Shaughnessy, *Macromolecules* **26**: 4888 (1993).
33. B. Friedman and B. O'Shaughnessy, *J. Phys. II (Paris)* **3**: 1657 (1993).
34. M. Doi, *Chem. Phys.* **11**: 115 (1975).
35. P.-G. de Gennes, *J. Chem. Phys.* **76**: 3316, 3322 (1982).
36. B. Friedman and B. O'Shaughnessy, *Europhys. Lett.* **23**: 1993, 667; *Macromolecules* **26**, 1993, 5726.
37. B. O'Shaughnessy, *Phys. Rev. Lett.* **71**, 1993, 3331; *Macromolecules* **27**, 1994, 3875.
38. J. des Cloizeaux, *J. Phys. (Paris)* **41**: 223 (1980).
39. T. A. Witten and J. J. Prentis, *J. Chem. Phys.* **77**: 4247 (1982).
40. B. Duplantier, *J. Stat. Phys.* **54**: 581 (1989).
41. L. Schafer, C. Vonferber, U. Lehr, and B. Duplantier, *Nuclear Phys. B* **374**: 473 (1992).
42. B. Friedman and B. O'Shaughnessy, *Int. J. Mod. Phys. B* **8**: 2555 (1994).
43. A. R. Khokhlov, *Makromol. Chem., Rapid Commun.* **2**: 633 (1981).
44. C. J. Durning and B. O'Shaughnessy, *J. Chem. Phys.* **88**: 7117 (1988).
45. B. O'Shaughnessy and U. Sawhney, *Phys. Rev. Lett.* **76**: 3444 (1996).
46. B. O'Shaughnessy and U. Sawhney, *Macromolecules* **29**: 7230 (1996).
47. G. H. Fredrickson, *Phys. Rev. Lett.* **76**: 3440 (1996).
48. G. H. Fredrickson and S. T. Milner, *Macromolecules* **29**: 7386 (1996).
49. B. O'Shaughnessy and D. Vavylonis (1998, Submitted to *J. Phys. II (Paris)*).
50. B. O'Shaughnessy and D. Vavylonis (1998, Submitted to *Macromolecules*).
51. J. D. Ferry, *Viscoelastic Properties of Polymers,* third ed., Wiley, New York, 1980.
52. G. Wilemski and M. Fixman, *J. Chem. Phys.* **58**: 4009 (1973).
53. D. J. Amit, *Field Theory, the Renormalization Group, and Critical Phenomena,* World Scientific, Singapore, 1984.
54. Y. Oono, *Advances in Chemical Physics,* I. Prigogine and S. A. Rice Eds., Vol. 61 Wiley, New York, 1985.
55. Doi M., *Chem. Phys.,* **11**: 107 (1975).

56. B. O'Shaughnessy, *J. Chem. Phys.* **94:** 4042 (1991).

57. B. O'Shaughnessy and D. Vavylonis, (1998, Submitted to *Macromolecules*).

58. M. von Smoluchowski, *J. Phys. Chem.* **92:** 192 (1917).

59. J. des Cloizeaux and G. Jannink, *Polymers in Solution, Their Modelling and Structure,* Clarendon Press, Oxford, 1990.

60. M. Muller, *Macromolecles* **30:** 6353 (1997).

61. M. Doi, *J. Phys. A* **9:** 1465, 1479 (1976).

62. E. B. Zhulina, O. V. Borisov, and L. Brombacher, *Macromolecules* **24:** 4679 (1991).

63. E. B. Zhulina and O. V. Borisov, *J. Colloid Interface Sci.* **144:** 507 (1991).

64. A. N. Semenov, *Sov. Phys. JETP* **61:** 733 (1985).

65. A. N. Semenov, *Macromolecules* **25:** 4967 (1992).

66. G. Oshanin, M. Moreau, and S. Burlatsky, *Adv. Colloid Interface Sci.* **49:** 1 (1994).

67. S. Zhu, Y. Tian, A. E. Hamielec, and D. R. Eaton, *Macromolecules* **23:**1144 (1990).

68. J. Shen, Y. Tian, G. Wang, and M. Yang, *Makromol. Chem.* **192:** 2669 (1991).

69. B. O'Shaughnessy, *Makromol. Chem., Theory Simul.* **4:** 481 (1995).

70. E. Karetekin, B. O'Shaughnessy, and N. J. Turro, (to be published).

71. N. J. Turro, *Modern Molecular Photochemistry,* University Science Books, Mill Valley, CA, 1991.

Index